AF361804

Numerical Simulation in Biomechanics and Biomedical Engineering-II

Numerical Simulation in Biomechanics and Biomedical Engineering-II

Editor

Mauro Malvè

MDPI • Basel • Beijing • Wuhan • Barcelona • Belgrade • Manchester • Tokyo • Cluj • Tianjin

Editor
Mauro Malvè
Department of Engineering,
Universidad Pública
de Navarra,
Pamplona, Spain

Editorial Office
MDPI
St. Alban-Anlage 66
4052 Basel, Switzerland

This is a reprint of articles from the Special Issue published online in the open access journal *Mathematics* (ISSN 2227-7390) (available at: https://www.mdpi.com/si/mathematics/Biomechanics_Biomedical_Eng_II).

For citation purposes, cite each article independently as indicated on the article page online and as indicated below:

LastName, A.A.; LastName, B.B.; LastName, C.C. Article Title. *Journal Name* **Year**, *Volume Number*, Page Range.

ISBN 978-3-0365-8100-2 (Hbk)
ISBN 978-3-0365-8101-9 (PDF)

Contents

About the Editor

Mauro Malvè

Mauro Malvè is an aeronautical engineer with a master's degree from the Politecnico di Milano (Italy). After spending a brief amount of time in the industry, in 2006, he obtained a PhD in Mechanical Engineering at the University of Karlsruhe (Germany) with a focus on computational cardiovascular biomechanics. From 2008 to 2012, he completed a Postdoctoral Fellowship at the Bioengineering Research Networking (CIBER-BBN) of the University of Zaragoza (Spain) centered on the fluid structure interaction analysis applied to large vessels and to the human upper airways. In 2013, he moved to Pamplona (Spain) to join the Engineering faculty of the Public University of Navarra (UPNA). He is currently an Associate Professor of Structural Mechanics at the Department of Engineering at UPNA and a scientific director of the research group IMAC (Applied and Computational Mechanical Engineering). His main research interests focus on computational fluid dynamics and on fluid–structure interaction analysis applied to human and veterinary medicine. More recently, he has been working extensively on the design and simulation of customizable, 3D-printable cardiovascular stents and respiratory prostheses. He is currently a member of Bioengineering Research Networking (CIBER-BBN) and the European Society of Biomechanics. He maintains collaborations with several institutions in Spain and worldwide, such as the University of Zaragoza (Spain), the University of Saskatchewan (Canada), and the École Vétérinaire Maisons Alfort, where he partook in three research stays in the last few years, and the University of Grenobles-Alpes (France). He has published over 40 scientific publications indexed with the Journal of Citations Report, and he has presented his research in over 60 national and international conferences.

Preface to "Numerical Simulation in Biomechanics and Biomedical Engineering-II"

In comparison with some of the traditional fields of engineering, biomedical engineering is a relatively new discipline that combines knowledge from several aspects of medicine, biology, mathematics and physics among others. In last two decades, it has received significant attention from the scientific community due to its direct relation to the health of humans, animals and even of the environment.

This second volume of the Special Issue entitled 'Numerical Modeling in Biomechanics and Biomedical Engineering' in *Mathematics* presents a collection of different applications of mathematical methods and computational modeling to biomechanics and biomedical engineering. Some of the trending topics included in the book are the structural analysis of the muscle skeletal system and bone tissue, the modeling of arterial biomechanics and its pathologies such as the atherosclerosis, discrete particle modeling for drug delivery or tumor targeting, virtual surgery such as arthroplasty or mandibular advancement, and the computational analysis of micro-vascular networks and complex arterial flows for the identification specific parameters of diseased vessels among other interesting subjects.

The Editor thanks all the contributors for presenting their results and achievements through their clinical and basic scientific research articles. This book has been written and printed thanks to their expertise, dedication, and enthusiasm.

Mauro Malvè
Editor

Review

Modelling the Upper Airways of Mandibular Advancement Surgery: A Systematic Review

Mohd Faruq Abdul Latif [1], Nik Nazri Nik Ghazali [2,*], M. F. Abdullah [3], Norliza Binti Ibrahim [4], Roziana M. Razi [5], Irfan Anjum Badruddin [6,*], Sarfaraz Kamangar [6], Mohamed Hussien [7,8], N. Ameer Ahammad [9] and Azeem Khan [10]

[1]   Mechanical Engineering Technology Department, Faculty of Mechanical and Manufacturing Engineering Technology, Universiti Teknikal Malaysia Melaka, Melaka 76100, Malaysia
[2]   Department of Mechanical Engineering, Faculty of Engineering, University of Malaya, Kuala Lumpur 50603, Malaysia
[3]   Oral and Maxillofacial Surgery Unit, School of Dental Sciences, Universiti Sains Malaysia, Kubang Kerian 16150, Malaysia
[4]   Department of Oral & Maxillofacial Clinical Science, Faculty of Dentistry, University of Malaya, Kuala Lumpur 50603, Malaysia
[5]   Department of Paediatric Dentistry and Orthodontics, Faculty of Dentistry, University of Malaya, Kuala Lumpur 50603, Malaysia
[6]   Mechanical Engineering Department, College of Engineering, King Khalid University, Abha 61421, Saudi Arabia
[7]   Department of Chemistry, Faculty of Science, King Khalid University, Abha 61413, Saudi Arabia
[8]   Pesticide Formulation Department, Central Agricultural Pesticide Laboratory, Agricultural Research Center, Dokki, Giza 12618, Egypt
[9]   Department of Mathematics, Faculty of Science, University of Tabuk, Tabuk 71491, Saudi Arabia
[10]   Department of Science, Mahboobia Panjetan Degree College, Affiliated to Kakatiya University, Warangal 506002, India
*   Correspondence: nik_nazri@um.edu.my (N.N.N.G.); irfan@kku.edu.sa (I.A.B.)

Citation: Abdul Latif, M.F.; Ghazali, N.N.N.; Abdullah, M.F.; Ibrahim, N.B.; Razi, R.M.; Badruddin, I.A.; Kamangar, S.; Hussien, M.; Ahammad, N.A.; Khan, A. Modelling the Upper Airways of Mandibular Advancement Surgery: A Systematic Review. *Mathematics* **2023**, *11*, 219. https://doi.org/10.3390/math11010219

Academic Editors: Mauro Malvè and Lihua Wang

Received: 10 October 2022
Revised: 18 December 2022
Accepted: 23 December 2022
Published: 1 January 2023

Abstract: Obstructive sleep apnea syndrome is a conceivably hazardous ailment. Most end up with non-reversible surgical techniques, such as the maxillomandibular advancement (MMA) procedure. MMA is an amazingly obtrusive treatment, regularly connected to complexities and facial change. Computational fluid dynamic (CFD) is broadly utilized as an instrument to comprehend the stream system inside the human upper airways (UA) completely. There are logical inconsistencies among the investigations into the utilizations of CFD for OSAS study. Thus, to adequately understand the requirement for OSAS CFD investigation, a systematic literature search was performed. This review features the necessary recommendations to accurately model the UA to fill in as an ideal predictive methodology before mandibular advancement surgery.

Keywords: OSA; mandibular advancement; CFD; sleep apnea

MSC: 92-08

1. Introduction

Obstructive sleep apnea syndrome (OSAS) is a potentially life-threatening illness [1–4]. Obstructive sleep apnea (OSA) is a traditional chronic syndrome implicating the adult population, with the highest occurrence reported among middle-aged men [5]. The ailment is characterized by repetitive episodes of a complete or incomplete collapse of the upper airway during sleep, with a consequent decrease of the airflow [6]. Over the decade, several OSA management methods have been developed [7,8], and among them by utilizing positive airway pressure (PAP) treatment [4,9], an oral appliance [10] and several non-reversible surgical methods such as mandibular advancement surgery (MAS) [11,12]. MMA is a surgical treatment that involves cutting the upper and lower jaws to realign them [13].

The improvement of the jaw structures passively persuades an anterior displacement of the soft palate and the tongue, widening the pharyngeal space [12,14]. However, it is noteworthy that MAS is a highly invasive treatment, often associated with complications and aesthetic change [15–17].

Consequently, the treatment should apply to selected patients when all other approaches and first-level surgery have failed or patients with established craniofacial deformities [18,19]. CFD is widely used to fully understand the flow mechanism inside the human airways [20–23]. However, there are contradictions among the studies of the utilization of CFD for OSA study. Therefore, a systematic review is needed to understand the fundamental requirement of OSA CFD analysis fully.

The formulation of this systematic review began with the following research question: How can we properly model the upper airways (UA) as a prediction approach before MAS? Table 1 presents the keyword search string used for this article—the review's process flow, as depicted in Figure 1.

Table 1. Keywords for the search string.

Database	Keywords
WoS	ALL = ((maxillomandibular OR mandibular OR mandible* OR jaw* OR maxilla) AND (advance$ OR improvement OR gain OR elevation) AND ("obstructive sleep apnea" OR snoring OR "sleep* disorder breath$" OR "pharyngeal airway resist$" OR "sleep apnea"))

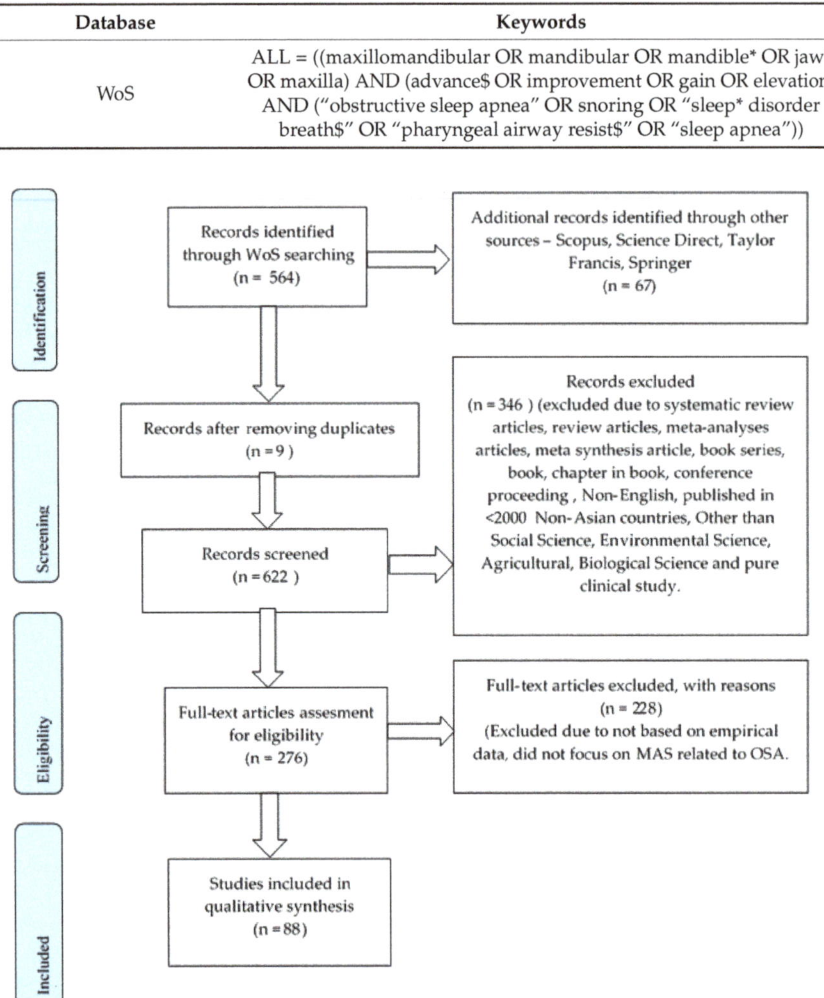

Figure 1. Flow chart of the review screening process.

2. Review Outcome Introduction

This paper reviews previous studies of OSAS from a CFD perspective focusing on MMA treatment. This review will detail previous CFD technology associated with OSA cases. In this paper, the author will be detailing the rigorous needs of the image processing technique, which is the essential pre-process of CFD modelling.

Furthermore, the UA boundary condition setup will be discussed in detail, starting from the meshing technique of the UA model. The discussion will detail the turbulence model used in the OSA literature, including the setup parameter of UA CFD modelling. Towards the end, the author will discuss related experimental validating processes to indicate the competency of CFD analysis as a tool for MMA treatment assessment.

2.1. Airways Imaging Technique

OSAS has become an interesting topic in research recently. Recent developments in OSA treatment procedures have heightened the need for advanced image techniques [24,25]. The imaging technique available in modern technology has two types: computerized axial tomography (CT) and magnetic resonance imaging (MRI).

One crucial theoretical topic that has engaged researchers for many years is how the UA trigger snoring [22]. The introduction image technique based on acoustics has opened a broad interest in the OSA prediction technique [26]. The MRI or CT approach is confined to a 2D stacking image of the subject's cross-section region, subjected to visualization, essential length, and volume measurement, as shown in Figure 2 [27–29]. The CT or MRI technique, however, has been widely used to evaluate the severity of OSA for surgery determination from the early 90s until today [30–32].

Until today, most of the clinical approaches to OSA treatment depended on the UA's generated cross-sectional image, which provides insight into the UA's narrowing gap and volume [33,34]. The medical decision is solely based on the judgment of the UA cross-section due to time limitation and virtual modeling capabilities limitation [35–37]. These results in non-responsive post-treatment occurred because not all OSA patients responded to surgical treatment positively [38,39]. A more detailed evaluation is needed to properly understand the behavior of OSA in the UA for a more accurate prediction of pretreatment.

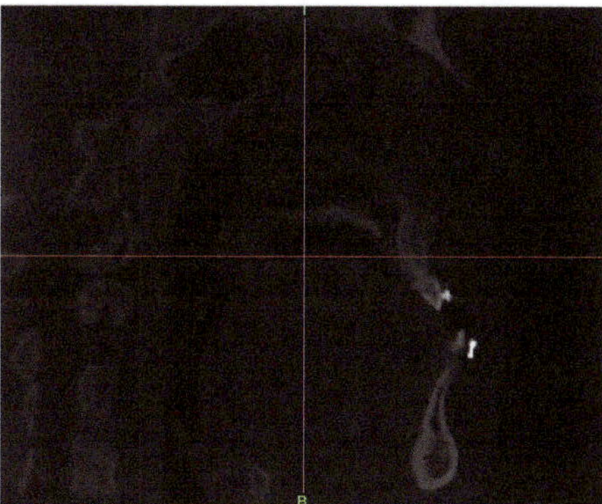

Figure 2. Example of CT-scan sagittal view of pretreatment of OSA. Adapted with permission from Ref. [28]. 2015, Butterfield.

2.2. Modelling of Human UA

Previously, studies of OSAS depended on statistical reviews of MRI and CT images and various clinical trials [40,41]. The introduction of CFD simulation has revolutionized OSA studies [42]. The knowledge of fluid dynamics is applied to understand the mechanism of OSA in the last decade [43]. The introduction of CFD has allowed studies of modelled UA to explain OSA theoretically [44]. The UA model was first studied based on the teaching model (Model C12, Carolina Biological Supply Company, Burlington, NC, USA) for medical school students [23] (Figure 3). This model assumes the human airways to be symmetrical. The study by Ted B. Martonen [23] scanned the silicon model into the computer as a 3D model. He then performed the simulation using different flow rates, demonstrating the possibility of applying CFD to model UA.

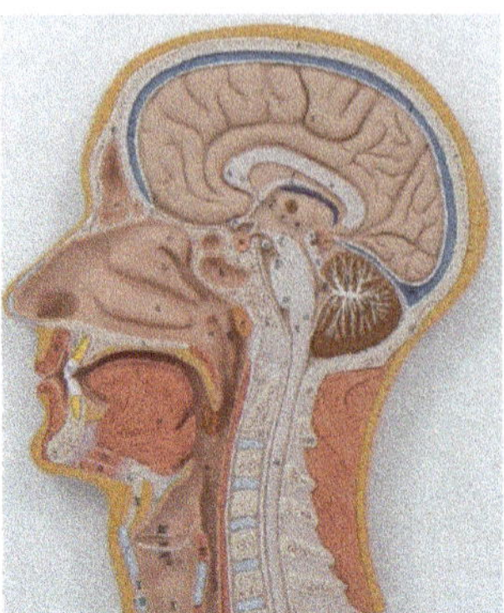

Figure 3. Example of human UA medical school teaching model used by Ted B. Martonen. Adapted with permission from Ref. [23]. 2002, Ted B. Martonen.

In 2003, Heenan et al. [45] developed a 3D model with realistic UA anatomy geometry (Figure 4). This model is an adaptation of the Weibel A model. Their study introduced a CFD method to study the airflow of the UA based on the model, which was less complicated than the actual human UA. Since then, CFD has been an increasingly important area in OSA study.

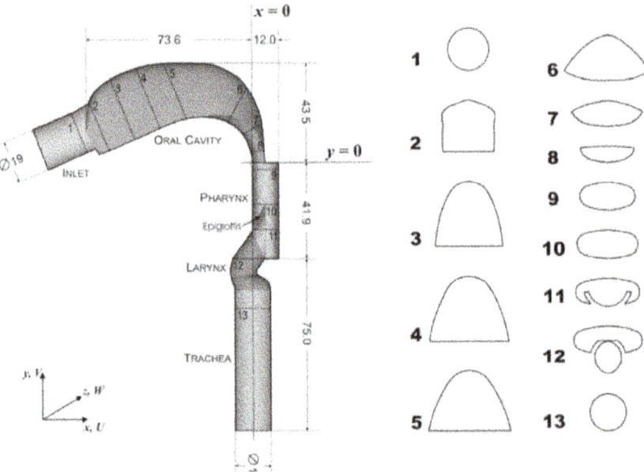

Figure 4. Idealize Weibel A human UA model. Reprinted with permission from Ref [45]. 2003, A. F. Heenan et al.

More literature has emerged that offers contradictory findings of the Weibel A model. Collins et al. [46] compared a geometrically accurate model sourced from the MRI model with the Weibel A model. The result shows a dissimilar comparison flow pattern because the Weibel A model has sharp edge geometry. In contrast, the accurate model has a smooth geometry, as shown in Figure 5. Thus, the simplified model is insufficient to accurately predict human UA's flow behavior. Although the result represented by Collins et al. [46] shows a significant variation in the flow pattern, the model developed by Heenan et al. [45] is a helpful reference for the CFD modelling of UA.

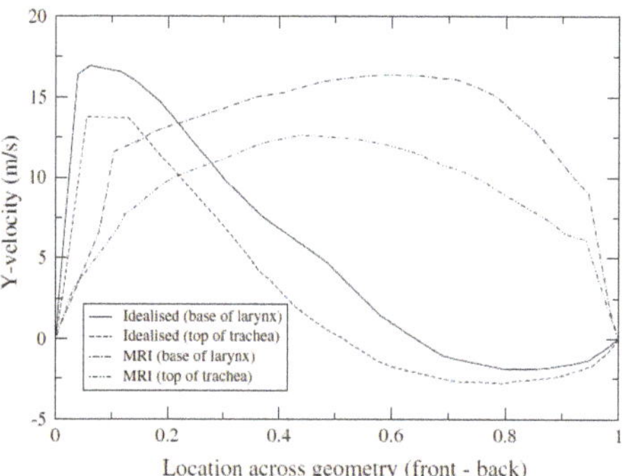

Figure 5. Comparison of the Idealize model with the MRI model. Reprinted with permission from Ref. [46]. 2007, T. P. Collins et al.

MRI and CT scan imaging techniques have revolutionized the human respiratory system [47]. This technique enables the construction of accurate geometry of human UA with the help of image processing software such as 3DVIEWNIX, AMIRA, MIMICS, and 3-MATICS [48]. Recent advances in the research of UA flow have emphasized the importance of anatomically accurate UA models. A significant volume of published studies describes the role of MRI or CT in describing the UA flow [49,50]. The first serious discussions and analyses of anatomically accurate UA models emerged during the early 2000s. Xu et al. [51] modelled three cases of OSAS in children aged three to five. The MRI slice image of the children UA was transformed into a 3D model using 3DVIEWNIX software. The study highlights a manual mask filtration technique applied to extract the UA, which removes some of the detail and small voids to simplify the geometry. The simple, clean geometry is necessary to have a good quality CAD model for CFD [52,53].

Most researchers report removing some details of voids and surface smoothing in the UA. It adds complexity to the CAD model, limiting the excellent meshing capabilities and weak CFD convergence [42,53]. Several studies that used MRI or CT show the potential of a functional imaging source for the 3D construction of the UA [25]. However, there are no published data on their sensitivity or specificity [41].

The excellent quality of CAD is a necessity for CFD application. However, the MRI or CT imaging technique requires tremendous effort to construct a good-quality CAD model. A specific technique for creating the UA geometry correct model has not yet been published. The technique is essential because it contributed to the CFD application's dependability and precision in comprehending OSAS [54].

2.3. Exclusion of the Nasal Cavity

Excluding the oral and nasal cavities simplifies the UA model in a significant portion of the UA research. This solved the issue of indefinitely characterizing the wall boundary of the UA. The oral and nasal cavities have less impact on UA flow. Under no circumstances does there seem to be evidence that the nasal cavity may fundamentally alter the pharyngeal flow characteristics, such as the slightest pressure or maximum velocity. The investigation by Zhao et al. [55] demonstrates that a missing nasal cavity will not radically fluctuate the UA's pressure drop and stream velocity profiles. This finding aligns with the examination by Persak et al. [56] and Shah et al. [57]. Calmet et al. [58] conducted a complete study of airways from the trachea up to the start of the nasal cavity (vestibule). In his study, he shows turbulence behavior in the middle of the nasal cavity. This is understandable due to the complexity of the nasal cavity funnel. It is hard to justify the influence of the nasal cavity on UA air flow purely based on a single data sample without comparison with other data in that investigation. Cheng et al. [59] also demonstrated full airways with a nasal cavity comparing pre-surgery and post-surgery data. In his reporting, he did not describe or demonstrate any changes within the nasal cavity, and the ensuing image on the pressure contour plot demonstrated no substantial change prior to entering the UA funnel. The same study by Ito et al. [10] also showed no significant change in the airflow. Cheng et al. [59] exhibited a full airways model study, including the nasal cavity. However, in that study, the nasal cavity was superimposed. Both pre-surgery and post-surgery provided similar inflow patterns. An investigation by other researchers also revealed the same result [60–62]. The finding clearly shows that excluding the nasal cavity is necessary, as it adds more complexity to the investigation.

2.4. The Meshing of UA Geometry Accurate Model

As stressed, good-quality geometry is vital for CFD application. A CFD prepossessing step meshing discretizes the geometry into a more distinctive element to make numerical calculation possible. Currently, there are several options for mesh generation available commercially. Typically, as documented in most publications on UA models, most of them use the built-in mesh-generating capability of the widely known ANSYS CFD software [63–66].

Due to the complexity of the geometry, researchers tend to utilize auto mesh generation for retaining the original geometry accurate model of the UA [20,67], resulting in the mesh density concentration on the tight curve or small surfaces. Researchers applied more exceptional mesh cells to solve the unevenly distributed mesh density issue, typically for a UA model consisting of between 500 thousand to 1.6 million cells [55,68]. Typically, for internal fluid flow assessments such as the UA model, a hybrid mesh with at least five inflation layers is used, since it represents the near-wall effect better [20]. Figure 6 shows the example of a hybrid meshing of the UA model.

A finer mesh does not necessarily result in solution convergence. Although convergence is possible with more excellent meshing, it always comes with the high cost of computing. It is undeniable that solution convergence relies on good quality meshing. Most of the UA model research does not report the mesh quality of their model. They depend on finer mesh due to geometry complexity for convergence, which costs unnecessary computing [43]. Researchers utilize the application of the smoothing technique in the hope of achieving better mesh quality [42]. Some of the studies of the UA model utilize third-party meshing software, such as MESHLAB [69], GAMBIT [70,71], and DEP MESH-WORK [72]. Still, most of them only report basic auto mesh generation applications [42]. Recent advancements in the simulation of UA models have increased the demand for improved meshing approaches that result in a higher mesh quality for UA models.

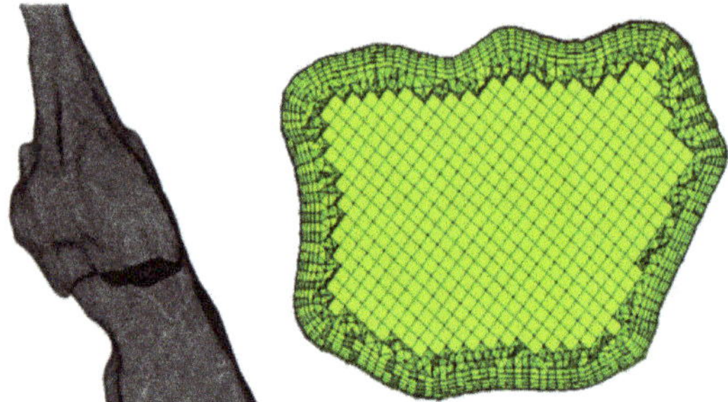

Figure 6. Hybrid meshing of a UA model Reprinted with permission from Ref. [43]. 2013, Moyin Zhao et al.

Finer meshing increases the accuracy of the result, which theoretically makes sense as it is close to the source model. However, finer meshing is computationally expensive [73]. In CFD, there is a well-adapted method to minimize unnecessary computing costs. The technique is grid sensitivity analysis [74]. The process determines the optimum mesh density while retaining the accuracy of the result. The study model simulates various element sizes, starting from a coarse to a finer mesh [75]. This method has been well adopted in the UA model by multiple researchers. Zhao et al. [55] show a plot of axial velocity along a vertical line from the inlet to the larynx wall, starting with a 200k mesh to a most excellent 1.8 million mesh. The result shows that 1.3 million mesh has a similar velocity profile as the 1.8 million mesh while saving 30% of the computing time.

Rahimi-Gorji et al. [73] demonstrate the execution of distinctive grid sizes comprising around 2.4, 3.4, 4.2, and 5.1 million cells and acquire pivotal velocity profiles at two cross areas. They have found that the expansion of grid size measure from 4.2 to 5.1 million cells does not modify the outcomes as it shows an agreeable velocity profile. Subsequently, the study uses a grid with a 4.2 million mesh size for the simulation. However, the author feels that comparing the pressure or velocity profile against the number of cells does not justify the acceptability of the boundary condition. A comparable validation test rig is necessary as a reference to determine the appropriate cell number for the modelling, as shown by Amatoury et al. [76]. A grid size agreeable with accurate experimental data is the ideal mesh size for modelling the UA.

2.5. Boundary Conditions

More literature has emerged that offers contradictory findings of the appropriate boundary condition for UA simulation. The boundary conditions used in UA CFD simulations should characterize and replicate those seen in human respiratory flow. The UA inspiratory flow originates from the nasal cavity and ends at the trachea. In a perfect situation, encompassing static pressure should be characterized at the nostrils and, correspondingly, release airflow at the lower larynx. Numerous published studies define human inspiratory airflow rates differently, as summarized in Table 2. A few examinations determined a time-dependent flow to copy the respiratory cycle.

Table 2. Compilation of methods used for CFD modelling of the UA.

Author	Geometry	Turbulence Model	Flowrate Lmin	Remarks
Rahimi-Gorji, Gorji and Gorji-Bandpy [73]	CT-scan	k-ω	10, 15, 30, 60	Particle deposition study
Xi, April Si, Dong and Zhong [77]	CT-scan	LES	15	Effects of glottis motion on airflow
Collins, Tabor and Young [46]	MRI	k-ω	72	Comparison of the idealized vs. accurate geometry model
Zubair, Riazuddin, Abdullah, Ismail, Shuaib and Ahmad [71]	CT-scan	laminar	15	Study of the effect of posture
Zhao, Barber, Cistulli, Sutherland and Rosengarten [43]	MRI	SST k-ω	10	Study of upper airway response to oral appliance treatment
Cheng, Koomullil, Ito, Shih, Sittitavornwong and Waite [59]	CT-scan	K-ϵ	42	Surgical assessment
Heenan, Pollard and Finlay [45]	reconstruct model	k-ϵ	15, 30, and 90	Study of the idealized airways
Suga, et al. [78]	CT-scan	k-ϵ	30	Study of the effect of oral appliance treatment
Gutmark, et al. [79]	MRI	SST k-ω	10	Biomechanics of the soft palate
Srivastav, Paul and Jain [63]	CT-scan	k-ϵ, k-ω	60	Capturing the wall turbulence
Bates, Schuh, McConnell, Williams, Lanier, Willmering, Woods, Fleck, Dumoulin and Amin [69]	MRI	LES	time-dependent 0–3 Ls	New method to generate dynamic boundary conditions for airway
Fletcher et al. [80]	CT-scan	LES	9.06 and 20.52	Genioglossal advancement (GGA)
De Backer, Vos, Gorlé, Germonpré, Partoens, Wuyts and Parizel [64]	CT-scan	Laminar and k-ϵ	23	Analyses in the lower airways
Srivastav, Paul and Jain [63]	CT-scan	K-ω and k-ϵ	60	Simulation of the human respiratory tract
Patel, Li, Krebs, Zhao and Malhotra [65]	CT-scan	laminar	4.67	Congenital nasal pyriform aperture stenosis (CNPAS)
	MRI	k-ϵ		

Table 2. *Cont.*

Author	Geometry	Turbulence Model	Flowrate Lmin	Remarks
Zhao, Barber, Cistulli, Sutherland and Rosengarten [55]	MRI	K-ω	10	MAS
Premaraj, Ju, Premaraj, Kim and Gu [31]	CT-scan	K-ω	18	Maxillary anterior guided orthotics (MAGO)
Cheng, Koomullil, Ito, Shih, Sittitavornwong and Waite [59]	CT-scan	k-ϵ	42	Full airways with the nasal cavity
Liu, Yan, Liu, Choy and Wei [60]	CT-scan	LES	16.8, 30, 60	Including the nasal cavity with an extension funnel
Powell et al. [81]	CT-scan	LES, K-ω SST	30	Patterns in pharyngeal airflow study

In contrast, others described a mean airflow stream rate [71,73,77]. Table 2 shows considerable variation in the airflow rate used in UA simulation. The variation is understandable because, ethically, it is difficult to have a human subject measure the actual patient breathing airflow rate. It is also because the human actual breathing flow rate varies from one person to another. Other boundary conditions were customary settings for all the reviewed studies, like a smooth and non-slip wall and a five percent turbulence intensity of the inlet flow [71].

2.6. Turbulence Modelling of UA Flow

The turbulence model adopted in UA modelling varied among these four commonly adapted models: k-epsilon, k-omega, k-omega SST, and large eddy simulation (LES) [63,81]. Turbulent flow features are modelled by solving the variables for kinetic energy (k) and dissipation rate (ε) or specific dissipation rate (ω)—a literature finding of the adopted turbulence model for UA modelling, as in Table 2. The decision to model a fluid dynamics problem, either laminar or turbulent, is one of the most challenging decisions a fluid dynamicist must take [82]. Therefore, a thorough understanding of the predicted flow field is critical for obtaining the desired outcomes. The estimated Reynolds numbers (Re) range indicates the UA flow to be laminar or transitional. The standard k–ω shear stress transport (k–ω SST) model was proven to be appropriate to simulate this complex flow.

The SST k–ω model has advantages in solving complex transitional flow, including the UA transitional flow [43,83]. The employment of the k–ω shear stress transport (SST) turbulence model provides an enhanced description of flows involving adverse pressure gradients and curved boundary layers [79].

The k–ε model with enhanced wall treatment is suitable for monitoring flow separation with a strong pressure gradient and flow recirculation. These are suitable to determine the flow parameters near the wall of running airflow [63].

Shown by literature, LES has become the preeminent turbulence model for the UA model simulation, as most of the studies show good agreement with the experimental result [81]. However, a fundamental disadvantage of this turbulence model application is that it demands a high computing expense incurred due to the length of time necessary to solve it. As a result, LES is unsuitable for time-demanding clinical applications. The K–ω SST model also agreed well with the experimental result and is almost comparable with the LES model [43,80]. This turbulence model offers much cheaper computing costs and is suitable for clinical application. Again, the author stressed that an excellent comparison with the physical model provides a justification of the ideal turbulence model for the UA study.

2.7. Location of the Inlet

The velocity profile of many studies demonstrates a brisk airstream at the smallest cross-sectional area at the velopharynx. This flow feature is identified as the 'pharyngeal jet' [46,78,81,83]. Hence, it may be coincidental that the degree of structural difference in UA anatomy systems is virtually tremendous, resulting in a diversity of UA flow patterns.

Before surgery, the highest increase in airflow velocity is observed at the pharyngeal airway when inhaling, and the biggest decrease in pressure is observed downstream, according to Liu's study. Negative pressure and airflow velocity in the whole airway equalized postoperatively. They discovered that velocity reduction at the most constricted portion of the pharyngeal airway correlates most strongly with surgical outcome. However, they emphasized that measurements of non-theoretical dynamic airflow during sleep would be excellent for further validating CFD models [62].

The study by Zhao et al. [55] of flow profiles for both inhaling and exhaling situations indicate that inhaling results in a 30% greater pressure loss and a higher flow velocity than exhaling. The airflow enters the UA stream with higher turbulent kinetic energy and lower total pressure, which expands the likelihood of UA collapse. In short, narrowing the UA increases the velocity of the territorial flow and decreases the pressure. This low-pressure peak may be associated with the severity of OSA, as a high-pressure gradient across the wall border would cause the UA structure to collapse [46,78,81,83]. Furthermore, 10 L/min, 15 L/min, and 30 L/min are the most prevalent input flowrate settings, according to our review [45,73,77–79,81,82].

2.8. Experimental Validation

Experimental validation in the UA model based on recent studies focused on verification checking for numerical studies [43,45,55,64,83,84]. Rapid prototyping is the most common method to create an identically exact UA model. [85]. Heenan et al. [45] have constructed an idealized Weibel A model representing the human oropharynx. The design slightly differs from the one described by Stapleton et al. [86]. The model is two times larger than the original, which describes double the adequate accuracy in setting up using the particle image velocity (PIV) method described in Figure 7. The flow rate of the double scale model was doubled to maintain the full-scale Reynolds number. The construction of the physical model uses the fused deposition modelling (FDM) method with surface smoothing using dichloromethane and an epoxy coating. The result shows a discrepancy between the experimental and CFD. Heenan et al., assume the differences is because of the error in the CFD simulation due to the weak boundary layer of the CFD model [45].

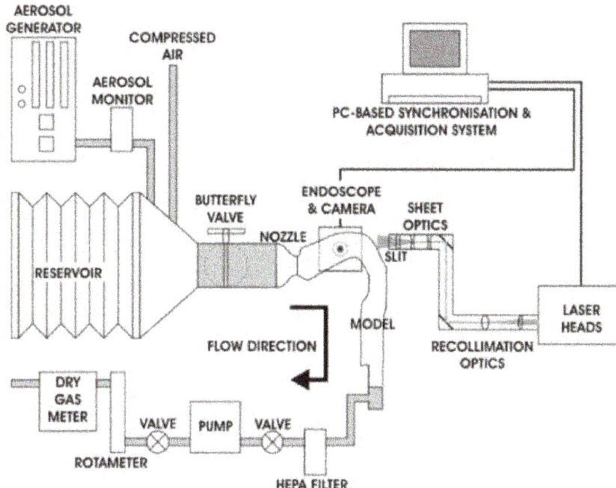

Figure 7. PIV experimental setup Reprinted with permission from Ref. [45]. 2003, A. F. Heenan et al.

Collins, Tabor and Young [46] argue that the flow character in the idealized geometry is affected by the shape of the geometries used. The absence of air curvature and surface irregularities in the Weibel A model makes it unusable for predicting the flow pattern. Thus Collins, Tabor and Young [46] developed a hybrid geometry based on the one described by

Stapleton, Guentsch, Hoskinson and Finlay [86] and Heenan, Pollard and Finlay [45], as shown in Figure 8. The result are compared with an MRI based model of a 21 years old male. Both were simulated using the CFD method.

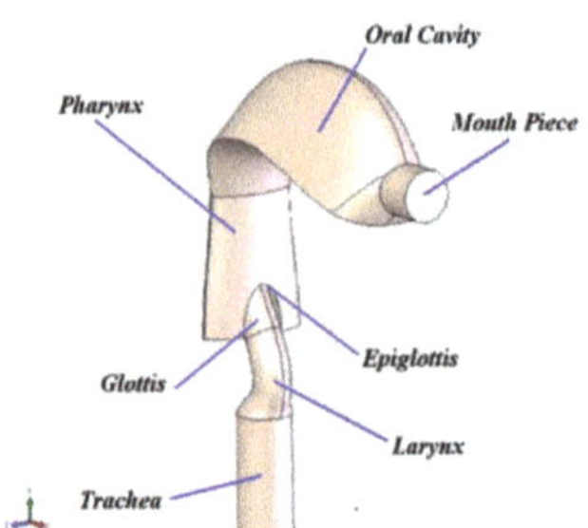

Figure 8. Hybrid idealized geometry. Reprinted with permission from Ref. [46] 2007, T. P. Collins et al.

The idealized hybrid model shows good agreement with the one represented by Heenan, Pollard and Finlay [45], as it is a similar geometry. However, the idealized geometry offers a slight advantage over the MRI geometry due to the recirculation zone created by the sharp edge or steps in the idealized geometry. Thus, it is concluded that the idealized geometry is inadequate at predicting the features of the flow of human UA.

Xu, et al. [87] cast a ¾ scale of a geometrically accurate model of UA using a silicon cast for result validation (Figure 9). The reason for using a scaled model is unknown. The study placed eight pressure sensors along the surface of the silicon UA model. The study also placed the same pressure point positions in the CFD model. However, the CFD model did not scale to the cast model. The results agreed to within 2.0 Pa in both inspiration and expiration phases at almost all locations. However, the maximum pressure drop that CFD predicted was 20 Pa higher than in the experiment. The scale difference might play an essential role in pressure distribution; thus, a comparable 1:1 scale model is desirable.

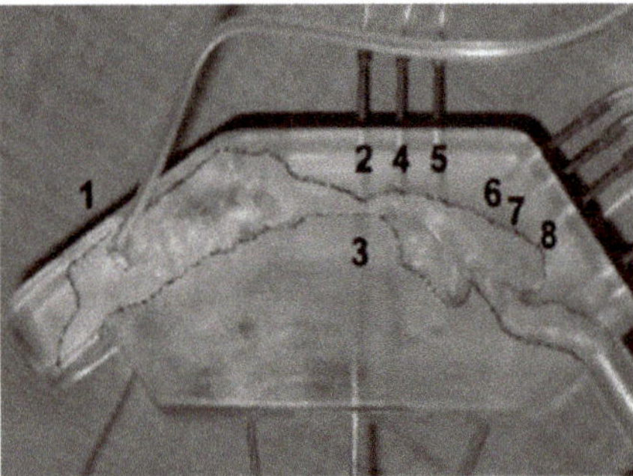

Figure 9. 85% scaled silicon cast UA model. Reprinted with permission from Ref. [51]. 2006, Chun Xu et al.

Latter, Zhao, Barber, Cistulli, Sutherland and Rosengarten [43] developed a 1:1 scale model of a geometry-accurate physical model. The model uses the rapid prototyping (RP) method using a transparent polymer. A comparison of pressure distribution along the surface of the UA model is shown in Figure 10. The finding shows good agreement between the CFD and the experimental result. This finding is significant as it can be used as a base model to verify one's UA model CFD result by following the exact boundary condition and turbulence model.

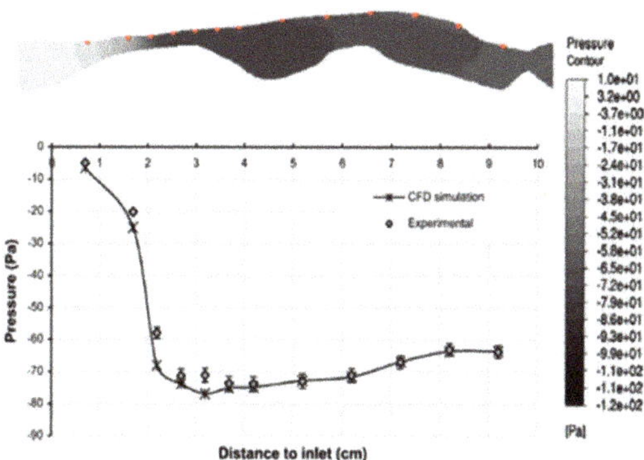

Figure 10. 1:1 scale of the UA model pressure distribution comparison between CFD and the experimental results. Reprinted with permission from Ref. [43] 2013, Moyin Zhao et al.

Due to the complexity and individuality of the geometrically accurate MRI or CT scan model, it is not easy to adjust the geometry to understand the geometrical influence on UA flow better. An idealized geometry should have a better shape agreement than the average-person UA and not generalize as the Weibel model. The model should have detailed geometry that another researcher efficiently replicates.

2.9. Numerical Modelling Issues of UA

A previous study by Collins has indicated the disadvantage and limitations of the Weibel model. In term of shape based on his study, it is impossible to mimic the organic shape of UA by mean of 3D CAD construction (Idealize model) [46]. The model is oversimplified and has a sharp edge that is not present in the actual human UA. In addition, the model's nasal cavity does not accurately represent the upper airways since it comprises a single large chamber rather than multiple layers of the air passage. The flow mechanism of such a design differs significantly from or completely contradicts the actual UA and specifies the flow characteristic entering the trachea, influencing the UA study's total flow [46]. The authors suggest that the 3D model is the most appropriate geometrical model for comprehending the fundamental flow behavior within the UA. In terms of CAD and meshing, the 3D model is simple to configure for numerical simulation due to its simple shape [45]. CT scans offer an actual or geometry-accurate model of the UA, which can be converted into a CAD model using specialized software [59,63,81]. In numerical modelling, this precise model provides real-world information regarding air movement within the UA. Due to the organic nature of this model, preparing the CAD for CFD analysis necessitated extensive attention to detail [52,53].

Using CT scans to represent UA has advanced the study of UA. However, it is not without flaws. The geometry accuracy can vary depending on the image quality or resolution of the CT scan. This image is difficult to comprehend. Most researchers employ MIMICS to produce the 3D mask of the UA. The masking process requires additional treatment, detailing, or processing, such as filling the unneeded void or removing the surface spike. Poorly processed masks could result in poor meshing and compromise the CFD output [70,71].

The laminar model is unsuitable for UA research because, according to Barber's research, the UA's Reynolds number (Re) ranges from 400 to 3000, indicating that the flow is either laminar or transitional [45,73]. The flow turbulence was modelled using the standard shear stress transfer (SST) k–ω model. Their early sensitivity research applying multiple turbulence models to the physical experimental results demonstrated the suitability of the SST k–ω model in solving the complex UA flow [43]. Therefore, only a few studies have reported the UA utilizing laminar flow [65,71,88]. Compared to experimental results, LES provides the most accurate model of UA due to its superior observation at the wall and in the center. However, LES simulations require powerful computing capabilities. Other turbulence models, such as k–ε, are utilized extensively for external flow but infrequently for inside flow [78]. The researcher's findings also indicate that k–ε has low precision for the UA investigation [88]. The most acceptable and cost-effective turbulence model is k–ω, which offers comparable accuracy to the LES model [81].

Meshing of UA requires hybrid meshing for internal flow so that flow mechanisms close to the wall can be captured accurately. This hybrid meshing requires a high-quality CAD without elements that degrade mesh quality. For the solution to converge, high-quality mesh is required [73].

3. Conclusions

A complete evaluation is required to fully understand OSA's UA behavior and make more accurate pretreatment predictions. The simplified model is insufficient for effectively predicting the flow behavior of the human UA. It is also agreeable that the inhalation airflow shows higher airflow, which defines the air inlet's location in the upper airway model. The nasal cavity adds more complexity to the investigation; thus, removing it is necessary. The highlighted technique of removing some details of voids and surface smoothing in the modelling of the upper airway reduces the complexity of the CAD model. Complex geometry limits excellent meshing capabilities and results in weak CFD convergence. This technique is a good reference; however, it needs more published data on sensitivity or specificity. Additionally, this article emphasized the importance of a precise geometry construction technique for the UA geometry-accurate model, which is critical for the CFD application's reliability and accuracy in comprehending OSAS. It is undeniable that good quality meshing is required for solution convergence, which has increased the demand for detailed advanced meshing approaches.

A comparable 1:1 scale physical model is necessary to determine the appropriate cell number or element size for CFD modelling. Through grid sensitivity analysis, a grid size agreeable with accurate experimental data is the ideal mesh size reference for modelling the UA. However, the data reading in the physical validation should be pressure-based, as it shows good agreement with the CFD validation.

The k–ω SST model is the most economical, as it has advantages in solving complex transitional flow and provides an enhanced description of flows involving adverse pressure gradients and curved boundary layers. However, it is vital to have an excellent comparison with the physical model to justify the ideal turbulence model for the UA study.

An idealized geometry focused on the UA from the pharynx until the larynx was greatly needed. It should have a better shape agreement than the average-person UA and not be as oversimplified as the Weibel model. The idealized model should have detailed geometry that another researcher efficiently replicates. Thus, it is concluded that the idealized geometry is inadequate at predicting the features of the flow of human UA.

Author Contributions: Conceptualization, M.F.A.L, N.N.N.G. and M.F.A.; methodology, M.F.A.L, N.N.N.G., M.F.A., N.B.I. and R.M.R.; software, I.A.B., S.K., M.H., N.A.A. and A.K.; validation, I.A.B., S.K., M.H., N.A.A. and A.K.; formal analysis, I.A.B., S.K., M.H., N.A.A. and A.K.; investigation, M.F.A.L., N.N.N.G., M.F.A., N.B.I. and R.M.R.; resources, I.A.B., S.K., M.H., N.A.A. and A.K.; data curation, I.A.B., S.K., M.H., N.A.A. and A.K.; writing—original draft preparation, M.F.A.L, N.N.N.G. and M.F.A.; writing—review and editing, N.B.I., R.M.R., I.A.B., S.K. and M.H.; visualization, N.N.N.G., N.B.I. and R.M.R.; supervision, N.N.N.G. and N.B.I.; project administration, N.B.I. and R.M.R.; funding acquisition, N.B.I., S.K. and M.H. All authors have read and agreed to the published version of the manuscript.

Funding: King Khalid University under grant number RGP 2/101/43.

Data Availability Statement: Data available in manuscript itself.

Acknowledgments: The authors extend their appreciation to the Deanship of Scientific Research at King Khalid University for funding this work through the Large Groups Project under grant number RGP 2/101/43.

Conflicts of Interest: The authors declare no conflict of interest.

References

1. Lin, C.-H.; Chin, W.-C.; Huang, Y.-S.; Wang, P.-F.; Li, K.K.; Pirelli, P.; Chen, Y.-H.; Guilleminault, C. Objective and subjective long term outcome of maxillomandibular advancement in obstructive sleep apnea. *Sleep Med.* **2020**, *74*, 289–296. [CrossRef] [PubMed]
2. Boyd, S.B.; Walters, A.S.; Waite, P.; Harding, S.M.; Song, Y. Long-term effectiveness and safety of maxillomandibular advancement for treatment of obstructive sleep apnea. *J. Clin. Sleep Med.* **2015**, *11*, 699–708. [CrossRef] [PubMed]
3. Richard, W.; Kox, D.; den Herder, C.; van Tinteren, H.; de Vries, N. One stage multilevel surgery (uvulopalatopharyngoplasty, hyoid suspension, radiofrequent ablation of the tongue base with/without genioglossus advancement), in obstructive sleep apnea syndrome. *Eur. Arch. Oto-Rhino-Laryngol.* **2007**, *264*, 439–444. [CrossRef]
4. Uniken Venema, J.A.; Doff, M.H.; Joffe-Sokolova, D.; Wijkstra, P.J.; van der Hoeven, J.H.; Stegenga, B.; Hoekema, A. Long-term obstructive sleep apnea therapy: A 10-year follow-up of mandibular advancement device and continuous positive airway pressure. *J. Clin. Sleep Med.* **2020**, *16*, 353–359. [CrossRef]
5. Tsui, W.; Yang, Y.; McGrath, C.; Leung, Y. Improvement in quality of life after skeletal advancement surgery in patients with moderate-to-severe obstructive sleep apnoea: A longitudinal study. *Int. J. Oral Maxillofac. Surg.* **2020**, *49*, 333–341. [CrossRef] [PubMed]
6. Huang, Y.; White, D.P.; Malhotra, A. The impact of anatomic manipulations on pharyngeal collapse: Results from a computational model of the normal human upper airway. *Chest* **2005**, *128*, 1324–1330. [CrossRef]
7. Aarts, M.C.; Rovers, M.M.; van der Heijden, G.J.; Grolman, W. The value of a mandibular repositioning appliance for the treatment of nonapneic snoring. *Otolaryngol. -Head Neck Surg.* **2011**, *144*, 170–173. [CrossRef]
8. Vicini, C.; Dallan, I.; Campanini, A.; De Vito, A.; Barbanti, F.; Giorgiomarrano, G.; Bosi, M.; Plazzi, G.; Provini, F.; Lugaresi, E. Surgery vs ventilation in adult severe obstructive sleep apnea syndrome. *Am. J. Otolaryngol.* **2010**, *31*, 14–20. [CrossRef]
9. Doff, M.H.; Hoekema, A.; Wijkstra, P.J.; van der Hoeven, J.H.; Huddleston Slater, J.J.; de Bont, L.G.; Stegenga, B. Oral appliance versus continuous positive airway pressure in obstructive sleep apnea syndrome: A 2-year follow-up. *Sleep* **2013**, *36*, 1289–1296. [CrossRef]
10. Ito, S.; Otake, H.; Tsuiki, S.; Miyao, E.; Noda, A. Obstructive sleep apnea syndrome in a pubescent boy of short stature was improved with an orthodontic mandibular advancement oral appliance: A case report. *J. Clin. Sleep Med.* **2015**, *11*, 75–76. [CrossRef]
11. Nishanth, R.; Sinha, R.; Paul, D.; Uppada, U.K.; Rama Krishna, B.; Tiwari, P. Evaluation of changes in the pharyngeal airway space as a sequele to mandibular advancement surgery: A cephalometric study. *J. Maxillofac. Oral Surg.* **2020**, *19*, 407–413. [CrossRef] [PubMed]
12. Prinsell, J.R. Maxillomandibular advancement surgery for obstructive sleep apnea syndrome. *J. Am. Dent. Assoc.* **2002**, *133*, 1489–1497. [CrossRef] [PubMed]
13. Doff, M.H.; Jansma, J.; Schepers, R.H.; Hoekema, A. Maxillomandibular advancement surgery as alternative to continuous positive airway pressure in morbidly severe obstructive sleep apnea: A case report. *CRANIO®* **2013**, *31*, 246–251. [CrossRef] [PubMed]
14. Gokce, S.M.; Gorgulu, S.; Gokce, H.S.; Bengi, A.O.; Karacayli, U.; Ors, F. Evaluation of pharyngeal airway space changes after bimaxillary orthognathic surgery with a 3-dimensional simulation and modeling program. *Am. J. Orthod. Dentofac. Orthop.* **2014**, *146*, 477–492. [CrossRef]
15. Liu, S.-R.; Yi, H.-L.; Guan, J.; Chen, B.; Wu, H.-M.; Yin, S.-K. Changes in facial appearance after maxillomandibular advancement for severe obstructive sleep apnoea hypopnoea syndrome in Chinese patients: A subjective and objective evaluation. *Int. J. Oral Maxillofac. Surg.* **2012**, *41*, 1112–1119. [CrossRef]
16. Torres, H.M.; Valladares-Neto, J.; Torres, É.M.; Freitas, R.Z.; Silva, M.A.G. Effect of genioplasty on the pharyngeal airway space following maxillomandibular advancement surgery. *J. Oral Maxillofac. Surg.* **2017**, *75*, 189.e1–189.e12. [CrossRef]
17. Rashmikant, U.; Chand, P.; Singh, S.; Singh, R.; Arya, D.; Kant, S.; Agarwal, S. Cephalometric evaluation of mandibular advancement at different horizontal jaw positions in obstructive sleep apnoea patients: A pilot study. *Aust. Dent. J.* **2013**, *58*, 293–300. [CrossRef]
18. Mhlaba, J.M.; Chen, M.L.; Bandla, H.P.; Baroody, F.M.; Reid, R.R. Predictive soft tissue airway volume analysis in mandibular distraction: Pushing the envelope in surgical planning for obstructive sleep apnea. *J. Craniofacial Surg.* **2016**, *27*, 181–184. [CrossRef]
19. Denolf, P.L.; Vanderveken, O.M.; Marklund, M.E.; Braem, M.J. The status of cephalometry in the prediction of non-CPAP treatment outcome in obstructive sleep apnea patients. *Sleep Med. Rev.* **2016**, *27*, 56–73. [CrossRef]
20. Aasgrav, E.; Johnsen, S.G.; Simonsen, A.J.; Müller, B. CFD simulations of turbulent flow in the human upper airways. *Arxiv Med. Phys.* **2017**, *1706*, 02565.
21. Yu, C.-C.; Hsiao, H.-D.; Lee, L.-C.; Yao, C.-M.; Chen, N.-H.; Wang, C.-J.; Chen, Y.-R. Computational fluid dynamic study on obstructive sleep apnea syndrome treated with maxillomandibular advancement. *J. Craniofacial Surg.* **2009**, *20*, 426–430. [CrossRef] [PubMed]

22. Sittitavornwong, S.; Waite, P.D.; Shih, A.M.; Cheng, G.C.; Koomullil, R.; Ito, Y.; Cure, J.K.; Harding, S.M.; Litaker, M. Computational fluid dynamic analysis of the posterior airway space after maxillomandibular advancement for obstructive sleep apnea syndrome. *J. Oral Maxillofac. Surg.* **2013**, *71*, 1397–1405. [CrossRef] [PubMed]

23. Martonen, T.B.; Quan, L.; Zhang, Z.; Musante, C. Flow simulation in the human upper respiratory tract. *Cell Biochem. Biophys.* **2002**, *37*, 27–36. [CrossRef] [PubMed]

24. Chang, Y.; Koenig, L.J.; Pruszynski, J.E.; Bradley, T.G.; Bosio, J.A.; Liu, D. Dimensional changes of upper airway after rapid maxillary expansion: A prospective cone-beam computed tomography study. *Am. J. Orthod. Dentofac. Orthop.* **2013**, *143*, 462–470. [CrossRef]

25. Alsufyani, N.A.; Noga, M.L.; Witmans, M.; Cheng, I.; El-Hakim, H.; Major, P.W. Using cone beam CT to assess the upper airway after surgery in children with sleep disordered breathing symptoms and maxillary-mandibular disproportions: A clinical pilot. *J. Otolaryngol. -Head Neck Surg.* **2017**, *46*, 31. [CrossRef] [PubMed]

26. Chang, C.-S.; Wallace, C.G.; Hsiao, Y.-C.; Hsieh, Y.-J.; Wang, Y.-C.; Chen, N.-H.; Liao, Y.-F.; Liou, E.J.-W.; Chen, P.K.-T.; Chen, J.-P. Airway changes after cleft orthognathic surgery evaluated by three-dimensional computed tomography and overnight polysomnographic study. *Sci. Rep.* **2017**, *7*, 1–9. [CrossRef]

27. Susarla, S.M.; Abramson, Z.R.; Dodson, T.B.; Kaban, L.B. Upper airway length decreases after maxillomandibular advancement in patients with obstructive sleep apnea. *J. Oral Maxillofac. Surg.* **2011**, *69*, 2872–2878. [CrossRef]

28. Butterfield, K.J.; Marks, P.L.; McLean, L.; Newton, J. Linear and volumetric airway changes after maxillomandibular advancement for obstructive sleep apnea. *J. Oral Maxillofac. Surg.* **2015**, *73*, 1133–1142. [CrossRef]

29. Ip, M.; Tan, K.; Peh, W.; Lam, K. Effect of Sandostatin® LAR® on sleep apnoea in acromegaly: Correlation with computerized tomographic cephalometry and hormonal activity. *Clin. Endocrinol.* **2001**, *55*, 477–483. [CrossRef]

30. Ito, Y.; Cheng, G.C.; Shih, A.M.; Koomullil, R.P.; Soni, B.K.; Sittitavornwong, S.; Waite, P.D. Patient-specific geometry modeling and mesh generation for simulating obstructive sleep apnea syndrome cases by maxillomandibular advancement. *Math. Comput. Simul.* **2011**, *81*, 1876–1891. [CrossRef]

31. Premaraj, T.S.; Ju, S.; Premaraj, S.; Kim, S.K.; Gu, L. Computational fluid dynamics modeling of pharyngeal airway resistance based on cone-beam computed tomography. *J. Mech. Med. Biol.* **2019**, *19*, 1950045. [CrossRef]

32. Hsieh, Y.-J.; Liao, Y.-F.; Chen, N.-H.; Chen, Y.-R. Changes in the calibre of the upper airway and the surrounding structures after maxillomandibular advancement for obstructive sleep apnoea. *Br. J. Oral Maxillofac. Surg.* **2014**, *52*, 445–451. [CrossRef] [PubMed]

33. Chang, J.E.; Min, S.W.; Kim, C.S.; Kwon, Y.S.; Hwang, J.Y. Effects of the jaw-thrust manoeuvre in the semi-sitting position on securing a clear airway during fibreoptic intubation. *Anaesthesia* **2015**, *70*, 933–938. [CrossRef] [PubMed]

34. Butterfield, K.J.; Marks, P.L.; McLean, L.; Newton, J. Pharyngeal airway morphology in healthy individuals and in obstructive sleep apnea patients treated with maxillomandibular advancement: A comparative study. *Oral Surg. Oral Med. Oral Pathol. Oral Radiol.* **2015**, *119*, 285–292. [CrossRef] [PubMed]

35. Souza, F.J.F.d.B.; Evangelista, A.R.; Silva, J.V.; Périco, G.V.; Madeira, K. Cervical computed tomography in patients with obstructive sleep apnea: Influence of head elevation on the assessment of upper airway volume. *J. Bras. Pneumol.* **2016**, *42*, 55–60. [CrossRef] [PubMed]

36. Li, K.K.; Guilleminault, C.; Riley, R.W.; Powell, N.B. Obstructive sleep apnea and maxillomandibular advancement: An assessment of airway changes using radiographic and nasopharyngoscopic examinations. *J. Oral Maxillofac. Surg.* **2002**, *60*, 526–530. [CrossRef]

37. Lin, C.H.; Liao, Y.F.; Chen, N.H.; Lo, L.J.; Chen, Y.R. Three-dimensional computed tomography in obstructive sleep apneics treated by maxillomandibular advancement. *Laryngoscope* **2011**, *121*, 1336–1347. [CrossRef]

38. Lee, W.H.; Hong, S.-N.; Kim, H.J.; Rhee, C.-S.; Lee, C.H.; Yoon, I.-Y.; Kim, J.-W. A comparison of different success definitions in non-continuous positive airway pressure treatment for obstructive sleep apnea using cardiopulmonary coupling. *J. Clin. Sleep Med.* **2016**, *12*, 35–41. [CrossRef]

39. Ristow, O.; Rückschloß, T.; Berger, M.; Grötz, T.; Kargus, S.; Krisam, J.; Seeberger, R.; Engel, M.; Hoffmann, J.; Freudlsperger, C. Short-and long-term changes of the pharyngeal airway after surgical mandibular advancement in class II patients—A three-dimensional retrospective study. *J. Cranio-Maxillofac. Surg.* **2018**, *46*, 56–62. [CrossRef]

40. Gottsauner-Wolf, S.; Laimer, J.; Bruckmoser, E. Posterior airway changes following orthognathic surgery in obstructive sleep apnea. *J. Oral Maxillofac. Surg.* **2018**, *76*, 1093.e1–1093.e21. [CrossRef]

41. Slaats, M.A.; Van Hoorenbeeck, K.; Van Eyck, A.; Vos, W.G.; De Backer, J.W.; Boudewyns, A.; De Backer, W.; Verhulst, S.L. Upper airway imaging in pediatric obstructive sleep apnea syndrome. *Sleep Med. Rev.* **2015**, *21*, 59–71. [CrossRef] [PubMed]

42. Mylavarapu, G.; Mihaescu, M.; Fuchs, L.; Papatziamos, G.; Gutmark, E. Planning human upper airway surgery using computational fluid dynamics. *J. Biomech.* **2013**, *46*, 1979–1986. [CrossRef] [PubMed]

43. Zhao, M.; Barber, T.; Cistulli, P.; Sutherland, K.; Rosengarten, G. Computational fluid dynamics for the assessment of upper airway response to oral appliance treatment in obstructive sleep apnea. *J. Biomech.* **2013**, *46*, 142–150. [CrossRef] [PubMed]

44. Achilles, N.; Pasch, N.; Lintermann, A.; Schröder, W.; Mösges, R. Computational fluid dynamics: A suitable assessment tool for demonstrating the antiobstructive effect of drugs in the therapy of allergic rhinitis. *Acta Otorhinolaryngol. Ital.* **2013**, *33*, 36–42.

45. Heenan, A.M.; Pollard, E.; Finlay, A.W. Experimental measurements and computational modeling of the flow field in a idealized human oropharynx. *Exp. Fluids* **2003**, *35*, 70–84. [CrossRef]

46. Collins, T.; Tabor, G.; Young, P. A computational fluid dynamics study of inspiratory flow in orotracheal geometries. *Med. Biol. Eng. Comput.* **2007**, *45*, 829–836. [CrossRef]
47. Barrera, J.E. Virtual surgical planning improves surgical outcome measures in obstructive sleep apnea surgery. *Laryngoscope* **2014**, *124*, 1259–1266. [CrossRef]
48. Sutherland, K.; Chan, A.S.; Cistulli, P.A. Three-dimensional assessment of anatomical balance and oral appliance treatment outcome in obstructive sleep apnoea. *Sleep Breath.* **2016**, *20*, 903–910. [CrossRef]
49. Schneider, D.; Kämmerer, P.W.; Schön, G.; Bschorer, R. A three-dimensional comparison of the pharyngeal airway after mandibular distraction osteogenesis and bilateral sagittal split osteotomy. *J. Cranio-Maxillofac. Surg.* **2015**, *43*, 1632–1637. [CrossRef]
50. Spinelli, G.; Agostini, T.; Arcuri, F.; Conti, M.; Raffaini, M. Three-dimensional airways reconstruction in syndromic pedriatric patients following mandibular distraction osteogenesis. *J. Craniofacial Surg.* **2015**, *26*, 650–654. [CrossRef]
51. Xu, C.; Sin, S.; McDonough, J.M.; Udupa, J.K.; Guez, A.; Arens, R.; Wootton, D.M. Computational fluid dynamics modeling of the upper airway of children with obstructive sleep apnea syndrome in steady flow. *J. Biomech.* **2006**, *39*, 2043–2054. [CrossRef] [PubMed]
52. Huynh, J.; Kim, K.B.; McQuilling, M. Pharyngeal airflow analysis in obstructive sleep apnea patients pre-and post-maxillomandibular advancement surgery. *J. Fluids Eng.* **2009**, *131*, 091101. [CrossRef]
53. Kita, S.; Oshima, M.; Shimazaki, K.; Iwai, T.; Omura, S.; Ono, T. Computational Fluid Dynamic Study of Nasal Respiratory Function Before and After Bimaxillary Orthognathic Surgery With Bone Trimming at the Inferior Edge of the Pyriform Aperture. *J. Oral. Maxil. Surg.* **2016**, *74*, 2241–2251. [CrossRef] [PubMed]
54. Fan, Y.; Cheung, L.; Chong, M.; Chow, K.; Liu, C. Computational study on obstructive sleep apnea syndrome using patient–specific Models. In Proceedings of the World Congress on Engineering, London, UK, 6–8 July 2011.
55. Zhao, M.; Barber, T.; Cistulli, P.A.; Sutherland, K.; Rosengarten, G. Simulation of upper airway occlusion without and with mandibular advancement in obstructive sleep apnea using fluid-structure interaction. *J. Biomech.* **2013**, *46*, 2586–2592. [CrossRef]
56. Persak, S.C.; Sin, S.; McDonough, J.M.; Arens, R.; Wootton, D.M. Noninvasive estimation of pharyngeal airway resistance and compliance in children based on volume-gated dynamic MRI and computational fluid dynamics. *J. Appl. Physiol.* **2011**, *111*, 1819–1827. [CrossRef]
57. Shah, D.H.; Kim, K.B.; McQuilling, M.W.; Movahed, R.; Shah, A.H.; Kim, Y.I. Computational fluid dynamics for the assessment of upper airway changes in skeletal Class III patients treated with mandibular setback surgery. *Angle Orthod.* **2016**, *86*, 976–982. [CrossRef]
58. Calmet, H.; Gambaruto, A.M.; Bates, A.J.; Vázquez, M.; Houzeaux, G.; Doorly, D.J. Large-scale CFD simulations of the transitional and turbulent regime for the large human airways during rapid inhalation. *Comput. Biol. Med.* **2016**, *69*, 166–180. [CrossRef]
59. Cheng, G.C.; Koomullil, R.P.; Ito, Y.; Shih, A.M.; Sittitavornwong, S.; Waite, P.D. Assessment of surgical effects on patients with obstructive sleep apnea syndrome using computational fluid dynamics simulations. *Math. Comput. Simul.* **2014**, *106*, 44–59. [CrossRef]
60. Liu, X.; Yan, W.; Liu, Y.; Choy, Y.S.; Wei, Y. Numerical investigation of flow characteristics in the obstructed realistic human upper airway. *Comput. Math. Methods Med.* **2016**, *2016*, 3181654. [CrossRef]
61. Kim, T.; Kim, H.-H.; ok Hong, S.; Baek, S.-H.; Kim, K.-W.; Suh, S.-H.; Choi, J.-Y. Change in the upper airway of patients with obstructive sleep apnea syndrome using computational fluid dynamics analysis: Conventional maxillomandibular advancement versus modified maxillomandibular advancement with anterior segmental setback osteotomy. *J. Craniofacial Surg.* **2015**, *26*, e765–e770. [CrossRef]
62. Liu, S.Y.-C.; Huon, L.-K.; Iwasaki, T.; Yoon, A.; Riley, R.; Powell, N.; Torre, C.; Capasso, R. Efficacy of maxillomandibular advancement examined with drug-induced sleep endoscopy and computational fluid dynamics airflow modeling. *Otolaryngol. -Head Neck Surg.* **2016**, *154*, 189–195. [CrossRef] [PubMed]
63. Srivastav, V.K.; Paul, A.R.; Jain, A. Capturing the wall turbulence in CFD simulation of human respiratory tract. *Math. Comput. Simul.* **2019**, *160*, 23–38. [CrossRef]
64. De Backer, J.; Vos, W.; Gorlé, C.; Germonpré, P.; Partoens, B.; Wuyts, F.; Parizel, P.M.; De Backer, W. Flow analyses in the lower airways: Patient-specific model and boundary conditions. *Med. Eng. Phys.* **2008**, *30*, 872–879. [CrossRef] [PubMed]
65. Patel, T.R.; Li, C.; Krebs, J.; Zhao, K.; Malhotra, P. Modeling congenital nasal pyriform aperture stenosis using computational fluid dynamics. *Int. J. Pediatr. Otorhinolaryngol* **2018**, *109*, 180–184. [CrossRef] [PubMed]
66. Sul, B.; Wallqvist, A.; Morris, M.J.; Reifman, J.; Rakesh, V. A computational study of the respiratory airflow characteristics in normal and obstructed human airways. *Comput. Biol. Med.* **2014**, *52*, 130–143. [CrossRef] [PubMed]
67. Chang, K.K.; Kim, K.B.; McQuilling, M.W.; Movahed, R. Fluid structure interaction simulations of the upper airway in obstructive sleep apnea patients before and after maxillomandibular advancement surgery. *Am. J. Orthod. Dentofac. Orthop.* **2018**, *153*, 895–904. [CrossRef]
68. Habbeche, A.; Saoudi, B.; Jaouadi, B.; Haberra, S.; Kerouaz, B.; Boudelaa, M.; Badis, A.; Ladjama, A. Purification and biochemical characterization of a detergent-stable keratinase from a newly thermophilic actinomycete Actinomadura keratinilytica strain Cpt29 isolated from poultry compost. *J. Biosci. Bioeng.* **2014**, *117*, 413–421. [CrossRef]
69. Bates, A.J.; Schuh, A.; McConnell, K.; Williams, B.M.; Lanier, J.M.; Willmering, M.M.; Woods, J.C.; Fleck, R.J.; Dumoulin, C.L.; Amin, R.S. A novel method to generate dynamic boundary conditions for airway CFD by mapping upper airway movement with non-rigid registration of dynamic and static MRI. *Int. J. Numer. Methods Biomed. Eng.* **2018**, *34*, e3144. [CrossRef]

70. Luo, H.; Sin, S.; McDonough, J.M.; Isasi, C.R.; Arens, R.; Wootton, D.M. Computational fluid dynamics endpoints for assessment of adenotonsillectomy outcome in obese children with obstructive sleep apnea syndrome. *J. Biomech.* **2014**, *47*, 2498–2503. [CrossRef]
71. Zubair, M.; Riazuddin, V.N.; Abdullah, M.Z.; Ismail, R.; Shuaib, I.L.; Ahmad, K.A. Computational fluid dynamics study of the effect of posture on airflow characteristics inside the nasal cavity. *Asian Biomed.* **2013**, *7*, 835–840. [CrossRef]
72. Iwasaki, T.; Sato, H.; Suga, H.; Minami, A.; Yamamoto, Y.; Takemoto, Y.; Inada, E.; Saitoh, I.; Kakuno, E.; Kanomi, R. Herbst appliance effects on pharyngeal airway ventilation evaluated using computational fluid dynamics. *Angle Orthod.* **2017**, *87*, 397–403. [CrossRef]
73. Rahimi-Gorji, M.; Gorji, T.B.; Gorji-Bandpy, M. Details of regional particle deposition and airflow structures in a realistic model of human tracheobronchial airways: Two-phase flow simulation. *Comput. Biol. Med.* **2016**, *74*, 1–17. [CrossRef] [PubMed]
74. Faizal, W.; Ghazali, N.N.N.; Khor, C.; Zainon, M.Z.; Badruddin, I.A.; Kamangar, S.; Ibrahim, N.B.; Razi, R.M. Computational analysis of airflow in upper airway under light and heavy breathing conditions for a realistic patient having obstructive sleep apnea. *Comput. Model. Eng. Sci.* **2021**, *128*, 583–604. [CrossRef]
75. Zheng, Z.; Liu, H.; Xu, Q.; Wu, W.; Du, L.; Chen, H.; Zhang, Y.; Liu, D. Computational fluid dynamics simulation of the upper airway response to large incisor retraction in adult class I bimaxillary protrusion patients. *Sci. Rep.* **2017**, *7*, 45706. [CrossRef] [PubMed]
76. Amatoury, J.; Cheng, S.; Kairaitis, K.; Wheatley, J.R.; Amis, T.C.; Bilston, L.E. Development and validation of a computational finite element model of the rabbit upper airway: Simulations of mandibular advancement and tracheal displacement. *J. Appl. Physiol.* **2016**, *120*, 743–757. [CrossRef]
77. Xi, J.; April Si, X.; Dong, H.; Zhong, H. Effects of glottis motion on airflow and energy expenditure in a human upper airway model. *Eur. J. Mech. B/Fluids* **2018**, *72*, 23–37. [CrossRef]
78. Suga, H.; Iwasaki, T.; Mishima, K.; Nakano, H.; Ueyama, Y.; Yamasaki, Y. Evaluation of the effect of oral appliance treatment on upper-airway ventilation conditions in obstructive sleep apnea using computational fluid dynamics. *Cranio* **2019**, *39*, 209–217. [CrossRef]
79. Gutmark, E.J.; Wootton, D.M.; Sin, S.; Wagshul, M.E.; Subramaniam, D.R.; Arens, R. Biomechanics of the soft-palate in sleep apnea patients with polycystic ovarian syndrome. *J. Biomech.* **2018**, *76*, 8–15. [CrossRef]
80. Fletcher, A.; Choi, J.; Awadalla, M.; Potash, A.E.; Wallen, T.J.; Fletcher, S.; Chang, E.H. The effect of genioglossal advancement on airway flow using a computational flow dynamics model. *Laryngoscope* **2013**, *123*, 3227–3232. [CrossRef]
81. Powell, N.B.; Mihaescu, M.; Mylavarapu, G.; Weaver, E.M.; Guilleminault, C.; Gutmark, E. Patterns in pharyngeal airflow associated with sleep-disordered breathing. *Sleep Med.* **2011**, *12*, 966–974. [CrossRef]
82. Zhao, M.Y.; Barber, T.; Cistulli, P.; Sutherland, K.; Rosengarten, G. Predicting the treatment response of oral appliances for obstructive sleep apnea using computational fluid dynamics and fluid-structure interaction simulations. In Proceedings of the Asme International Mechanical Engineering Congress and Exposition, San Diego, CA, USA, 15–21 November 2013.
83. Mihaescu, M.; Mylavarapu, G.; Gutmark, E.J.; Powell, N.B. Large eddy simulation of the pharyngeal airflow associated with obstructive sleep apnea syndrome at pre and post-surgical treatment. *J. Biomech.* **2011**, *44*, 2221–2228. [CrossRef] [PubMed]
84. Zhao, M.; Barber, T.; Cistulli, P.; Sutherland, K.; Rosengarten, G. Computational fluid dynamic study of upper airway flow to predict the success of oral appliances in treating sleep apnea. In Proceedings of the 17th Australasian Fluid Mechanics Conference, Auckland, New Zealand, 5–9 December 2010; pp. 579–582.
85. Paxman, T.; Noga, M.; Finlay, W.H.; Martin, A.R. Experimental evaluation of pressure drop for flows of air and heliox through upper and central conducting airway replicas of 4-to 8-year-old children. *J. Biomech.* **2019**, *82*, 134–141. [CrossRef] [PubMed]
86. Stapleton, K.W.; Guentsch, E.; Hoskinson, M.K.; Finlay, W.H. On the suitability of k-ε turbulence modeling for aerosol deposition in the mouth and throat: A comparison with experiment. *J. Aerosol Sci.* **2000**, *31*, 739–749. [CrossRef]
87. Xu, C.; Brennick, M.J.; Dougherty, L.; Wootton, D.M. Modeling upper airway collapse by a finite element model with regional tissue properties. *Med. Eng. Phys.* **2009**, *31*, 1343–1348. [CrossRef]
88. De Backer, J.W.; Vanderveken, O.M.; Vos, W.G.; Devolder, A.; Verhulst, S.L.; Verbraecken, J.A.; Parizel, P.M.; Braem, M.J.; de Heyning, P.H.V.; De Backer, W.A. Functional imaging using computational fluid dynamics to predict treatment success of mandibular advancement devices in sleep-disordered breathing. *J. Biomech.* **2007**, *40*, 3708–3714. [CrossRef]

Article

The Correlation between Bone Density and Mechanical Variables in Bone Remodelling Models: Insights from a Case Study Corresponding to the Femur of a Healthy Adult

José Luis Calvo-Gallego [1,*], Fernando Gutiérrez-Millán [1], Joaquín Ojeda [1], María Ángeles Pérez [2] and Javier Martínez-Reina [1,*]

1 Departmento de Ingeniería Mecánica y Fabricación, Universidad de Sevilla, 41092 Seville, Spain
2 Multiscale in Mechanical and Biological Engineering, Instituto de Investigación en Ingeniería de Aragón (I3A), Instituto de Investigación Sanitaria Aragón (IIS Aragón), University of Zaragoza, 50018 Zaragoza, Spain
* Correspondence: joselucalvo@us.es (J.L.C.-G.); jmreina@us.es (J.M.-R.)

Abstract: Bone remodelling models (BRM) are often used to estimate the density distribution in bones from the loads they are subjected to. BRM define a relationship between a certain variable measuring the mechanical stimulus at each bone site and either the local density or the local variation of density. This agrees with the Mechanostat Theory, which establishes that overloaded bones increase their density, while disused bones tend to decrease their density. Many variables have been proposed as mechanical stimuli, with stress or strain energy density (SED) being some of the most common. Yet, no compelling reason has been given to justify the choice of any of these variables. This work proposes a set of variables derived from the local stress and strain tensors as candidates for mechanical stimuli; then, this work correlates them to the density in the femur of one individual. The stress and strain tensors were obtained from a FE model and the density was obtained from a CT-scan, both belonging to the same individual. The variables that best correlate with density are the stresses. Strains are quite uniform across the femur and very poorly correlated with density, as is the SED, which is, therefore, not a good variable to measure the mechanical stimulus.

Keywords: bone remodelling; mechanical stimulus; correlation; bone density distribution; strain energy density; absolute maximum principal stress; fluctuation of stresses

MSC: 92-10

Citation: Calvo-Gallego, J.L.; Gutiérrez-Millán, F.; Ojeda, J.; Pérez, M.Á.; Martínez-Reina, J. The Correlation between Bone Density and Mechanical Variables in Bone Remodelling Models: Insights from a Case Study Corresponding to the Femur of a Healthy Adult. *Mathematics* **2022**, *10*, 3367. https://doi.org/10.3390/math10183367

Academic Editor: Mauro Malvè

Received: 21 August 2022
Accepted: 14 September 2022
Published: 16 September 2022

Publisher's Note: MDPI stays neutral with regard to jurisdictional claims in published maps and institutional affiliations.

1. Introduction

Bone is a living tissue that can adapt its apparent density and internal microstructure (through the process called bone remodelling), and its shape and external dimensions (through bone modelling) as a response to different mechanical and biological stimuli. Regarding the former, it has been hypothesized that one of the goals of bone remodelling is to maintain bone as an optimal structure that supports the loads with the minimum weight [1]. Thus, bone density, and consequently, stiffness, are high in overloaded regions and low in regions with a low stress level. In other words, there is a direct relationship between density and stresses. Many bone remodelling models (BRM) have been proposed in the literature to quantify this relationship [1–5].

These BRM have been very often used to estimate the density distribution in bones from the loads they are subjected to, mainly in the human femur [6,7], but also in other bones such as the mandible [8]. This problem, that we will name here *Density Prediction Through Bone Remodelling* (DPTBR), is usually approached through the following iterative process:

1. Assign an initial uniform density distribution to the bone under study.
2. Apply the loads and boundary conditions to a Finite Element (FE) model of the bone and calculate the stresses/strains at every point of the mesh.

3. Apply the BRM to relate the stress/strain state with the new density [9] or with the change in density [1–5], and then update the density at each point.
4. Update the stiffness tensor at each point based on the new density. Go back to step 2 to start a new iteration.

This iterative process is repeated until convergence of the density distribution is achieved, which usually resembles the real one with good accuracy. The BRM can be applied following two approaches, as mentioned above in step 3. In an evolutionary model, the stress/strain state determines the local change in apparent density, $\dot{\rho}$:

$$\dot{\rho} = f(S) \qquad \text{Evolutionary model} \tag{1}$$

where S is a certain mechanical stimulus related to the stress/strain state. This approach is based on the Mechanostat Theory [10], which is behind most of the BRM. This theory hypothesizes that bone adapts itself to overloads by increasing its apparent density and adapts to disuse states by decreasing it. Overload and disuse are defined in the Mechanostat Theory by certain strain ranges. The theory also establishes the existence of a so-called "lazy zone", a strain range between disuse and overload, for which no evident change in apparent density is observed or, at least, it is not significant. Implementing this approach in DPTBR problems has a major drawback, as the uniqueness of the solution is not guaranteed. This path dependence occurs due to the implementation of the lazy zone [11], and the final density distribution depends on the initially assumed distribution.

Recently, we have proposed a non-evolutionary strategy to solve DPTBR problems [9]. Instead of calculating the rate of change in density as a function of the stimulus, we calculated a priori a relationship between the stimulus S and the density achieved at the equilibrium (see Equation (2)) and used that relationship to assign the density to an element as a function of the local stimulus. Since this stimulus can, in turn, change with density (through the stiffness), an iterative process is still required, but the number of iterations needed to achieve convergence is much lower. Notwithstanding, the uniqueness of the solution is not guaranteed in this case. The only advantage of this approach is its higher speed of convergence.

$$\rho = g(S) \qquad \text{Non-evolutionary model.} \tag{2}$$

On the other hand, the non-evolutionary approach is not suitable to predict the changes in density in a bone subjected to changes in its biomechanical environment. The evolutionary approach is preferable in this case, as it allows a real-time simulation of those changes. However, to this end, it is very important to start from a realistic initial density distribution, in order to avoid the path dependence of the solution previously mentioned.

The stimulus S is the variable that drives bone remodelling and must comprise biological and mechanical factors. Focusing on the latter, the amount of damage has been used as the stimulus (or a part of it) in targeted bone remodelling models [12–15]. This is based on the hypothesis that one goal of bone remodelling is to repair the microstructural damage accumulated in the bone matrix by daily activity. Other models simply apply the Mechanostat Theory to account for disuse and overload and the influence of these states on bone adaptation. To this end, these models have used a stimulus that measures the intensity of the loads, with the strain energy density (SED) being the most commonly used variable to account for that intensity (see [1–5], among many others). However, in the original Mechanostat Theory, the disuse, lazy zone and overload states were defined in terms of strain; thus, establishing the hypothesis that strain is the magnitude driving the bone response. In such case, after a change in load that could alter the level of strain, the bone microstructure (and consequently, stiffness) must be regulated to return to the homeostatic situation [12].

In this work, we will focus on the mechanical part of the stimulus—the main objective being to study which is the best variable among a series of candidates to account for the mechanical feedback in bone remodelling models. We will discuss which is the best mechanical stimulus, or equivalently, which is the best predictor of bone density, either for an evolutionary or a non-evolutionary approach. For this purpose, we have analysed the stress and strain distributions in a human femur using a FE model and the loads extracted from a gait analysis performed on the same individual, from whom we have also estimated the bone density distribution from a CT scan. Several mechanical variables calculated from the stress and strain tensors throughout the gait cycle have been proposed as candidates for the mechanical stimulus (or predictors) and will be correlated with the density distribution.

This paper is organised as follows: In Section 2.1, we provide a description of the gait analysis performed to estimate the loads acting on the femur during walking. In Section 2.2, we describe the image analysis of CT scans of a human femur, which were performed to obtain its bone density distribution. In Section 2.3, we briefly describe the methodology to estimate the stiffness tensor at every point of the femur based on the density distribution. In that section, we focus on the case that considers bone as an isotropic material, while through Sections 2.4–2.6, we develop the methodology to estimate that stiffness tensor when bone is considered anisotropic. In Section 2.7, we briefly describe how the gait cycle was simulated with a FE model. The correlation coefficients used to evaluate the relationship between bone density and the variables used as candidates for mechanical stimuli (or predictors) are defined in Section 2.8. In Section 2.9, we define those predictors. In Section 3, we provide the correlations between the bone apparent density and the proposed predictors. We discuss the results of these correlations in Section 4 and highlight the conclusions of this study in Section 5. Finally, the Appendices contain a more detailed description of the procedures used in the Section 2.

2. Materials and Methods

The procedure followed here is summarised in Figure 1 and is explained as follows: First, a gait analysis was performed to estimate the forces exerted by those muscles inserted in the femur and joint reactions at the knee and the hip in the subject under study. A CT scan of the subject's femur was taken and the greyscale value was related to the bone apparent density using a linear relationship. With this, a FE model of the femur was built and the loads, estimated through a gait analysis performed on the same subject, were applied to the FE model in order to obtain the distribution of stresses and strains throughout the gait cycle. A set of variables, defined in Sections 2.7 and 2.9, was assessed from the temporal evolutions of stresses and strains. Finally, the bone apparent density estimated from the CT scan was correlated with these variables.

2.1. Gait Analysis and Subject Data

The gait analysis was performed at the Motion Analysis Laboratory of the Department of Mechanical Engineering and Manufacturing of the Universidad de Sevilla. A Vicon®system of 12 infra-red cameras was used to record motion at a frequency of 100 Hz along with 2 AMTI force platforms to record the ground reaction forces at a frequency of 1000 Hz. The marker placement protocol employed was the modified Cleveland protocol [16].

The subject under study was an adult male of 27 years old, with no reported pathologies, 1.85 m tall and weighing 75 kg, who walked at a freely chosen forward speed. The participant signed an informed consent form prior to the recording of the measurements. The study protocol was approved by a medical ethics committee through the Andalusian Biomedical Research Ethics Platform (approval number 20151012181252).

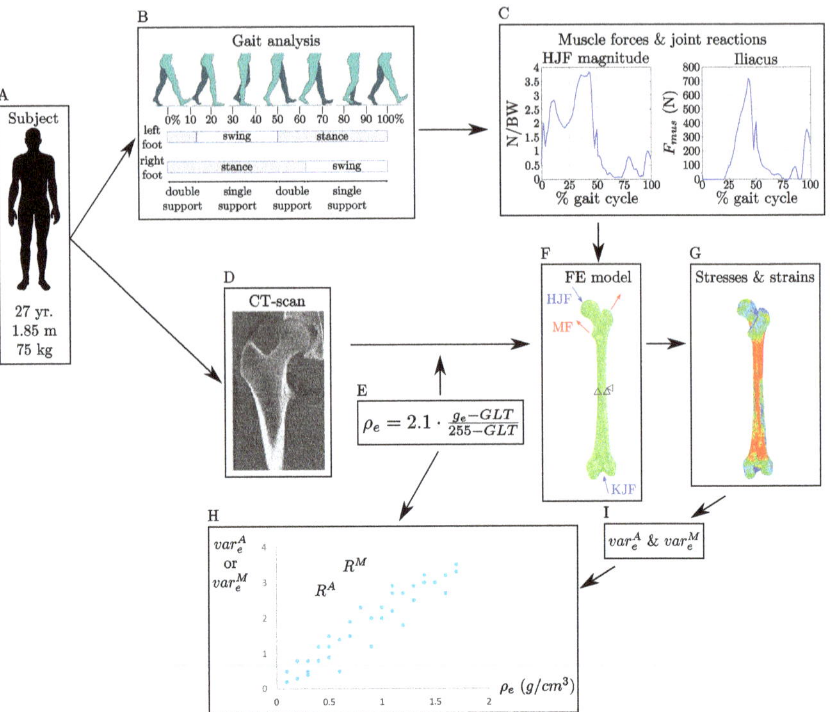

Figure 1. The summary of the procedure followed in this work. One male subject was selected for the study (**A**). A CT scan (**D**) was used to build a FE model of the subject's femur (**F**). The gait analysis (**B**) perfomed on the same subject was used (**C**) to estimate the muscle forces (MF) and joint reaction forces (HJF: hip joint force, KJF: knee joint force) applied to the FE model, together with isostatic boundary conditions. The mechanical properties of the bone were estimated from the density, which was estimated, in turn, from the CT scan (**E**). The FE simulation of the gait cycle yielded the temporal evolution of stresses and strains (**G**) throughout the cycle. A set of variables (**I**) were derived from those temporal evolutions. Finally, these variables were correlated (**H**) to the bone apparent density in order to evaluate a good predictor of density for bone remodelling models.

The recorded experimental data were processed using OpenSim software [17]. The biomechanical model implemented was the *Gait2392*, available in the OpenSim library. This model was developed to perform 3D gait analysis and consists of 23 degrees of freedom and 92 actuators that simulate the action of muscle forces. However, for the sake of simplicity, the subtalar and metatarsophalangeal joints were blocked in this work, so that the foot was considered a single biomechanical segment. The model was scaled to fit the subject's morphology from the recorded experimental data. The mass ratio of each segment was assumed constant during scaling. In this model, muscle attachment points were placed where OpenSim locate them by default, using numerical approximation [18] of cadavers' data [19,20].

The forces applied to the FE model were obtained in two steps. First, the inverse dynamic problem was solved using the kinetics experimental data recorded in the laboratory. From these results, the time evolutions of the muscle forces were calculated solving a force-sharing problem through a static optimisation algorithm that considered the dynamics of muscle contraction and activation [17,21]. The results were collected for those muscles inserted in the femur and are provided in the Supplementary Materials. Additionally, joint reaction forces at the hip and knee were estimated using the algorithm proposed by Steele et al. [22]. A detailed description of this procedure is included in Appendix B.

2.2. CT Scan of the Femur, FE Model and Density Distribution

A CT scanner (LightSpeed16, GE Medical Systems, Milwaukee, WI, USA; 120 kVp, reconstructed with bone Plus kernel, 1.25 mm slice thickness) was used to estimate bone density on the right femur of the same individual on which the gait analysis was conducted. The software Abaqus (version 2020, SIMULIA, Dassault Systémes, Madrid, España) was used to run FE analysis. The 3D geometry of the femur was reconstructed from the CT scan and meshed with 4-node linear tetrahedral elements (C3D4 in Abaqus element library). The final mesh consisted in 339,168 elements and 64,757 nodes. In order to check the convergence of the FE solution, we compared the results with a different mesh, built with 10-node quadratic tetrahedral elements (C3D10). This new mesh was obtained from the previous one by simply placing mid-side nodes in the edges of the former elements, resulting—obviously—in the same number of elements and 480,480 nodes.

To assign a value of density to each element of the mesh, a linear relationship between the greyscale value and the bone apparent density was used, as in [23–26]:

$$\rho_e = A + B \cdot g_e, \tag{3}$$

where g_e and ρ_e are the greyscale value and the density estimated for element e, respectively. Two pairs of greyscale–density values are needed to define the linear relationship (constants A and B). These points are usually obtained through calibration of the CT scan. However, this calibration was not available in our case, and therefore, we needed to make two assumptions. Thus, as the first pair, an apparent density value of 2.1 g/cm^3 [4] was assigned to the maximum greyscale value of the CT scan (255), i.e., to the densest cortical bone. The second point chosen was that corresponding to a null bone apparent density, which is sometimes called a grey level (or greyscale) threshold (GLT). It can be seen in Equation (1) that $\rho_e = 0$ for $g_e = GLT$. This lower limit is usually made to correspond to bone marrow, which was assumed here to be placed inside the diaphyseal canal. However, the greyscale value inside the canal varied in the range 70–90, thus, making it impossible to assign a unique and reliable value to GLT. For this reason, three cases were studied, assuming different values for GLT, namely: 70, 80 and 90. This uncertainty in the choice of GLT affects the slope of the linear relationship, which is then:

$$\rho_e = 2.1 \cdot \frac{g_e - GLT}{255 - GLT}. \tag{4}$$

The greyscale distribution was mapped from the CT scan to the FE mesh using the software *Bonemat*, and Equation (3) was used to convert the greyscale value into bone apparent density. Those elements with a greyscale value below GLT were assigned a very small density (0.001 g/cm^3), thus, yielding a negligible—although, not null—stiffness, something necessary to avoid convergence problems. Figure 2 shows the distributions of density obtained for the three GLT. It can be seen that those three GLT led to distributions of the estimated density which are similar, in general, with the exception of the proximal region and the thickness of the cortical layer in the diaphysis. As GLT rises, the volume identified as marrow increases, and this makes GLT = 90 produce a thinner cortical layer and a slightly underestimated density in the proximal region. Nonetheless, it can be noted in Figure 2 that the uncertainty introduced in the density distribution by GLT is not too important, in the range of 70–90. Moreover, we will show later that the conclusions of this study are the same regardless of the choice of GLT.

The external boundary of the bone was not perfectly defined in some slices of the CT scan where the cortex was too thin, since the average greyscale value between the background and the periosteum produced a cortex of intermediate density. For this reason, the model was covered with a layer of shell elements to simulate the cortex, with a thickness of 1 mm, a Young's modulus equal to 19 GPa [27] and a Poisson's ratio $\nu = 0.32$ [28].

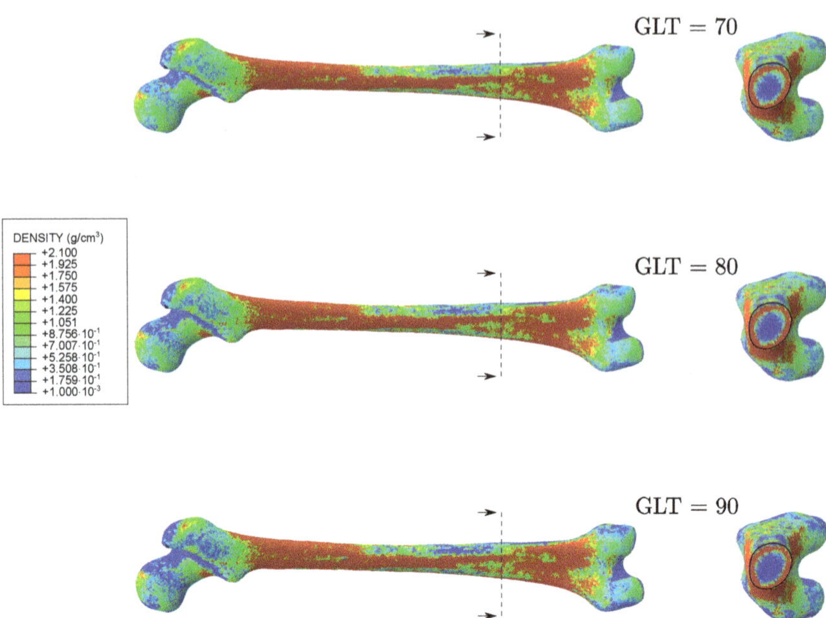

Figure 2. Distribution of bone density estimated from Equation (3) for three GLT: in the whole femur (**left**), in a diaphyseal cross-section (**right**).

2.3. Estimation of the Stiffness Tensor

Isotropic and anisotropic models were considered. In the former, the stiffness tensor of bone at each material point can be estimated through the density obtained from the CT scan, by using one of the numerous correlations between the elastic constants and the apparent density that can be found in the literature. For example, Jacobs proposed the following [28]:

$$E_{\text{Jacobs}}\,(\text{MPa}) = \begin{cases} 2014\,\rho^{2.5} & \text{if } \rho \leq 1.2\,\text{g/cm}^3 \\ 1763\,\rho^{3.2} & \text{if } \rho > 1.2\,\text{g/cm}^3 \end{cases} \tag{5a}$$

$$\nu = \begin{cases} 0.2 & \text{if } \rho \leq 1.2\,\text{g/cm}^3 \\ 0.32 & \text{if } \rho > 1.2\,\text{g/cm}^3 \end{cases}. \tag{5b}$$

Hernandez et al. [29] proposed another correlation for E, based on the ash fraction, α (a variable used to measure the mineral content of bone tissue), and on the bone volume fraction (or bone volume per total volume), BV/TV. If we express $BV/TV = \frac{\rho}{\hat{\rho}}$, with ρ and $\hat{\rho}$ being the apparent density and the tissue density, respectively, then the correlation proposed by Hernandez et al. reads:

$$E(\text{MPa}) = 84{,}370 \left(\frac{\rho}{\hat{\rho}}\right)^{2.58} \alpha^{2.74}. \tag{6}$$

If typical values of ash fraction $\alpha = 0.68$ and tissue density $\hat{\rho} = 2.31\,\text{g/cm}^3$ [29] are used, Equation (6) becomes:

$$E_{\text{Hernandez}}(\text{MPa}) = 3388\,\rho^{2.58}, \tag{7}$$

which produces an estimation of the Young's modulus up to 70% higher than Equation (5a) depending on the density (see Figure 3). Each correlation was used in the corresponding isotropic model, named, respectively, *IsoJ* (Equation (5a)) and *IsoH* (Equation (7)) after Jacobs and Hernandez. A constant Poisson's ratio $\nu = 0.3$ was used in conjunction with Equation (7) in model *IsoH*.

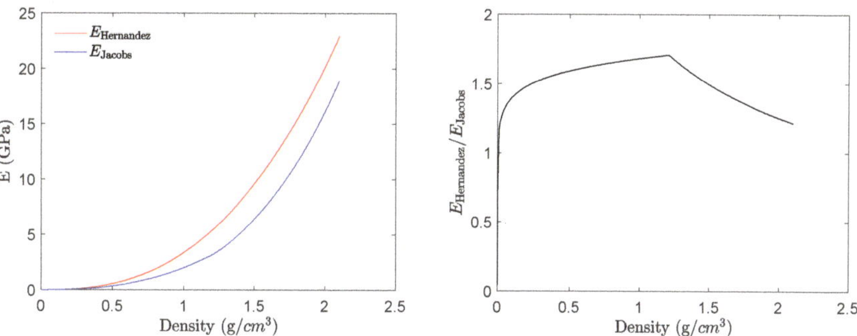

Figure 3. Comparison of the correlation provided by Jacobs [28] and the one derived from the correlation of Hernandez et al. [29] for the Young's modulus as a function of bone apparent density.

Bone is actually an anisotropic material with a dependence of the mechanical properties on the direction and the type of tissue. Particularly, cortical bone is usually modelled as a transversely isotropic material [30] and trabecular bone as an orthotropic material [31]. Consequently, its anisotropy must be taken into account in the estimation of the stiffness tensor, and so was carried out in the anisotropic model, named here *AnisoH* (note that the H refers to the fact that the anisotropic model relies on Hernandez correlation, as explained later in Section 2.4).

Some bone remodelling models have been proposed to predict not only the adaptation of bone apparent density but also of its anisotropy [4,5]. The model developed by Doblaré and García [4] was also used to predict the distribution of anisotropy in the same manner as DPTBR simulations are used to predict the distribution of bone density. In fact, these authors predicted both distributions simultaneously. The starting point was an isotropic material with a uniform distribution of density. Their BRM adapted bone density and anisotropy, the latter through the fabric tensor, and both variables were used to update the stiffness tensor. Convergence was deemed when both density and anisotropy remained constant between simulations. A brief summary of this model is provided next, in Section 2.4.

Given that DPTBR simulations were shown to be path-dependent [11], we estimated the density distribution from the CT scan and used the bone remodelling model developed by Doblaré and García [4] to estimate the anisotropy. This required a slight variation of the original model, explained in Section 2.5. Besides, another variation of the original model was used to account for time-varying loads, such as walking. This variation was introduced by Ojeda [32] and is presented in Section 2.6.

The objective of comparing the three models referred to above (*IsoJ*, *IsoH* and *AnisoH*) was to rule out that the assumption made for the constitutive model is forcing a certain correlation between the density and the predictors.

2.4. Anisotropic Bone Remodelling Model Based on Continuum Damage Mechanics—Model by Doblaré and García

A brief description of the anisotropic bone remodelling model developed by Doblaré and García [4] is given next. Consulting the original paper is advised for a more detailed description of the model. This model is an extension to the anisotropic case of the model developed by Beaupré et al. [3]. It applies the Mechanostat Theory by defining a stepwise linear relationship between the mechanical stimulus, Ψ_t, and the bone formation/resorption rate, $\dot{r}$ (see Figure 4). The reference value of the stimulus, Ψ_t^* defines a region of width 2w called lazy zone (analogous to the "adapted-window" of the Mechanostat Theory), where no net change of bone density is produced. The value of $\dot{r}$ determines the temporal evolution of apparent density, $\dot{\rho}$. In turn, apparent density determines the variation of the Young's modulus through correlations such as (5a) or (7).

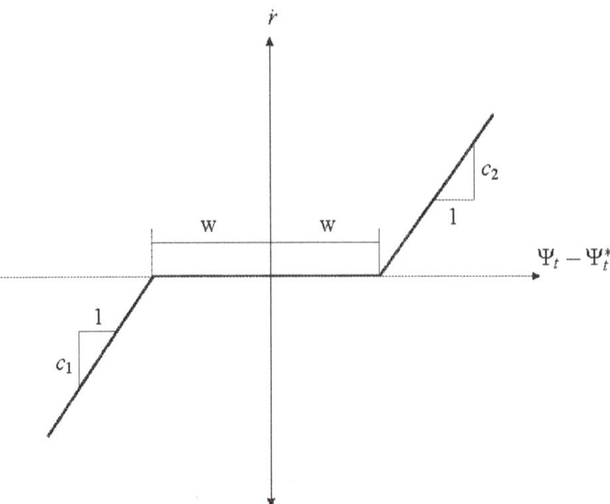

Figure 4. The Mechanostat Theory as interpreted by Beaupré et al. [3]. This law establishes a stepwise linear relationship between the bone formation/resorption rate, $\dot{r}$, and the difference between the mechanical stimulus, Ψ_t, and a reference value of that stimulus, Ψ_t^*.

The model developed by Doblaré and García was based on the theory of Continuum Damage Mechanics (CDM), where the stiffness tensor of the damaged material, $\mathbf{C}$, is obtained from the tensor of the the non-damaged material, $\mathbf{C_0}$, and damage:

$$\mathbf{C} = (1 - D)\mathbf{C_0}, \tag{8}$$

where D is the damage variable, null for an intact material and equal to 1 for a completely damaged or failed material. In the extension of CDM to the anisotropic case introduced by Cordebois and Sideroff [33], the scalar damage D is replaced with a damage tensor $\mathbf{D}$ and the resulting material is orthotropic, with the principal directions of orthotropy aligning with the principal axes of the damage tensor $\mathbf{D}$. Damage is understood as a measure of porosity and the directionality of that porosity and both are incorporated into the model jointly, by following the idea suggested by Cowin [34] for the fabric tensor, $\mathbf{H}$. Therefore, the undamaged material is an ideal situation of a perfectly isotropic bone with null porosity. The damage and fabric tensors are related by:

$$\mathbf{D} = \mathbf{1} - \mathbf{H}^2, \tag{9}$$

with $\mathbf{1}$ being the second order identity tensor. Equation (9) leads to the following relationship between the components of the compliance tensor in the principal directions and the eigenvalues of the fabric tensor, h_i:

$$\frac{1}{E_i} = \frac{1}{\hat{E}}\frac{1}{h_i^4}$$

$$-\frac{\nu_{ij}}{E_i} = -\frac{\hat{\nu}}{\hat{E}}\frac{1}{h_i^2 h_j^2}$$

$$\frac{1}{G_{ij}} = \frac{1+\hat{\nu}}{\hat{E}}\frac{1}{h_i^2 h_j^2}, \tag{10}$$

where $\hat{E}$ and $\hat{\nu}$ are, respectively, the Young's modulus and Poisson's ratio of the bone with no porosity. These values can be obtained through correlations (5) or (7) for an apparent density ρ equal to the density of the bone matrix, $\hat{\rho}$, which is assumed constant. In this particular model *AnisoH*, we assumed $\hat{\rho} = 2.1$ g/cm^3 and chose the Hernandez correlation (7), together with $\nu = 0.3$. The influence of porosity and directionality are factorised in this model by redefining the fabric tensor as follows:

$$\mathbf{H} = \left(\frac{\rho}{\hat{\rho}}\right)^{\beta/4} A^{1/4}\hat{\mathbf{H}}^{1/2}, \tag{11}$$

where $\hat{\mathbf{H}}$ is the normalised fabric tensor. This tensor is normalised by imposing $\det(\hat{\mathbf{H}}) = 1$ in order to account only for the directionality of the pores. Additionally, β is the exponent of the apparent density in the correlations (5a) or (7) and A is a parameter introduced to ensure that the formulation reproduces the isotropic bone remodelling model developed by Beaupré et al. [3] if it is applied to an isotropic case. A can here be considered a constant. As stated before, the quotient $\rho/\hat{\rho}$ in Equation (11) is equal to the bone volume fraction, $BV/TV = 1 - p$, with p as the porosity. The mechanical stimulus is defined in this model through the tensor, $\mathbf{Y}$:

$$\mathbf{Y} = 2\left\{2\hat{G}\,\mathrm{sym}\left[(\mathbf{H}\boldsymbol{\varepsilon}\mathbf{H})(\mathbf{H}\boldsymbol{\varepsilon})\right] + \hat{\lambda}\,\mathrm{tr}(\mathbf{H}^2\boldsymbol{\varepsilon})\,\mathrm{sym}(\mathbf{H}\boldsymbol{\varepsilon})\right\}, \tag{12}$$

where $\hat{G}$ and $\hat{\lambda}$ are the Lamé constants corresponding to the cortical bone with no porosity and $\mathrm{tr}(\bullet)$ and $\mathrm{sym}(\bullet)$ represent the trace and symmetric part of a tensor, respectively. Doblaré and García defined another tensor, $\mathbf{J}$, to quantify the relative influence of the spherical and deviatoric parts of the stimulus:

$$\mathbf{J} = \frac{1-\omega}{3}\,\mathrm{tr}(\mathbf{Y})\mathbf{1} + \omega\,\mathrm{dev}(\mathbf{Y}), \tag{13}$$

where $\mathbf{1}$ is the identity tensor, $\mathrm{dev}(\bullet)$ represents the deviatoric part of a tensor and the anisotropy factor, ω, weights the importance of the anisotropy of the stimulus in the model. This factor ranges from $\omega = 0$, which means that the model only depends on the isotropic component of the stimulus, to $\omega = 1$, which produces the maximum level of anisotropy. The same value used by Doblaré and García [4] was used here ($\omega = 0.1$). Two functions g_r and g_f are proposed to establish the remodelling criteria. These functions depend on the stimulus $\mathbf{J}$ and are allowed to distinguish the formation, resorption and lazy zones, as carried out in Figure 4. For that reason, those functions also depend on the reference value of the stimulus, Ψ_t^*, and the width of the lazy zone, through w. Their expressions are quite complex and can be consulted in [4]. The remodelling criteria are given by the following conditions:

$$\begin{array}{lll} g_f(\mathbf{J},\Psi_t^*,\mathrm{w}) \leq 0 & g_r(\mathbf{J},\Psi_t^*,\mathrm{w}) > 0 & \text{resorption;} \\ g_r(\mathbf{J},\Psi_t^*,\mathrm{w}) \leq 0 & g_f(\mathbf{J},\Psi_t^*,\mathrm{w}) > 0 & \text{formation;} \\ g_f(\mathbf{J},\Psi_t^*,\mathrm{w}) \leq 0 & g_r(\mathbf{J},\Psi_t^*,\mathrm{w}) \leq 0 & \text{lazy zone.} \end{array} \tag{14}$$

Based on the fulfilment of the corresponding criterion, the variation of the fabric tensor $\mathbf{H}$, that accounts for the variation of anisotropy (through tensor $\hat{\mathbf{H}}$) and porosity (see Equation (11)), is provided by:

$$\dot{\mathbf{H}} = k_1 \frac{\hat{\rho}}{\rho} J^{-3} \hat{\omega} \quad \text{resorption;}$$

$$\dot{\mathbf{H}} = k_2 \frac{\hat{\rho}}{\rho} J \hat{\omega} \quad \text{formation;} \tag{15}$$

$$\dot{\mathbf{H}} = 0 \quad \text{lazy zone;}$$

where the tensor $\hat{\omega}$ is introduced to simplify the expression, as follows:

$$\hat{\omega} = \frac{1 - 2\omega}{3} \mathbf{1} \otimes \mathbf{1} + \omega \mathbf{I} \tag{16}$$

with $\mathbf{I}$ being the fourth-order identity tensor. The factors k_1 and k_2 in Equation (15) depended on several parameters in the original model. One of those parameters is $\dot{r}$, so that the amount of formed or resorbed tissue modifies the fabric tensor through porosity.

2.5. Modification of the BRM to Maintain Density Constant

In the original model, Doblaré and García used $\dot{\mathbf{H}}$ to assess the variation of porosity and anisotropy, but in this work, since the density is known from the CT scan, we have forced it to remain constant and that is the reason why k_1 and k_2 can be assumed as constants. In such case, by deriving Equation (11):

$$\dot{\mathbf{H}} = \frac{1}{2} \left(\frac{\rho}{\hat{\rho}} \right)^{\beta/4} A^{1/4} \hat{\mathbf{H}}^{-1/2} \dot{\hat{\mathbf{H}}} \tag{17}$$

and given that $\hat{\mathbf{H}}$ must remain normalised ($\det(\hat{\mathbf{H}}) = 1$), we finally adopted:

$$\dot{\hat{\mathbf{H}}} = c\, \hat{\mathbf{H}}^{1/2} \dot{\mathbf{H}}, \tag{18}$$

where c is a constant. This expression can be used in an Euler forward integration algorithm to yield:

$$\hat{\mathbf{H}}(t_{j+1}) = \hat{\mathbf{H}}(t_j) + c\, \hat{\mathbf{H}}^{1/2}(t_j)\, \dot{\mathbf{H}}(t_j)\, \Delta t, \tag{19}$$

where t_{j+1} and t_j are two consecutive integration steps, the time step $\Delta t = 1$ is chosen and c is the constant necessary to enforce the condition $\det \hat{\mathbf{H}}(t_{j+1}) = 1$. The simulation started from an initially isotropic material ($\hat{\mathbf{H}}(t_1)$ equal to the identity tensor) and was stopped when the norm of the fabric tensor averaged for all the elements was almost invariable between iterations, i.e.:

$$\frac{\sum_{e=1}^{n} \hat{\mathbf{H}}_e(t_{j+1}) : \hat{\mathbf{H}}_e(t_{j+1}) - \sum_{e=1}^{n} \hat{\mathbf{H}}_e(t_j) : \hat{\mathbf{H}}_e(t_j)}{\sum_{e=1}^{n} \hat{\mathbf{H}}_e(t_j) : \hat{\mathbf{H}}_e(t_j)} < 0.001. \tag{20}$$

2.6. Modification of the BRM to Consider Time-Varying Loads

Doblaré and García [4] and Beaupré et al. [6] applied their models to estimate the bone density distribution in a human femur by applying the normal walking loads. These authors considered three instants of the gait cycle and treated those instants as independent loads. This procedure does not seem very plausible as they are not independent but part of the same load. For that reason, the procedure was modified by Ojeda [32] to treat the gait cycle as a single load. Moreover, the particularity of time-varying loads is taken into account with this modification.

As stated by Carter et al. [35], bone remodelling depends on the maximum stresses that the bone withstands throughout its load history. Thus, the peaks of mechanical stimulus reached in time-varying loads would control the bone remodelling response, and it is important to note that these peaks can be reached at different instants at each bone site. Let us consider, for example, the temporal evolution of the mechanical stimulus shown in Figure 5, with three different activities. The remodelling response is assumed to depend on the maximum stimulus representative of each activity (black dots in Figure 5). The cycles are grouped by the level of stimulus and an average cycle must be chosen as representative of a certain activity. In Figure 5, A1, A2 and A3 represent, respectively, a high, medium and low intensity load. The mechanical stimulus must be obtained by superimposing the effect of all activities [35], but let us consider for a moment that only one of those loads is applied. In such case, A1 would stimulate formation, A3 resorption and A2 would produce no net change of bone mass.

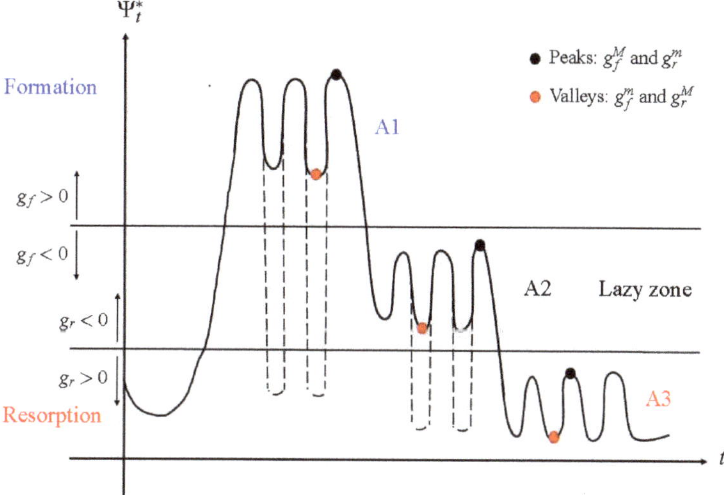

Figure 5. Remodelling criteria with different loads. The horizontal lines limit the zones of formation (top line) and resorption (bottom line). If the peak is over the top line, then $g_f^M > 0$ and $g_r^m < 0$ and formation occurs. If the peak is below the bottom line, then $g_f^M < 0$ and $g_r^m > 0$ and resorption occurs. If the peak lies between both lines, $g_f^M < 0$ and $g_r^m < 0$ and neither formation nor resorption occurs (lazy zone).

The peaks of the stimulus coincide with the maximum of the formation criterion function, g_f, which is proportional to the stimulus. Those peaks are termed here, g_f^M. Since the resorption criterion function, g_r, is inversely proportional to the stimulus [4], the local minimum of this function, g_r^m, also coincides with the peaks of the stimulus (it must be noted that g_f^M and g_r^m do not necessarily coincide. They are simply reached at the same instant). Analogously, the valleys of the stimulus coincide with the minimum of g_f and the maximum of g_r, termed here g_f^m and g_r^M, respectively (red dots in Figure 5).

The procedure to analyse the bone remodelling process for time-varying loads begins with the calculation of the stimulus (and thus, of g_f and g_r) at every point of the FE mesh throughout the cycle, in order to capture the peaks g_f^M and g_r^m for each element. Based on the ideas of Carter et al. [35], Ojeda assumed that only the peaks are important from the bone remodelling perspective. Therefore, the activities plotted in solid and dashed lines in Figure 5 would lead to the same bone remodelling response, regardless of the valleys.

The criterion must identify the peaks and place them in one of the three regions: formation, resorption or lazy zone. Thus, the remodelling criteria (14) are replaced by:

$$
\begin{aligned}
g_f^M &> 0 && \text{formation;} \\
g_r^m &\leq 0 \quad g_f^M \leq 0 && \text{lazy zone;} \\
g_f^M &\leq 0 \quad g_r^m > 0 && \text{resorption.}
\end{aligned}
\tag{21}
$$

It is important to highlight that g_f^M and g_r^m are not necessarily reached at the same instant for all the elements. This is the reason why a detailed description of the cycle is required in the modification proposed by Ojeda, as the remodelling response at each element could be driven by the stress-state reached at a different time point. For the same reason, the simplification consisting in considering three instants of the cycle as independent loads is not valid.

2.7. FE Simulation of the Gait Cycle—Temporal Evolution of Stresses and Strains in the Femur

The temporal evolution of stresses and strains in the femur during the gait cycle can be obtained by solving an elastic problem. Let $\Omega \subset \mathbb{R}^3$ be an open bounded domain and $\Gamma = \partial\Omega$ be its boundary, assumed to be Lipschitz continuous and divided into two disjoint parts Γ_D and Γ_N where Dirichlet and Neumann boundary conditions are applied, respectively. We denote, by $\mathbf{x} = x_i\,\mathbf{e}_i$, a generic point of Ω and $\mathbf{n}(\mathbf{x}) = n_i\,\mathbf{e}_i$ as the outward unit normal vector to Γ at a point $\mathbf{x}$. The Einstein summation notation was adopted and a Cartesian basis $\mathbf{e}_i$ ($i = 1, 2, 3$) can be used without loss of generality.

Let $\mathbf{u} = u_i\,\mathbf{e}_i$, $\boldsymbol{\sigma} = \sigma_{ij}\,\mathbf{e}_i \otimes \mathbf{e}_j$ and $\boldsymbol{\varepsilon}(\mathbf{u}) = \varepsilon_{ij}\,\mathbf{e}_i \otimes \mathbf{e}_j$ denote the displacement field, the stress tensor and the linearised strain tensor, respectively. $\mathbf{e}_i \otimes \mathbf{e}_j$ ($i, j = 1, 2, 3$) represents the Cartesian tensorial basis. Let $\mathbf{b} = b_i\,\mathbf{e}_i$ denote the known vector of body forces, $\mathbf{t} = t_i\,\mathbf{e}_i$ the known vector of surface traction forces at Γ_N and $\boldsymbol{\delta} = \delta_i\,\mathbf{e}_i$ the known displacements at Γ_D.

The Momentum Conservation Principle states:

$$
\sigma_{ij,j} + b_i = \rho\,\ddot{u}_i,
\tag{22}
$$

where $,j$ denotes the partial derivative $\partial/\partial x_j$ and $(\ddot{\bullet})$ denotes the second derivative with respect to time. The right-hand side of Equation (22) represents the inertial force per unit volume. Linearised strains are related to displacements by:

$$
\varepsilon_{ij}(\mathbf{u}) = \frac{1}{2}\left(u_{i,j} + u_{j,i}\right) \qquad i,j = 1, 2, 3
\tag{23}
$$

and finally, strains and stresses are related by the constitutive equation, which for linear elastic materials reads:

$$
\sigma_{ij} = C_{ijkl}\,\varepsilon_{kl} \qquad i, j, k, l = 1, 2, 3,
\tag{24}
$$

where C_{ijkl} are the components of the fourth-rank stiffness tensor $\mathbf{C}$. In the case of an isotropic material, this tensor is completely defined by the Young's modulus, E, and the Poisson's ratio, ν, which are expressed as the functions of the density in the case of bone (see Equation (5)). In general anisotropic materials, this tensor has 21 independent elastic constants. In our anisotropic case, bone is assumed to be an orthotropic material, and the compliance tensor (inverse of the stiffness tensor) is given as a function of the fabric tensor (recall Equation (10)). The elastic problem is completed with the Dirichlet and Neumann boundary conditions, respectively, applied at Γ_D and Γ_N:

$$
\mathbf{u} = \boldsymbol{\delta} \quad \text{at } \Gamma_D
\tag{25a}
$$

$$
\sigma_{ij}\,n_j = t_i \quad \text{at } \Gamma_N.
\tag{25b}
$$

This boundary value problem rarely has an analytical solution, and hence, it is usually solved by means of the FEM, as carried out here.

At this point, the FE model of the femur had a complete definition of density, and consequently, of stiffness. Next, the muscle loads and joint reaction forces, estimated in the gait analysis, were applied as external forces. Muscle loads were applied as concentrated loads at the insertion points and with the direction defined by the insertion and origin points (taken from OpenSim [22]). The joint reaction forces were applied as concentrated loads on the corresponding articular surfaces, i.e., the hip reaction force on the surface of the femoral head and the knee reaction force on the surface of the epicondyles or of the epicondylar fossa. In both cases, the node of application at each instant was calculated as follows: A line was defined as passing through the corresponding joint centre (defined in OpenSim [22]) and with the direction of the reaction force at that instant. The reaction force was applied at the node closest to the intersection of the articular surface with that line. Furthermore, the minimum number of displacement boundary conditions (isostatic) was applied to restrain the rigid body motion of the FE model. The loads were varied over time, as a result from the gait analysis, but a quasi-static analysis was performed by disregarding the inertial forces in the FE simulation ($\rho\,\ddot{u}_i$ in Equation (22)). Nonetheless, we must note that the loads estimated with OpenSim arise from enforcing the equilibrium of all the external forces, including the inertial ones. For that reason, these inertial forces were indirectly considered in the simulations. The FE model, including the loads and the boundary conditions, is provided in the Supplementary Materials.

This pseudo-static analysis provided the temporal evolution of stresses and strains at each element e of the mesh (in fact, it is obtained at each integration point in full integration elements. In our case, C3D4 elements have only one integration point, which can be, therefore, identified with the element. In the case of C3D10 elements, the variables were evaluated at the centroid of the element) and at every instant i of the gait cycle. Several variables were derived from the stress and strain tensors (see Section 2.9). For a certain variable proposed as stimulus S, its value was calculated at each instant i and for each element, e, thus, yielding S_e^i. Then, the following maximum, minimum and amplitude were defined to represent its evolution throughout the gait cycle:

$$S_e^M = \operatorname*{Max}_i(S_e^i) \tag{26a}$$

$$S_e^m = \operatorname*{Min}_i(S_e^i) \tag{26b}$$

$$S_e^A = S_e^M - S_e^m. \tag{26c}$$

Note that the definition of S_e^M is related to the aforementioned hypothesis of Carter, according to which the local remodelling response would depend on the peak of stress that a bone site withstands throughout its load history [35]. As stated before, the peaks of stress do not necessarily occur at the same time for all bone sites. Therefore, considering only the loads of a single instant of the cycle is not enough to analyse the remodelling response of the whole bone. Even considering three instants of the cycle, as carried out in [4,6], is not enough. A detailed description of the gait cycle is required and the modification of the BRM presented in Section 2.6 is related to this idea. The variable, S_e^M, generalises this concept of the peak of stress to the peak of stimulus. As an alternative to the maximum of the stimulus throughout the cycle, the amplitude is considered in S_e^A.

2.8. Correlation Coefficients

The Spearman and Pearson correlation coefficients were used to assess the statistical dependence between the bone density and the variables defined later in Section 2.9. Based on the definitions made in Equation (26), the following coefficients were calculated, taking each element as a point of the sample:

- R_j^M, between the maximum throughout the cycle S_e^M and the apparent density ρ_e.

- R_j^A, between the amplitude throughout the cycle S_e^A and the apparent density ρ_e.

where j stands for P (Pearson) or S (Spearman). Weighted coefficients were calculated to account for the relative importance of a sample point based on the volume of the element. The expressions of the weighted correlation coefficients are given in Appendix A. The Spearman correlation coefficient is a non-parametric measure of rank correlation and it assesses how well the relationship between two variables can be described by a monotonic function. In particular, it evaluates if there is a concordance between the highest density and the highest values of a predictor variable. The Pearson coefficient is a parametric measure of correlation between two variables that assesses if they are related by a specific function. In particular, the linear, quadratic and power correlation coefficients were calculated for those variables that yielded a high Spearman coefficient, in order to confirm a strong correlation and to evaluate the best function relating the variable to bone density. Despite that strain energy density (SED) did not yield a high Spearman coefficient, it was also correlated with bone density through the Pearson coefficient and using those three functions, for reasons that will be explained later.

It is important to note that concentrated loads or displacement boundary conditions can produce spurious stress concentrations in the FE model. For this reason, the elements closest to the nodes where loads or displacement boundary conditions were applied up to a distance of two elements in all directions were removed from the correlations.

2.9. Evolution of Stresses and Strains—Definition of Predictor Variables

The set of predictor variables analysed in this work includes some variables that measure the magnitude of the stress or the strain tensor and the SED that accounts for the magnitude of both tensors in a single variable. The principal stresses are named here, $\sigma_1 \geq \sigma_2 \geq \sigma_3$, and analogously, $\varepsilon_1 \geq \varepsilon_2 \geq \varepsilon_3$. The maximum and minimum principal stresses (σ_1, σ_3) and strains (ε_1, ε_3) are proposed as predictor variables. The maximum tensile stress is defined as:

$$\sigma_t = \begin{cases} \sigma_1 & \text{if } \sigma_1 \geq 0 \\ 0 & \text{if } \sigma_1 < 0 \end{cases} \tag{27}$$

and the maximum compressive stress as:

$$\sigma_c = \begin{cases} -\sigma_3 & \text{if } \sigma_3 < 0 \\ 0 & \text{if } \sigma_3 \geq 0 \end{cases}. \tag{28}$$

The fluctuations of stresses throughout the cycle can be measured by the variable:

$$\sigma_f = \sigma_1^M - \sigma_3^m, \tag{29}$$

where the superscripts M and m follow the definition given in Equation (26). The absolute maximum principal stress (AMPσ) and strain (AMPε) are defined analogously as:

$$AMP\sigma = \text{Max}\{|\sigma_1|, |\sigma_3|\} \tag{30a}$$

$$AMP\varepsilon = \text{Max}\{|\varepsilon_1|, |\varepsilon_3|\}. \tag{30b}$$

The von Mises (σ_{vonMises}) and Tresca (σ_{Tresca}) stresses (in metallic materials, these variables are used in yielding criteria, which are not applicable to bone, but they can be regarded as well as a measure of the stress intensity) as well as the hydrostatic stress (σ_o) and volumetric strain (ε_o) are also proposed as predictor variables:

$$\sigma_{\text{vonMises}} = \sqrt{\frac{(\sigma_1 - \sigma_2)^2 + (\sigma_1 - \sigma_3)^2 + (\sigma_2 - \sigma_3)^2}{2}} \tag{31a}$$

$$\sigma_{\text{Tresca}} = \frac{\sigma_1 - \sigma_3}{2} \tag{31b}$$

$$\sigma_o = \frac{\sigma_1 + \sigma_2 + \sigma_3}{3} \tag{31c}$$

$$\varepsilon_o = \varepsilon_1 + \varepsilon_2 + \varepsilon_3. \tag{31d}$$

Finally, the SED, which has been extensively used as a mechanical stimulus in bone remodelling algorithms, is also proposed as a predictor. SED is given by the following expression in terms of the stress ($\boldsymbol{\sigma}$) and strain ($\boldsymbol{\varepsilon}$) tensors or their components (σ_{ij}, ε_{ij}):

$$\text{SED} = \frac{1}{2}\boldsymbol{\sigma} : \boldsymbol{\varepsilon} = \frac{1}{2}\sigma_{ij}\,\varepsilon_{ij}. \tag{32}$$

Beaupré et al. [3] defined the mechanical stimulus that drives bone remodelling, Ψ, as a combination of two factors: the SED and the number of cycles of each activity. Furthermore, these authors proposed to superimpose the effect of all the activities, i, performed by the individual during one day, by weighting the SED of each activity, SED_i, with the corresponding number of cycles, n_i. The interested reader can consult the details in [3]. In the following, we will assume that only the most representative activity is carried out daily and that the number of cycles is constant. In that case, there is a linear relationship between Ψ and SED [3]:

$$\Psi = k \cdot \text{SED}. \tag{33}$$

In other words, we can identify Ψ and SED for correlation purposes. More importantly, Beaupré et al. proposed to take into account the porosity of the tissue to redefine the mechanical stimulus. Thus, Ψ represents the mechanical stimulus at the continuum (or macroscopic) level. SED can be calculated through FEM and if the constant k in Equation (33) is known, the mechanical stimulus at the continuum level, Ψ, can also be evaluated for each element of the mesh. Beaupré et al. hypothesized that this mechanical stimulus must be sensed by the existing tissue within the element in order to produce a remodelling response. Therefore, Ψ can be distributed among the tissue existing in the element through porosity, analogously to what localisation procedures do in multiscale approaches, i.e., allowing to move from the macro to the micro scale. To this end, those authors proposed the following mechanical stimulus at the tissue (or microscopic) level [3]:

$$\Psi_t = \frac{\Psi}{(1-p)^n}, \tag{34}$$

where $p \in [0,1]$ is the porosity and $n=2$ is the exponent they used [3]. In this way, if the porosity of one element is close to 1, the mechanical stimulus must be distributed among the little existing tissue and this will be heavily overloaded. Later, we will analyse the effect of the exponent n. As stated before, the linear, quadratic and power Pearson coefficients were also evaluated for the SED. The rationale for this is based on the theoretical dependence of SED upon density, which can be deduced from previous works found in the literature [2,35]. In the particular case of a uniaxial stress-state, the stress tensor is $\boldsymbol{\sigma} = \sigma\,\boldsymbol{e}_i \otimes \boldsymbol{e}_i$, with $\boldsymbol{e}_i$ being the loading direction and σ the applied stress. In such case, and assuming a linearly elastic and isotropic behaviour for bone, the SED can be calculated through Equation (32) as:

$$\text{SED}^* = \frac{1}{2}\sigma\varepsilon = \frac{1}{2}E\,\varepsilon^2, \tag{35}$$

where $\boldsymbol{\varepsilon}$ is the strain tensor and ε is the strain in the loading direction. The asterisk has been added to highlight that this expression corresponds to a particular case. Additionally, a typical power correlation between the Young's modulus and the apparent bone density can be assumed, for example Equations (5a) and (7), which would read:

$$E = B\rho^\beta, \tag{36}$$

where B and β are constants. In this case, Ψ can be rewritten using Equations (33) and (35) as:

$$\Psi^* = k\frac{1}{2}B\,\varepsilon^2\,\rho^\beta = K\rho^\beta, \tag{37}$$

where the constants preceding the factor ρ^β were grouped in a new constant K. The right-hand side of Equation (37) is only valid if bone is subjected to a constant strain, which would be in accordance with the Mechanostat Theory. Under all these assumptions, Ψ could be related to the bone apparent density through a power law. Recalling Equation (34) and given that $(1-p)$ is equal to the bone volume fraction, which is proportional to the bone density, the mechanical stimulus at the tissue level, Ψ_t, could also be related to bone density through a power law. For this reason, we will check if such a power correlation between apparent density and Ψ (or equivalently SED) or Ψ_t is suitable.

3. Results

The weighted Spearman correlation coefficients between bone density and the predictors proposed here are shown in Table 1 for the constitutive model *AnisoH*. As stated before, three different values were used for the grey level threshold (GLT) used in Equation (4), thus, leading to three different FE models (see Figure 2). The correlation coefficients are given in the three cases for comparison.

The fact that the Spearman correlation coefficient is non-parametric makes it more appropriate to evaluate the correlation between density and the variables, as it implies no assumption on the type of relationship. It simply establishes if there is a concordance between those points having the highest density and those having the highest values of a certain predictor.

Table 1. Weighted Spearman correlation coefficients obtained with the constitutive model, *AnisoH*, using C3D4 elements and for different values of GLT: for the maximum variable throughout the cycle, R_s^M, and for the amplitude throughout the cycle, R_s^A. The predictors analysed are: the maximum principal stress (σ_1), the minimum principal stress (σ_3), the absolute maximum principal stress (AMPσ), the corresponding strains (ε_1, ε_3 and AMPε), the strain energy density (SED or mechanical stimulus at the continuum level Ψ), the mechanical stimulus at the tissue level (Ψ_t), the maximum tensile and compressive stresses (σ_t and σ_c, respectively), the fluctuation of stresses (σ_f), the von Mises and Tresca stresses, the hydrostatic stress (σ_o) and the volumetric strain (ε_o). Values higher than 0.9 are highlighted in boldface; negative values are in red.

Predictor	GLT = 70		GLT = 80		GLT = 90	
	R_s^M	R_s^A	R_s^M	R_s^A	R_s^M	R_s^A
σ_1	0.871	0.880	**0.905**	**0.912**	**0.936**	**0.942**
σ_3	−0.040	0.872	−0.047	0.897	−0.045	**0.926**
σ_t	0.871	0.871	**0.905**	**0.905**	**0.936**	**0.936**
σ_c	0.868	0.868	0.894	0.894	**0.923**	**0.923**
σ_f	**0.904**	**0.904**	**0.927**	**0.927**	**0.951**	**0.951**
AMPσ	0.896	0.896	**0.921**	**0.920**	**0.945**	**0.944**
σ_{vonMises}	0.898	0.898	**0.922**	**0.922**	**0.948**	**0.948**
σ_{Tresca}	0.898	0.898	**0.922**	**0.922**	**0.948**	**0.947**
σ_o	0.701	0.886	0.739	**0.911**	0.782	**0.938**
ε_1	−0.344	−0.340	−0.431	−0.429	−0.503	−0.502
ε_3	0.410	−0.355	0.432	−0.445	0.457	−0.504
AMPε	−0.328	−0.325	−0.427	−0.425	−0.499	−0.498
Ψ	0.698	0.698	0.721	0.721	0.772	0.772
Ψ_t for n=2	−0.048	−0.048	−0.092	−0.092	−0.097	−0.099
ε_o	−0.202	−0.303	−0.255	−0.468	−0.327	−0.571

It can be seen that most of the stress magnitudes are highly correlated with the density except for the peak of the minimum principal stress, σ_3^M, for obvious reasons, as the sign of σ_3 (usually negative) is considered in the calculus of this peak. Therefore, σ_3^M usually corresponds to the lowest absolute value throughout the cycle. The amplitude, σ_3^A, is better correlated with the density as it usually measures the range of the compressive stress.

The strain magnitudes are poorly correlated with the density, in some cases with negative coefficients. SED (or Ψ) is only moderately correlated with the density and not correlated at all if it is corrected to account for the porosity (Ψ_t).

Regarding the influence of GLT, it can be seen that though the values of R are different, the same trend is observed in the three cases. In fact, if we ordered the predictors based on R, the same order would result for the three GLT.

The weighted Pearson correlation coefficients are shown in Table 2 for the constitutive model, *AnisoH*. Only some of the variables previously analysed in Table 1—those that are considered more interesting—are studied here; in particular, some of the stress variables that had a higher Spearman coefficient together with the SED at the continuum and at the tissue level for $n = 1$ and $n = 2$ (see Equation (34)). The Pearson coefficients are parametric and presuppose a certain relationship between the variables being correlated. Thus, we have tried linear, quadratic and power functions (see Appendix A for details).

Compared to the Spearman, the Pearson correlations have worsened notably as we are forcing them to fit a certain function which is probably not the most appropriate to relate the density with the predictor. Among the three types of functions tested, the quadratic is slightly better, followed by the power and the linear function. The low correlation between SED and density stands out—something that does not improve in the case of SED at the tissue level—for which even negative correlation coefficients were obtained, as in the case of the Spearman coefficients.

Table 2. Weighted Pearson correlation coefficients obtained with the constitutive model, *AnisoH*, using C3D4 elements and for different values of GLT: for the maximum variable throughout the cycle, R_P^M, and for the amplitude throughout the cycle, R_P^A. The predictors analysed in this case are: the absolute maximum principal stress (AMPσ), the maximum tensile and compressive stresses (σ_t and σ_c, respectively), the fluctuation of stresses (σ_f) and the mechanical stimulus at the continuum level (Ψ) and at the tissue level (Ψ_t) for two values of n. Values higher than 0.8 are highlighted in boldface; negative values are in red.

Predictor	Type	GLT = 70		GLT = 80		GLT = 90	
		R_P^M	R_P^A	R_P^M	R_P^A	R_P^M	R_P^A
AMPσ	Linear	0.773	0.772	0.793	0.791	**0.808**	**0.806**
σ_t	Linear	0.746	0.745	0.761	0.760	0.770	0.769
σ_c	Linear	0.671	0.672	0.684	0.684	0.695	0.695
σ_f	Linear	**0.810**	**0.810**	**0.825**	**0.825**	**0.838**	**0.838**
Ψ	Linear	0.012	0.012	0.012	0.012	0.012	0.012
Ψ_t for n = 1	Linear	−0.016	−0.016	−0.016	−0.016	−0.013	−0.013
Ψ_t for n = 2	Linear	−0.023	−0.023	−0.026	−0.026	−0.023	−0.023
AMPσ	Quadratic	0.784	0.782	**0.804**	**0.803**	**0.818**	**0.817**
σ_t	Quadratic	0.762	0.761	0.778	0.778	0.787	0.786
σ_c	Quadratic	0.678	0.679	0.690	0.691	0.700	0.700
σ_f	Quadratic	**0.823**	**0.823**	**0.839**	**0.839**	**0.850**	**0.850**
Ψ	Quadratic	0.026	0.026	0.026	0.026	0.029	0.029
Ψ_t for n = 1	Quadratic	0.017	0.017	0.017	0.017	0.018	0.018
Ψ_t for n = 2	Quadratic	0.021	0.021	0.026	0.026	0.023	0.023
AMPσ	Power	0.780	0.778	0.799	0.798	**0.813**	**0.811**
σ_t	Power	0.761	0.760	0.777	0.776	0.785	0.784
σ_c	Power	0.674	0.675	0.683	0.684	0.690	0.691
σ_f	Power	**0.820**	**0.820**	**0.835**	**0.835**	**0.845**	**0.845**
Ψ	Power	0.006	0.006	0.005	0.005	0.005	0.005
Ψ_t for n = 1	Power	−0.016	−0.016	−0.017	−0.017	−0.015	−0.015
Ψ_t for n = 2	Power	−0.019	−0.019	−0.022	−0.020	−0.015	−0.015

Tables 3 and 4 compare, respectively, the Spearman and Pearson coefficients obtained using the three constitutive models analysed in this work: *AnisoH*, *IsoH* and *IsoJ*. As indicated above, the effect of GLT was not important, and hence, only one case (GLT = 80) was studied. It can be noted that the effect of the constitutive model is negligible on the Spearman correlations and small on the Pearson correlations, at least for the three cases tested here. The biggest difference is obtained for the power fit between the *IsoH* and *IsoJ* models, which, in turn, follow a different power correlation between the density and the Young's modulus, but is not greater than 0.03.

Table 3. Weighted Spearman correlation coefficients obtained for GLT = 80, constitutive models *AnisoH*, *IsoH* and *IsoJ* and using C3D4 elements: for the maximum variable throughout the cycle, R_S^M, and for the amplitude throughout the cycle, R_S^A. The predictors analysed are: the maximum principal stress (σ_1), the minimum principal stress (σ_3), the absolute maximum principal stress (AMPσ), the corresponding strains (ε_1, ε_3 and AMPε), the strain energy density (SED or mechanical stimulus at the continuum level Ψ), the mechanical stimulus at the tissue level (Ψ_t), the maximum tensile and compressive stresses (σ_t and σ_c, respectively), the fluctuation of stresses (σ_f), the von Mises and Tresca stresses, the hydrostatic stress (σ_o) and the volumetric strain (ε_o). Values higher than 0.9 are highlighted in boldface; negative values are in red.

Predictor	AnisoH		IsoH		IsoJ	
	R_S^M	R_S^A	R_S^M	R_S^A	R_S^M	R_S^A
σ_1	**0.905**	**0.912**	**0.906**	**0.913**	**0.905**	**0.913**
σ_3	−0.047	0.897	−0.058	0.899	−0.059	**0.895**
σ_t	**0.905**	**0.905**	**0.906**	**0.906**	**0.905**	**0.905**
σ_c	0.894	0.894	0.895	0.895	0.892	0.892
σ_f	**0.927**	**0.927**	**0.928**	**0.928**	**0.925**	**0.925**
AMPσ	**0.921**	**0.920**	**0.922**	**0.921**	**0.920**	**0.920**
σ_{vonMises}	**0.922**	**0.922**	**0.924**	**0.924**	**0.921**	**0.920**
σ_{Tresca}	**0.922**	**0.922**	**0.923**	**0.923**	**0.920**	**0.920**
σ_o	0.739	**0.911**	0.742	**0.913**	0.739	**0.911**
ε_1	−0.431	−0.429	−0.431	−0.429	−0.443	−0.441
ε_3	0.432	−0.445	0.431	−0.443	0.429	−0.458
AMPε	−0.427	−0.425	−0.425	−0.423	−0.440	−0.438
Ψ	0.721	0.721	0.722	0.721	0.708	0.708
Ψ_t for n = 2	−0.092	−0.092	−0.092	−0.092	−0.114	−0.114
ε_o	−0.255	−0.468	−0.254	−0.464	−0.247	−0.489

We have also analysed the spatial distribution of the correlations, in particular of the Pearson coefficients (power fit), by assessing separately the correlations for the elements of the proximal, distal and diaphyseal thirds (see Table 5). The aim of this comparison was to investigate if there are regions of the femur where the density is better to the predictors. Given the limited influence of GLT and the constitutive model, we only show the case GLT = 80 and *IsoH*. Besides, we only compare some of the variables that show a higher correlation (σ_f, AMPσ) and only the coefficients for the maximum variable throughout the cycle, R_S^M. The other stress variables follow the same trend, as well as R_S^A. The comparison of the strain variables is meaningless since they are not correlated with density, as shown previously. It can be noted that the correlation coefficients are high in the diaphysis, significantly worse in the proximal and especially worse in the distal third, influenced by the simplified way the joint reaction forces were modelled. They were applied as concentrated nodal forces, as explained in Section 2.7, rather than as a load distributed over the articular surface, as it actually occurs. This simplification affects the stresses near the articular region and, therefore, the correlations. The hip joint force can be more plausibly applied as a concentrated nodal force since the pressure on that joint spans a narrower region than that on the knee joint. Probably, this makes the correlations be slightly better in the proximal third than in the distal one.

Table 4. Weighted Pearson correlation coefficients obtained for GLT = 80, constitutive models *AnisoH, IsoH* and *IsoJ* and using C3D4 elements: for the maximum variable throughout the cycle, R_P^M, and for the amplitude throughout the cycle, R_P^A. The predictors analysed in this case are: the absolute maximum principal stress (AMPσ), the maximum tensile and compressive stresses (σ_t and σ_c, respectively), the fluctuation of stresses (σ_f), the mechanical stimulus at the continuum level (Ψ) and at the tissue level (Ψ_t) for two values of n. Values higher than 0.8 are highlighted in boldface; negative values are in red.

Predictor	Type of Correlation	AnisoH R_P^M	R_P^A	IsoH R_P^M	R_P^A	IsoJ R_P^M	R_P^A
AMPσ	Linear	0.793	0.791	**0.802**	**0.800**	**0.814**	**0.813**
σ_t	Linear	0.761	0.760	0.767	0.766	0.771	0.770
σ_c	Linear	0.684	0.684	0.692	0.692	0.707	0.708
σ_f	Linear	**0.825**	**0.825**	**0.832**	**0.832**	**0.840**	**0.840**
Ψ	Linear	0.012	0.012	0.012	0.012	0.015	0.015
Ψ_t for n = 1	Linear	−0.016	−0.016	−0.016	−0.016	−0.017	−0.017
Ψ_t for n = 2	Linear	−0.026	−0.026	−0.026	−0.026	−0.029	−0.029
AMPσ	Quadratic	0.804	0.803	**0.814**	**0.812**	**0.834**	**0.833**
σ_t	Quadratic	0.778	0.778	0.785	0.784	0.797	0.796
σ_c	Quadratic	0.690	0.691	0.698	0.699	0.719	0.720
σ_f	Quadratic	**0.839**	**0.839**	**0.846**	**0.846**	**0.861**	**0.861**
Ψ	Quadratic	0.026	0.026	0.026	0.026	0.030	0.030
Ψ_t for n = 1	Quadratic	0.017	0.017	0.017	0.017	0.020	0.020
Ψ_t for n = 2	Quadratic	0.026	0.026	0.026	0.026	0.029	0.029
AMPσ	Power	0.799	0.798	**0.809**	**0.808**	**0.833**	**0.832**
σ_t	Power	0.777	0.776	0.783	0.783	0.797	0.796
σ_c	Power	0.683	0.684	0.692	0.693	0.717	0.718
σ_f	Power	**0.835**	**0.835**	**0.842**	**0.842**	**0.861**	**0.861**
Ψ	Power	0.005	0.005	0.006	0.006	0.008	0.008
Ψ_t for n = 1	Power	−0.017	−0.017	−0.017	−0.017	−0.018	−0.018
Ψ_t for n = 2	Power	−0.022	−0.022	−0.022	−0.022	−0.023	−0.023

Table 5. Weighted Pearson correlation coefficients (power fit) obtained for the maximum variable throughout the cycle, R_S^M, for GLT = 80, constitutive model *IsoH* and using C3D4 elements. The predictors analysed are: the fluctuation of stresses (σ_f) and the absolute maximum principal stress (AMPσ). Values higher than 0.8 are highlighted in boldface.

	Proximal	Diaphysis	Distal	Global
σ_f	0.718	**0.879**	0.660	**0.844**
AMPσ	0.694	**0.862**	0.640	**0.813**

The influence of the mesh (C3D4 vs. C3D10) was analysed by comparing the correlation coefficients in Table 6; in particular, the Spearman and the power Pearson coefficients in the case GLT = 80 and using the constitutive model *IsoH*. The Pearson coefficients improved moderately with the use of quadratic elements (C3D10), but only for the good predictors, i.e., those variables that are highly correlated with density. The rest of variables, such as Ψ and the strain magnitudes (not shown), did not improve their correlations. The other types of fit (linear and quadratic) also improved with C3D10, although to a lesser extent. It is noteworthy that the Spearman coefficients were almost identical in both meshes.

Table 6. Influence of the FE mesh (C3D4 vs. C3D10) on the weighted Spearman and Pearson (power) correlation coefficients obtained for GLT = 80 and constitutive model *IsoH*: for the maximum variable throughout the cycle, R_P^M, and for the amplitude throughout the cycle, R_P^A. The same predictors analysed for Spearman and Pearson coefficients are compared here. Values higher than 0.9 (in Spearman) or 0.8 (in Pearson) are highlighted in boldface; negative values are in red.

Predictor	Type of Correlation	C3D4		C3D10	
		R_P^M	R_P^A	R_P^M	R_P^A
σ_1	Spearman	**0.906**	**0.913**	**0.904**	**0.912**
σ_3	Spearman	−0.058	0.899	−0.236	0.896
σ_t	Spearman	**0.906**	**0.906**	**0.904**	**0.904**
σ_c	Spearman	0.895	0.895	0.893	0.894
σ_f	Spearman	**0.928**	**0.928**	**0.930**	**0.930**
AMPσ	Spearman	**0.922**	**0.921**	**0.923**	**0.922**
σ_{vonMises}	Spearman	**0.924**	**0.924**	**0.925**	**0.925**
σ_{Tresca}	Spearman	**0.923**	**0.923**	**0.925**	**0.924**
σ_0	Spearman	0.742	**0.913**	0.734	**0.911**
ε_1	Spearman	−0.431	−0.429	−0.499	−0.496
ε_3	Spearman	0.431	−0.443	0.484	−0.495
AMPε	Spearman	−0.425	−0.423	−0.482	−0.480
Ψ	Spearman	0.722	0.721	0.709	0.709
Ψ_t for n = 2	Spearman	−0.092	−0.092	−0.182	−0.182
ε_0	Spearman	−0.254	−0.464	−0.287	−0.499
AMPσ	Pearson (Power)	**0.809**	**0.808**	**0.874**	**0.874**
σ_t	Pearson (Power)	0.783	0.783	**0.818**	**0.818**
σ_c	Pearson (Power)	0.692	0.693	0.741	0.742
σ_f	Pearson (Power)	**0.842**	**0.842**	**0.902**	**0.902**
Ψ	Pearson (Power)	0.006	0.006	0.011	0.011
Ψ_t for n = 1	Pearson (Power)	−0.017	−0.017	−0.019	−0.019
Ψ_t for n = 2	Pearson (Power)	−0.022	−0.022	−0.025	−0.025

Histograms

There are so many points involved in the correlations that the plots of density against the different variables are very difficult to distinguish. Instead, histograms are used to show the percentage of volume occupied by those elements whose values of AMPσ or AMPε are within a given range. The elements of cortical ($\rho > 1.2$ g/cm^3) and trabecular bone ($\rho \leq 1.2$ g/cm^3) have been separated into two different histograms and the results for the three constitutive models were plotted jointly (see Figure 6). Thus, for example, in the model *AnisoH*, the elements of trabecular bone whose AMPε is in the range [200,600] $\mu\varepsilon$ occupy 30% of the total volume of trabecular bone ($\rho \leq 1.2$ g/cm^3).

It can be seen that while the strain range is similar for cortical and trabecular bone, the stresses are completely different, with trabecular bone having stresses several orders of magnitude lower than cortical bone. In general, strains are found in a very narrow range, especially in the cortical bone, for which 87% of the volume has AMPε in the range of 200–600 $\mu\varepsilon$. This is not so evident in trabecular bone, though 51% is still within the range of 200–1000 $\mu\varepsilon$ and 65% is in the range of 200–1400 $\mu\varepsilon$. This is still a narrow range, as overload strains are up to 4000 $\mu\varepsilon$ [36,37].

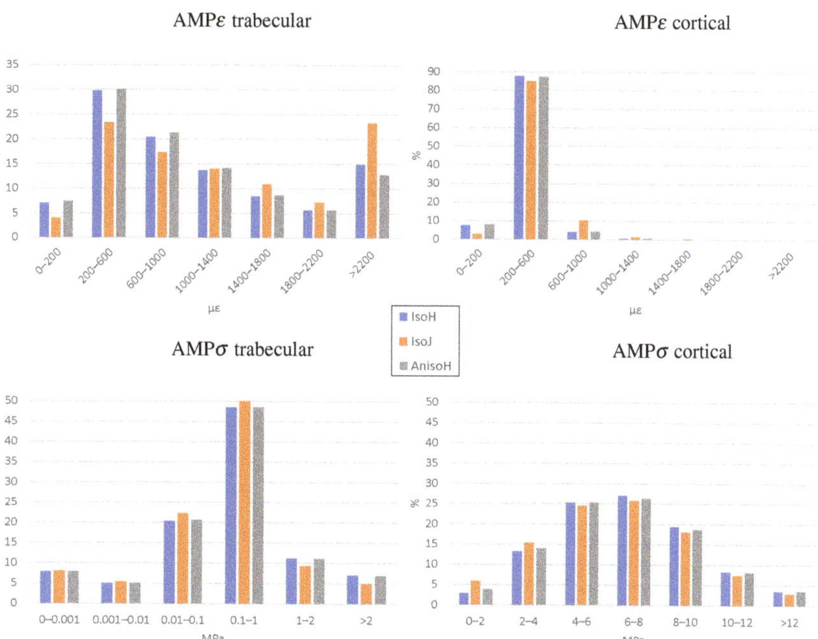

Figure 6. Distribution of the absolute maximum principal stress (AMPσ) and strain (AMPε) for the elements of trabecular ($\rho \leq 1.2$ g/cm^3) and cortical bone ($\rho < 1.2$ g/cm^3). The Y axis shows the percentage of the total volume occupied by the elements, whose AMPσ or AMPε are in a given range.

4. Discussion

A linear function (Equation (3)) was used here to estimate the bone apparent density from the grey level of the CT scan [23–26]. As in these previous works, the CT scan could not be calibrated, and thus, a threshold was set by identifying the grey level of the diaphyseal canal with bone marrow, that is, with null bone density. Notwithstanding, the greyscale value inside the canal was found to vary in the range of 70–90 and, thus, three different grey level thresholds (GLT) were used and their effect was investigated. The same trend was observed in R for the three values of GLT. Therefore, we can state that the choice of GLT had no effect on the correlation coefficients (see Tables 1 and 2), and hence, on the overall conclusions of this study. For that reason, the rest of results were compared only for the intermediate GLT = 80.

In general, the variables derived from the strain tensor has a low correlation with density and this can be due to the fact that the strains are concentrated in a narrow range, as deduced from the histograms. On the contrary, the stress-related variables are distributed over a much wider range and they are very closely related to density as the correlation coefficients showed. The high correlation found between density and the von Mises, Tresca, absolute maximum principal stress (AMPσ) and the fluctuation of stresses (σ_f) are noteworthy.

On the other hand, SED is only modestly correlated with density, probably because it depends on the strains, which are very poorly correlated with ρ. Since SED also depends on the stress level, the histograms of SED (not shown) are as spread as the histograms of stress; however, in view of the correlations, its variation does not seem to be as coupled to density as the mere variation of stress is. However, if SED seems a modest predictor of density, the mechanical stimulus at the tissue level proposed by Beaupré (Ψ_t) is even worse, as it yields negative correlation coefficients.

The Spearman coefficient is probably the most simple indicator that two variables are correlated by a monotonic function, since it does not assume any specific relationship as does the Pearson coefficient. If R_s is positive and high for a certain variable, this means that density increases with that variable, and that a large part of that increase can be explained by the variable, without assessing what specific relationship exists. Hence, that variable can serve as a good predictor of density in a bone remodelling model, though the particular relationship between the density and the specific variable should be further investigated. In this regard, the linear, quadratic or power functions tried for the Pearson coefficients worked only moderately well. Only σ_f and AMPσ showed a value of R_P slightly over 0.8 (up to 0.861 for σ_f in the *IsoJ* model). This means that $R^2 > \sim 0.64$, i.e., around 64% of the variation of the density, can be explained by σ_f (up to 74% in the *IsoJ* model).

Only some of the variables appearing in Table 1 were chosen for assessing its Pearson coefficient, those having a high Spearman coefficient plus SED and its related variables (Ψ_t), for its common use as a mechanical stimulus in many models of bone remodelling. Obviously, if the Spearman coefficients of SED and its related variables were low, the respective Pearson coefficients could not improve them, but it is worth noting the very low values of R_P obtained for SED compared to R_S. This would mean that Equation (37) was not very appropriate, probably because of the many assumptions involved in it, viz.: uniaxial stress-state, isotropic material and power correlation between Young's modulus and density.

The distribution of two variables, one representing the stress-state (AMPσ) and one representing the strain-state (AMPε), was analysed by means of histograms (see Figure 6) that distinguish between cortical ($\rho > 1.2$ g/cm^3) and trabecular bone ($\rho \leq 1.2$ g/cm^3) These histograms showed that the strains are concentrated in a relatively narrow range in the case of trabecular bone. Around 65% of the volume of trabecular bone was found to be within the range AMP$\varepsilon \in [200–1400]$ $\mu\varepsilon$ in the *AnisoH* model, 62% in *IsoH* and 53% in the *IsoJ*. The range was particularly narrow in the case of cortical bone, as about 85% of its total volume was found in the range AMP$\varepsilon \in [200–600]$ $\mu\varepsilon$ for all constitutive models.

It is noteworthy that the range of strains obtained here was low compared with the normal strains indicated by the Mechanostat Theory (between 800 and 1200 $\mu\varepsilon$ [10]) Nonetheless, other authors have established a different range of normal strains in the so-called "adapted-window" of the Mechanostat, between 200 and 1500 $\mu\varepsilon$ [36,37]. Yet, the strains seem relatively low for models, *AnisoH* and *IsoH*, which are likely overestimating the bone stiffness. The correlation, (5a), used by model *IsoJ*, is probably more adequate in view of the strains it produces.

In contrast to strains, the stresses are more uniformly distributed and over a much wider range, with the stresses of trabecular bone being one or two orders of magnitude lower than those of cortical bone. This strong relationship between bone density and stress is confirmed by the high Spearman correlation coefficients of most stress variables (see Table 1).

The distributions and correlations of strain and stress variables would confirm that bone is adapted to withstand a constant strain, or at least a strain within a narrow range, in accordance with the Mechanostat Theory, while the local stress-state seems to determine the bone density. This would suggest the use of a strain variable as the mechanical stimulus S in evolutionary BRM (see Equation (1)), such that bone density changes if the strain is out of the normal or adapted range. On the contrary, a stress variable would be more appropriate for the stimulus in a non-evolutionary BRM (see Equation (2)), such that the stress would determine the bone density at a given location. In the case of cyclic loads, such as the one applied here, a variable measuring the fluctuations of stresses throughout the cycle seems the more appropriate mechanical stimulus among those tested here, though the specific relationship between density and stress is yet to be determined. In no case does the SED seem a suitable variable to be used as the mechanical stimulus in BRM.

We have compared three constitutive models (*AnisoH*, *IsoH* and *IsoJ*) in order to check whether the correlation between predictors and density, $S - \rho$, is being forced by the relationship between stiffness and density, $E = F(\rho)$, in particular for those predictors derived from the stress tensor (for the sake of generality, we should write $\mathbf{C} = F(\rho)$, with $\mathbf{C}$ being the stiffness tensor, in order to account for anisotropic materials. However, the rationale provided is independent of this distinction and we will continue with $E = F(\rho)$). It is well known from the literature that F is a monotonically increasing function and given that the stiffer elements tend to withstand higher stress levels, we should expect a function such as $\sigma = H(E)$ to be monotonically increasing as well. Thus, if we write:

$$\sigma = H\big(F(\rho)\big), \tag{38}$$

we can expect a monotonically increasing function relating σ and ρ, in other words, a positive correlation $S - \rho$ for the predictors derived from the stress tensor. Therefore, it could seem that the function $E = F(\rho)$ is forcing $S - \rho$ and it is so to some extent, especially for the Spearman correlation coefficients. However, if $E = F(\rho)$ were the only determinants of the correlation $S - \rho$, this correlation should depend on the constitutive model and it does not—significantly, at least. Indeed, the increase in stiffness from *IsoJ* to *IsoH* is quite remarkable (up to 70%, see Figure 3, left) and yet the Pearson correlations do not change much from one model to the other (see Table 4). Moreover, since the ratio of stiffness is non-uniform (see Figure 3, right), a redistribution of stresses could be expected that would cause the stress patterns of the two models not to coincide. The same could be said of the comparison between models *AnisoH* and *IsoH*. Such redistribution of stresses would, therefore, affect the Spearman correlation coefficients or the AMPσ histograms, but both are almost identical for the three models (see Table 3 and Figure 6). Hence, we can conclude that the assumed constitutive model has no significant effect on the correlations.

However, there is another argument to support that $S - \rho$ is not completely forced by the constitutive behaviour, but only to some extent. Certainly, Equation (38) is incomplete as the stress-state also depends on the loads and boundary conditions:

$$\sigma = H\big(F(\rho), \mathbf{F}\big), \tag{39}$$

where $\mathbf{F}$ stands for the set of loads applied to the domain, including the body forces, the surface tractions applied on the boundaries (i.e., Neumann boundary conditions) and the reaction forces (and consequently the Dirichlet boundary conditions). Apart from that, the predictors, S, are obtained as a function of the stress tensor, that we can denote as $S = G(\sigma)$, thus leading to:

$$S = G\Big(H\big(F(\rho), \mathbf{F}\big)\Big). \tag{40}$$

In view of Equation (40) the predictors, S, need not be predetermined only by the function $F(\rho)$, not even the Spearman correlation coefficients, as the function G is not necessarily monotonic. This is confirmed by the different Spearman coefficients obtained for different predictors (see Tables 1 and 3). It should be noted that the same concepts that are behind the G functions can be applied to the predictors derived from the strain tensor or the SED-based predictors.

An example can serve to illustrate the key role played by the loads on the correlations $S - \rho$. We have obtained these correlations for every instant of the gait cycle separately. The stress and strain tensors obtained at each instant i were used to assess S_e^i (recall Section 2.7) and these variables were correlated with ρ_e to give R_p^i, a weighted Pearson correlation coefficient for each instant i. The worst coefficient between AMPσ and density throughout the cycle was $R_p^i = 0.48$ (for the quadratic fit and in the case GLT = 80, constitutive model *IsoH* and C3D4 elements), which is far from $R_p^M = 0.804$. The former is a very poor correlation. $(R_p^i)^2 = 0.23$, i.e., only about 25% of the variance of density can be explained by AMPσ^i and this is probably because the loads at that instant are not representative of the gait cycle. In fact, the loads at a single instant cannot be representative of the entire cycle on a general basis.

In summary, the function $E = F(\rho)$ is contributing to obtain the correlations $S - \rho$, at least for stress-based predictors, but only to some extent. There are other factors, not influenced by that function, that play an important role in the correlations; an accurate estimation of the loads is crucial, taking into account the variation of stresses throughout the cycle is also important and an appropriate choice of S (funtion G) is key to predicting density.

The type of element (C3D4 vs. C3D10) had a relative importance on the results, as the linear elements (C3D4) are stiffer than the quadratic ones (C3D10). The latter yielded moderately higher stresses and strains, which were slightly better correlated with density through a power law than those obtained with C3D4 elements (see Table 6). The linear and quadratic fits (not shown) also improved with C3D10, but to a lesser extent. Notwithstanding, the C3D4 mesh had a sufficient element density and was accurate enough as evidenced by the fact that the stress and strain patterns are almost identical in both meshes. Consequently, the Spearman correlation coefficients derived from both are also almost identical. For that reason, the use of C3D4 elements was justified, at least for the mesh density employed in our model.

Among the limitations of this study we must note that only one activity (walking) has been considered, among the many activities that a person can carry out in a normal day. Other activities can load the femur in a different way than walking, thus, affecting the local bone density. This could partially explain the variance of density not explained by our correlations. However, walking is by far the most common activity affecting the lower limb and the most common activity performed by the subject under study. Hence, little influence of other activities can be expected.

The joint loads were applied in a simplified way, as nodal-concentrated loads instead of loads distributed over the articular surface. This approximation is especially significant in the knee reaction force and this could explain the lower correlation coefficients found in the distal third of the femur. Another limitation is that the CT scan could not be calibrated, but we have shown that this fact did not influence the conclusions drawn. Finally, it must be noted that only one individual, a male healthy subject, was studied. As future work, it would be interesting to repeat the study in other cases, including different bones, age, gender, health status, bone pathologies, etc., in order to confirm whether these variables or others alter the dependence of bone density on the proposed predictors.

5. Conclusions

In view of the results of this study we can conclude the following:

- Stresses are highly variable in the femur and very different in cortical and trabecular sites.
- Stresses are strongly correlated with density and seem to determine it. Therefore, certain variables derived from the stress tensor could be good candidates for the mechanical stimulus in non-evolutionary BRM that can be used in DPTBR problems.
- In the case of cyclic loads, the fluctuation of stresses throughout the cycle is the best predictor of density among those tested in this work.
- In contrast to the above, strains are relatively uniform across the femur and remain within the "adapted-window" established by the Mechanostat Theory.
- Strains are very poorly correlated with density and all indications are that bone tends to keep them approximately constant. Therefore, variables derived from the strain tensor could be good candidates as mechanical stimulus in evolutionary BRM, in which density would be forced to change in order to maintain strains within the "adapted-window".
- Strain energy density does not seem to be a good variable to measure the mechanical stimulus in any case, and even less so if this variable is evaluated at the tissue level by using the porosity correction factor.

Supplementary Materials: The following supporting information can be downloaded at: https://www.mdpi.com/article/10.3390/math10183367/s1.

Author Contributions: Conceptualization, J.M.-R., J.O., J.L.C.-G. and M.Á.P.; methodology, J.M.-R., J.O. and J.L.C.-G.; software, F.G.-M., J.L.C.-G., J.M.-R. and J.O.; formal analysis, F.G.-M., J.M.-R. and J.O.; writing—original draft preparation, J.M.-R., J.O. and J.L.C.-G; writing—review and editing, J.L.C.-G. and M.Á.P.; visualization, J.L.C.-G., J.O. and J.M.-R.; funding acquisition, J.M.-R. All authors have read and agreed to the published version of the manuscript.

Funding: Funding was provided by the Fondo Europeo de Desarrollo Regional (FEDER), the Consejería de Economía, Conocimiento, Empresas y Universidad de la Junta de Andalucía, dentro del Programa Operativo FEDER 2014–2020 and the Universidad de Sevilla to the project P18-RT-3611 entitled "Efecto combinado del ejercicio físico y el denosumab en el tratamiento de la osteoporosis. Diseño de un tratamiento farmacológico específico de paciente" for which this paper was prepared.

Institutional Review Board Statement: The study was conducted in accordance with the Declaration of Helsinki, and approved by the Andalusian Biomedical Research Ethics Platform (approval number 20151012181252).

Informed Consent Statement: Informed consent was obtained from all subjects involved in the study.

Data Availability Statement: The finite element model, including boundary conditions and loads, used in this study is available in the Supplementary Materials. Other data presented in this study are not publicly available because they contain patient information.

Conflicts of Interest: The authors declare no conflicts of interest. The funders had no role in the design of the study; in the collection, analyses, or interpretation of data; in the writing of the manuscript; or in the decision to publish the results.

Abbreviations

The following abbreviations are used in this manuscript:

BRM	Bone remodelling model(s)
SED	Strain energy density
FE(M)	Finite Element (Method)
CT	Computer tomography
DPTBR	Density prediction through bone remodelling
GLT	Grey level threshold
AMPε	Absolute maximum principal strain
AMPσ	Absolute maximum principal stress
MF	Muscle force(s)
HJF	Hip joint force
KJF	Knee joint force

Appendix A. Weighted Pearson and Spearman Correlation Coefficients

In the case of a linear correlation, the Pearson coefficient between the bone apparent density ρ and the predictor var is weighted with the volume of the finite elements, which are the sample points. The weighted Pearson coefficient is defined as the weighted covariance divided by the weighted variances of both variables and determines the proportion of variance explained by the weighted linear fit of the point cloud (ρ_e , var_e):

$$R_P = \frac{\sum_{e=1}^{n}[V_e\,(\rho_e - \bar{\rho})(var_e - \overline{var})]}{\sqrt{\sum_{e=1}^{n}[V_e\,(\rho_e - \bar{\rho})^2]\ \cdot\ \sum_{e=1}^{n}[V_e\,(var_e - \overline{var})^2]}} \tag{A1}$$

with the weighted averages being $\overline{\bullet}$, for example, in the case of the predictor *var*:

$$\overline{var} = \frac{\sum\limits_{e=1}^{n} var_e \, V_e}{\sum\limits_{j=1}^{n} V_e} \tag{A2}$$

and with an analogous expresion for the weighted average of density, $\bar{\rho}$. V_e is the volume of the element e, used to weight the variable var_e.

The weighted Spearman correlation coefficient, R_S, is obtained through Equations (A1) and (A2) by simply replacing the pairs $(\rho_e \, , var_e)$ with the pairs composed of their respective ranks, i.e., $\left(\mathrm{rank}(\rho_e) \, , \mathrm{rank}(var_e)\right)$.

The weighted Pearson correlation coefficient was also obtained for both a quadratic and a power fit of the point cloud $(\rho_e \, , var_e)$. In the first case, the following function was fitted to the point cloud in the least squares sense, after being weighted with the element volume, using the *polyfitweighted* library of MATLAB:

$$\tilde{var}(\rho) = A\rho^2 + B\rho + C. \tag{A3}$$

In least squares fitting the Pearson coefficient is defined as the quotient between the covariance of the fitted and raw variables divided by the variances of both. In weighted fittings Equation (A1) is modified as follows:

$$R_P = \frac{\sum\limits_{e=1}^{n} [V_e \, (\tilde{var}(\rho_e) - \overline{\tilde{var}})(var_e - \overline{var})]}{\sqrt{\sum\limits_{e=1}^{n} [V_e \, (\tilde{var}(\rho_e) - \overline{\tilde{var}})^2] \cdot \sum\limits_{e=1}^{n} [V_e \, (var_e - \overline{var})^2]}}. \tag{A4}$$

In the case of a power fit, the function to be fitted $(var = K\rho^\beta)$ is transformed into a linear function by taking logarithms of both sides:

$$\log var = K + \beta \log \rho. \tag{A5}$$

Then, the Pearson coefficient of a linear fit (Equation (A1)) is used by replacing the pairs $(\rho_e \, , var_e)$ with $(\log \rho_e \, , \log var_e)$.

Appendix B. Inverse Dynamic Problem Methodology

The forces applied to the FE model were obtained solving an inverse dynamics problem, which is briefly described next (a more detailed description can be found for example in [38,39]). All the simulations were run in OpenSim. First, the inverse kinematics problem was solved. The input data in this problem are the trajectories of the markers, $\mathbf{x}^{\mathrm{ext}}$, obtained experimentally for the gait cycle by using a Vicon®system. The output is the temporal evolution of the generalized coordinates, $\mathbf{q}$. The inverse kinematics problem was solved using a least square pose estimator:

$$\min_{\mathbf{q}} \left(\sum_{i \in markers} w_i \, \|\mathbf{x}_i^{\mathrm{exp}} - \mathbf{x}_i(\mathbf{q})\|^2 \right), \tag{A6}$$

where $\mathbf{x}_i(\mathbf{q})$ is the position of the virtual marker i, which depends on the coordinates values, and w_i is a marker weight taken from [17,21]. The results of the inverse kinematics problem were used as input to solve the inverse dynamics problem, by means of the classical equations of motion:

$$\mathbf{M}(\mathbf{q})\ddot{\mathbf{q}} + \mathbf{C}(\mathbf{q}, \dot{\mathbf{q}}) + \mathbf{G}(\mathbf{q}) = \mathbf{0}, \tag{A7}$$

where $\mathbf{q}$, $\dot{\mathbf{q}}$ and $\ddot{\mathbf{q}}$ are the vectors of generalized positions, velocities and accelerations, respectively; $\mathbf{M}$ is the mass matrix; $\mathbf{C}$ is the vector of quadratic velocities; $\mathbf{G}$ is the vector of external forces including gravitational forces and ground reaction forces and $\boldsymbol{\varnothing}$ is the vector of generalized forces regarding motor tasks. Finally, muscle forces ($\mathbf{F_{mus}}$) were estimated solving the following optimization problem:

$$\min\left(J = \sum_{m=1}^{n} (a_m)^p \right)$$
$$s.t. \quad \sum_{m=1}^{n} [a_m \cdot f(F_m^0, l_m, v_m)] \cdot r_{m,j} = \tau_j \tag{A8}$$

where n is the number of muscles in the model; a_m is the activation level of muscle m at a given time step; F_m^0 its maximum isometric force; l_m its length; v_m its shortening velocity; $f(F_m^0, l_m, v_m)$ its force-length-velocity relation; $r_{m,j}$ its moment arm about the jth joint axis and τ_j is the generalized force acting about the jth joint axis. The cost function J of Equation (A8) minimizes the sum of muscle activation squared (p = 2) as in [17]. As an example, Figure A1 shows the temporal evolution of the forces exerted by the iliacus and the lateral gastrocnemius obtained following this procedure. The temporal evolutions of all the muscles considered in the analysis are included in the Supplementary Materials.

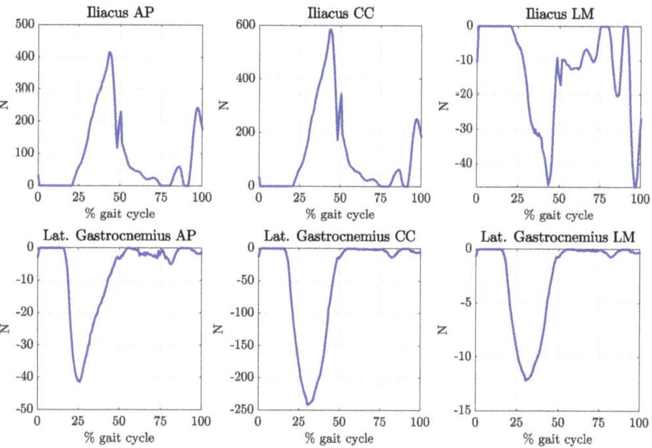

Figure A1. Temporal evolution of the components of the forces exerted by the iliacus and the lateral gastrocnemius throughout the gait cycle. AP: antero-posterior. CC: Cranio-caudal. LM: Lateral-medial.

Finally, the bone-on-bone forces on the hip and the knee were calculated using the algorithm proposed by Steele [22].

$$\mathbf{R_i} = \mathbf{M_i}(\mathbf{y})\ddot{\mathbf{y}}_i - \left(\sum \mathbf{F_{muscles}} + \sum \mathbf{F_{external}} + \mathbf{R_{i-1}} \right) \tag{A9}$$

In this equation, vector $\mathbf{R_i}$ contains the resultant forces and moments at joint i or bone-on-bone forces. The body distal to joint i, B_i, is treated as an independent body with known kinematics in a global reference frame. Thus, $\ddot{\mathbf{y}}_i$ represents the six dimensional vector of known angular and linear accelerations of B_i, while $\mathbf{M_i}(\mathbf{q})$ is the 6×6 mass matrix of B_i. $\mathbf{F_{external}}$ and $\mathbf{F_{muscles}}$ represent the previously calculated forces and moments produced by external loads and musculotendon actuators, respectively. $\mathbf{R_{(i-1)}}$ represents the joint reaction load applied at the distal joint and was calculated in the previous recursive step. Our aim was to study bone-on-bone forces only at the hip and the knee joints, thus, the generic name $\mathbf{R_i}$ will be replaced by **HJF** and **KJF** (acronyms for hip and knee joint forces, respectively) and expressed in the local frame attached to the femur. Figure A2 shows the temporal evolution of **HJF** and **KJF** throughout the gait cycle. These data are included in the Supplementary Materials.

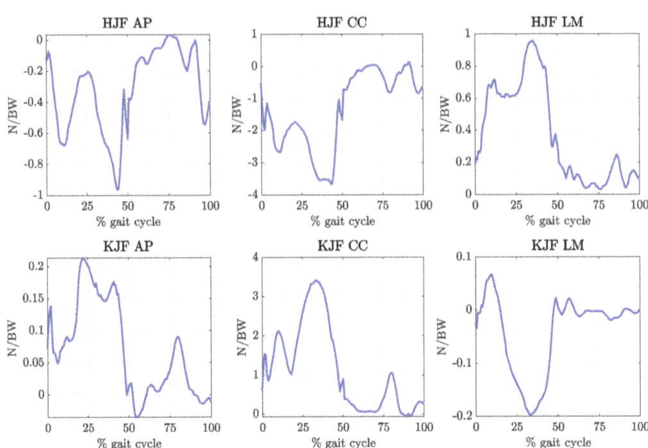

Figure A2. Temporal evolution of the HJF and KJF throughout the gait cycle. AP: Antero-posterior. CC: Cranio-caudal. LM: Lateral-medial. BW: Body weight.

References

1. Jacobs, C.R.; Simo, J.C.; Beaupre, G.S.; Carter, D.R. Adaptive bone remodeling incorporating simultaneous density and anisotropy considerations. *J. Biomech.* **1997**, *30*, 603–613. [CrossRef]
2. Huiskes, R.; Weinans, H.; Grootenboer, H.J.; Dalstra, M.; Fudala, B.; Sloof, T.J. Adaptive bone-remodeling theory applied to prosthetic-design analysis. *J. Biomech.* **1987**, *20*, 1135–1150. [CrossRef]
3. Beaupré, G.S.; Orr, T.E.; Carter, D.R. An approach for time-dependent bone modeling and remodeling–Theoretical development. *J. Orthop. Res.* **1990**, *8*, 651–661. [CrossRef] [PubMed]
4. Doblaré, M.; García, J.M. Anisotropic bone remodelling modelbased on a continuum damage-repair theory. *J. Biomech.* **2002**, *35*, 1–17. [CrossRef]
5. Martínez-Reina, J.; García-Aznar, J.M.; Domínguez, J.; Doblaré, M. A bone remodelling model including the directional activity of BMUs. *Biomech. Model. Mechanobiol.* **2009**, *8*, 111–127. [CrossRef]
6. Beaupré, G.S.; Orr, T.E.; Carter, D.R. An approach for time-dependent bone modeling and remodeling–Application: A preliminary remodeling simulation. *J. Orthop. Res.* **1990**, *8*, 662–670. [CrossRef]
7. Doblaré, M.; García, J.M. Application of an anisotropic bone-remodelling model based on a damage-repair theory to the analysis of the proximal femur before and after total hip replacement. *J. Biomech.* **2001**, *34*, 1157–1170. [CrossRef]
8. Reina, J.; García-Aznar, J.M.; Domínguez, J.; Doblaré, M. Numerical estimation of bone density and elastic constants distribution in a human mandible. *J. Biomech.* **2007**, *40*, 828–836. [CrossRef]
9. Calvo-Gallego, J.L.; Pivonka, P.; García-Aznar, J.M.; Martínez-Reina, J. A novel algorithm to resolve lack of convergence and checkerboard instability in bone adaptation simulations using non-local averaging. *Int. J. Numer Methods Biomed. Eng.* **2021**, *37*, e3419. [CrossRef]
10. Frost, H.M. Bone's mechanostat: A 2003 update. *Anat. Rec. A Discov. Mol. Cell Evol. Biol.* **2003**, *275*, 1081–1101. [CrossRef]
11. Martínez-Reina, J.; Ojeda, J.; Mayo, J. On the use of bone remodelling models to estimate the density distribution of bones. Uniqueness of the solution. *PLoS ONE* **2016**, *11*, e0148603. [CrossRef] [PubMed]
12. Prendergast, P.J.; Taylor, D. Prediction of bone adaptation using damage accumulation. *J. Biomech.* **1994**, *27*, 1067–1076. [CrossRef]
13. Hazelwood, S.J.; Martin, R.B.; Rashid, M.M.; Rodrigo, J.J. A mechanistic model for internal bone remodeling exhibits different dynamic responses in disuse and overload. *J. Biomech.* **2001**, *34*, 299–308. [CrossRef]
14. García, J.M.; Rueberg, T.; Doblaré, M. A bone remodelling model coupling micro-damage growth and repair by 3D BMU-activity. *Biomech. Model. Mechanobiol.* **2005**, *4*, 147–167. [CrossRef] [PubMed]
15. Martínez-Reina, J.; Reina, I.; Domínguez, J.; García-Aznar, J.M. A bone remodelling model including the effect of damage on the steering of BMUs. *J. Mech. Behav. Biomed. Mater.* **2014**, *32*, 99–112. [CrossRef] [PubMed]
16. Sutherland, D.H. The evolution of clinical gait analysis part II: Kinematics. *Gait Posture* **2002**, *16*, 159–179. s0966-6362(02)00004-8. [CrossRef]

17. Delp, S.L.; Anderson, F.C.; Arnold, A.S.; Loan, P.; Habib, A.; John, C.T.; Guendelman, E.; Thelen, D.G. OpenSim: Open-source software to create and analyze dynamic simulations of movement. *IEEE Trans. Biomed. Eng.* **2007**, *54*, 1940–1950. [CrossRef] [PubMed]
18. Van der Helm, F.C.; Veenbaas, R. Modelling the mechanical effect of muscles with large attachment sites: Application to the shoulder mechanism. *J. Biomech.* **1991**, *24*, 1151–1163. [CrossRef]
19. Brand, R.A.; Pedersen, D.R.; Friederich, J.A. The sensitivity of muscle force predictions to changes in physiologic cross-sectional area. *J. Biomech.* **1986**, *19*, 589–596. [CrossRef]
20. Brand, R.A.; Pedersen, D.R.; Davy, D.T.; Kotzar, G.M.; Heiple, K.G.; Goldberg, V.M. Comparison of hip force calculations and measurements in the same patient. *J. Arthroplast.* **1994**, *9*, 45–51. [CrossRef]
21. Martín-Sosa, E.; Martínez-Reina, J.; Mayo, J.; Ojeda, J. Influence of musculotendon geometry variability in muscle forces and hip bone-on-bone forces during walking. *PLoS ONE* **2019**, *14*, e0222491. [CrossRef] [PubMed]
22. Steele, K.M.; Demers, M.S.; Schwartz, M.H.; Delp, S.L. Compressive tibiofemoral force during crouch gait. *Gait Posture* **2012**, *35*, 556–560. [CrossRef] [PubMed]
23. Kerner, J.; Huiskes, R.; van Lenthe, G.H.; Weinans, H.; van Rietbergen, B.; Engh, C.A.; Amis, A.A. Correlation between pre-operative periprosthetic bone density and post-operative bone loss in THA can be explained by strain-adaptive remodelling. *J. Biomech.* **1999**, *32*, 695–703. [CrossRef]
24. Weinans, H.; Sumner, D.R.; Igloria, R.; Natarajan, R.N. Sensitivity of periprosthetic stress-shielding to load and the bone density-modulus relationship in subject-specific finite element models. *J. Biomech.* **2000**, *33*, 809–817. [CrossRef]
25. Wang, X.; Zauel, R.R.; Fyhrie, D.P. Postfailure modulus strongly affects microcracking and mechanical property change in human iliac cancellous bone: A study using a 2D nonlinear finite element method. *J. Biomech.* **2008**, *41*, 2654–2658. [CrossRef] [PubMed]
26. Pomwenger, W.; Entacher, K.; Resch, H.; Schuller-Götzburg, P. Need for CT-based bone density modelling in finite element analysis of a shoulder arthroplasty revealed through a novel method for result analysis. *Biomed. Eng.* **2014**, *59*, 421–430. [CrossRef]
27. Cuppone, M.; Seedhom, B.B.; Berry, E.; Ostell, A.E. The longitudinal Young's modulus of cortical bone in the midshaft of human femur and its correlation with CT scanning data. *Calcif. Tiss. Int.* **2004**, *74*, 302–309. [CrossRef]
28. Jacobs, C.R. Numerical Simulation of Bone Adaptation to Mechanical Loading. Ph.D. Thesis, Stanford University, Stanford, CA, USA, 1994.
29. Hernandez, C.J.; Beaupre, G.S.; Keller, T.S.; Carter, D.R. The influence of bone volume fraction and ash fraction on bone strength and modulus. *Bone* **2001**, *29*, 74–78. [CrossRef]
30. Martínez-Reina, J.; Domínguez, J.; García-Aznar, J.M. Effect of porosity and mineral content on the elastic constants of cortical bone: A multiscale approach. *Biomech. Model. Mechanobiol.* **2011**, *10*, 309–322. [CrossRef]
31. Cowin, S. On the strength anisotropy of bone and wood. *J. Appl. Mech.* **1979**, *46*, 832–838. [CrossRef]
32. Ojeda, J. Aplicación de las Técnicas MBS al Sistema Locomotor Humano. Ph.D. Thesis, Universidad de Sevilla, Sevilla, Spain, 2012.
33. Cordebois, J.P.; Sideroff, F. Damage Induced Elastic Anisotropy In *Mechanical Behavior of Anisotropic Solids*; Springer: Dordrecht, The Netherlands, 1982; pp. 761–774.
34. Cowin, S.C.; Sadegh, A.M.; Luo, G.M. An evolutionary Wolff's law for trabecular architecture. *J. Biomech. Eng.* **1992**, *114*, 129–136. [CrossRef] [PubMed]
35. Carter, D.R.; Fyhrie, D.P.; Whalen, R.T. Trabecular bone density and loading history: Regulation of connective tissue biology by mechanical energy. *J. Biomech.* **1987**, *20*, 785–794. [CrossRef]
36. Davies, H.M.S. The Adaptive Response of the Equine Metacarpus to Locomotory Stress. Ph.D. Thesis, University of Melbourne, Melbourne, Australia, 1995.
37. Nagaraja, S. Microstructural Stresses and Strains Associated with Trabecular Bone Microdamage. Ph.D. Thesis, Georgia Institute of Technology, Atlanta, GA, USA, 2006.
38. Mayo, J.; Ojeda, J. Influence of the kinematic constraints on dynamic residuals in inverse dynamic analysis during human gait without using force plates. *Multibody Syst. Dyn.* **2020**, *50*, 305–321. [CrossRef]
39. Ojeda, J.; Martínez-Reina, J.; Mayo, J. The effect of kinematic constraints in the inverse dynamics problem in biomechanics. *Multibody Syst. Dyn.* **2016**, *37*, 291–309. [CrossRef]

 *mathematics*

Article

Modified Whiteside's Line-Based Transepicondylar Axis for Imageless Total Knee Arthroplasty

Muhammad Sohail, Jaehyun Park, Jun Young Kim, Heung Soo Kim * and Jaehun Lee *

Department of Mechanical, Robotics and Energy Engineering, Dongguk University-Seoul, Seoul 04620, Korea
* Correspondence: heungsoo@dgu.edu (H.S.K.); jaehun@dgu.edu (J.L.)

Abstract: One of the aims of successful total knee arthroplasty (TKA) is to restore the natural range of motion of the infected joint. The operated leg motion highly depends on the coordinate systems that have been used to prepare the bone surfaces for an implant. Assigning a perfect coordinate system to the knee joint is a considerable challenge. Various commercially available knee arthroplasty devices use different methods to assign the coordinate system at the distal femur. Transepicondylar axis (TEA) and Whiteside's line are commonly used anatomical axes for defining a femoral coordinate system (FCS). However, choosing a perfect TEA for FCS is trickier, even for experienced surgeons, and a small error in marking Whiteside's line leads to a misaligned knee joint. This work proposes a modified Whiteside's line method for the selection of TEA. The Whiteside's line, along with the knee center and femur head center, define two independent central planes. Multiple prominent points on the lateral and medial sides of epicondyles are marked. Based on the lengths of perpendicular distances between the multiple points and central planes, the most prominent epicondyle points are chosen to define an optimal TEA. Compared to conventional techniques, the modified Whiteside's line defines a repeatable TEA

Keywords: imageless navigator; knee alignment device; total knee arthroplasty; repeatable transepicondylar axis; femoral coordinate system; modified Whiteside's line

MSC: 92B05

Citation: Sohail, M.; Park, J.; Kim, J.Y.; Kim, H.S.; Lee, J. Modified Whiteside's Line-Based Transepicondylar Axis for Imageless Total Knee Arthroplasty. *Mathematics* **2022**, *10*, 3670. https://doi.org/10.3390/math10193670

Academic Editor: Mauro Malvè

Received: 7 September 2022
Accepted: 1 October 2022
Published: 7 October 2022

Publisher's Note: MDPI stays neutral with regard to jurisdictional claims in published maps and institutional affiliations.

1. Introduction

The knee joint is the largest joint in the human body. Osteoarthritis-induced wear and tear of the articulating surfaces of the knee causes discomfort to the patient [1]. Arthroplasty is considered to be the permanent solution to treat the most severe knee arthritis [2]. Surgeons perform total knee arthroplasty (TKA) relying either on an imaging technique like ultrasound [3] or an imageless computer-based technique [4]. The computer-assisted TKA is becoming popular, because most of the alignment work is done by computers; hence, the transplanted joint is expected to be accurately aligned, with quicker recovery time, due to minimal incision [5]. The computer-assisted TKA device consists of software that takes some anatomical points as input. The marking system of these anatomical points usually relies on an infrared, laser, or electromagnetic pulses emitter, and reflectors paired with an optical camera to obtain the real-time position of the marked point [6–8].

The software constructs a virtual coordinate system for the knee to undergo arthroplasty, and based on the coordinate system, it guides the surgeon to prepare the bone surface for implant. The coordinate system can be assigned by different methodologies, for example, the transverse axis at the distal femur for assigning a femoral coordinate system (FCS) can be obtained by fitting a circle, a sphere, or a cylinder in the articulating surfaces of the distal femur, and using the center of the fitted geometry, the transverse axis is assigned [9,10]. Another approach is to join the most prominent points on the lateral and medial epicondyles to construct an anatomical transepicondylar axis (aTEA) [11]. Similarly, a surgical transepicondylar axis (sTEA) can be defined by joining the lateral epicondyle and medial sulcus [12]. In some cases, the posterior condylar axis (PCA) being in front of the surgeon can be easily used to define the transverse axis [13]. PCA and sTEA are parallel to each other, and perpendicular to the anterior posterior axis (AP axis) or Whiteside's line [14,15]. Whiteside's line is defined by using the deepest part of the patella groove (PG) anteriorly and the center of intercondylar notch (N) posteriorly [16]. sTEA is considered to be a standard to set the rotational alignment for TKA [17,18]. However, some studies have compared the sTEA and aTEA for their ability to be chosen as a transverse axis for FCS. For example, Tanavalee et al. studied the CT scans of 55 osteoarthritic knees considering both sTEA and aTEA, and they concluded that aTEA is near perpendicular to the AP axis and more reliable for rotation alignment compared to sTEA [19]. In the same year, Yoshino et al. examined 48 patients eligible for TKA and found that the medial sulcus was detectable in only one-fifth of the severe osteoarthritis cases; for less severe cases, it was detectable in half of the cases [20]. Hence, it was concluded that the chances of detecting the medial sulcus decrease with an increase in the severity of osteoarthritis.

Deterioration of PCA in severe osteoarthritis makes it less favorable for the selection of the transverse axis [20]. Whiteside's line being a smaller landmark leads to rotational error of up to $10°$, even for small uncertainties [21]. It is a well-known fact that alignment error $>3°$ from the natural alignment of the knee leads to quick wear and discomfort of the patient, and finite element analysis can accurately predict initial stability of an implant; however, the computational cost increases [15,22,23]. The sTEA, being an inconsistent landmark [24], leaves only aTEA to be a reliable choice for assigning the FCS and setting the rotational alignment of the implant [19,25]. Malrotation alignment leads to instability of implant, the discomfort of the patient, patella maltracking, and quick wear of inserted polyethylene [26]. So, the TKA is sometimes required to be revised. Fehring et al. summarized the 15-year data of revised TKA cases and found that 41% of the cases showed conditions related to rotational malalignment [27]. Another similar case study of 1632 revised TKA cases showed that 42% of the cases were linked with wear of the insertion and instability of the implant [28]. Dalury et al. also investigated the reasons why TKAs are revised, based on the analysis, 48% of the cases were linked with the symptoms attached to rotational malalignment [29]. Choosing a perfect aTEA is not always possible, which leads to rotational malalignment. Stoeckl et al. invited a team of four experienced surgeons to mark six cadaveric human legs. Skin and soft tissues were removed, and surgeons had to pick the most prominent point on lateral and medial epicondyles using an optical navigation system under perfect laboratory conditions. Each surgeon marked aTEA for three consecutive days; 144 points were marked by all the surgeons. Excluding extreme values, the selected points were distributed in an area of 298 mm^2 on the medial and 278 mm^2 on the lateral side of the bone. Using the extreme values of the marked area, a maximum of $8°$ of internal rotation was calculated ($3°$ allowed) [30]. Another study investigating the reproducibility of aTEA, where eight surgeons marked lateral and medial epicondyles on Thiel-embalmed cadaver specimens, shows the distribution of lateral and medial epicondyles on an area of 116 and 102 mm^2, respectively [31]. A team of another five surgeons studied the effect of errors in registering anatomical points on five cadavers for five days. Key anatomical points, including lateral and medial epicondyles, were intentionally marked wrong. The wrong registration of lateral and medial epicondyles led to an error in rotational alignment ranging from $11.1°$ external to $6.3°$ internal rotation [32]. Similarly, there are various studies in which the errors are either intentionally added, or induced due to human error, by the repetitive marking of points during a course of time. The objective of such studies is to show the effect of errors on the rotational alignment of TKA [32–36].

To reduce the errors caused by choosing a single point, bone morphing or selection of multiple points on the landmark have been reported in the literature. For example, Liu et al. marked a group of points and fitted an algorithm to choose the optimal point for defining TEA [37]. Perrin et al. studied the reproducibility of implant positioning in TKA using both morphing and conventional single-point selection techniques, and found the bone morphing technique to be more repeatable [38]. A system using bone morphing relies on a database gathered by CT scans from a large number of patients supplied by the device manufacturers. Once multiple anatomical points are registered by the surgeon, the algorithm finds the bone model that best fits the marked points [39]. Instead of model fitting, statistical shape models of the bone can be fitted in a marked cloud of points. The bone morphing requires the surgeon to carefully mark the entire distal femur. This is a time-consuming job, and error at this stage can lead to the failure of TKA [40]. This is because bone morphing requires model fitting, which increases the computational cost, and hence, the cost of TKA.

From the above literature review, it can be concluded that, since the beginning of computer-based TKA, researchers have relied on an individual anatomical axis to set the transverse axis of FCS. However, the reported outcomes are not repeatable and certain. Thus, it is necessary to investigate the repeatability and accuracy of a transverse axis defined by the combination of more than one anatomical axis. Currently, the aTEA is the most reliable transverse axis for FCS; however, it is not a repeatable axis. This work proposes a modified Whiteside's line, which combines the Whiteside's line and aTEA to define a repeatable transverse axis. Moreover, this method can also be added to the existing TKA devices just by their software upgradation. The rest of the paper is organized as follows: Section 2 reviews the generalized method to define FCS, the tibial coordinate system (TCS), and presents the proposed modified Whiteside's line. Section 3 describes the CAD and experimental setup. Section 4 presents and interprets the results, while Section 5 concludes the work.

2. Methodology

The imageless TKA device consists of two components: (a) a sensing system containing a pair of optical sensors coupled with an infrared camera, to capture anatomical points marked by the surgeon; and (b) an algorithm to utilize the surgeon-marked points for the construction of a coordinate system to prepare the bone surfaces for implant placement. The coordinate systems developed virtually at the distal femur and proximal tibia are referred to as FCS and TCS, respectively. The nomenclature of anatomical points to perform an imageless TKA is marked on open-source bone models, as shown in Figure 1 [41]. In the following section, a generalized procedure adopted to create FCS and TCS is explained first. Later, the modified Whiteside's line technique is developed.

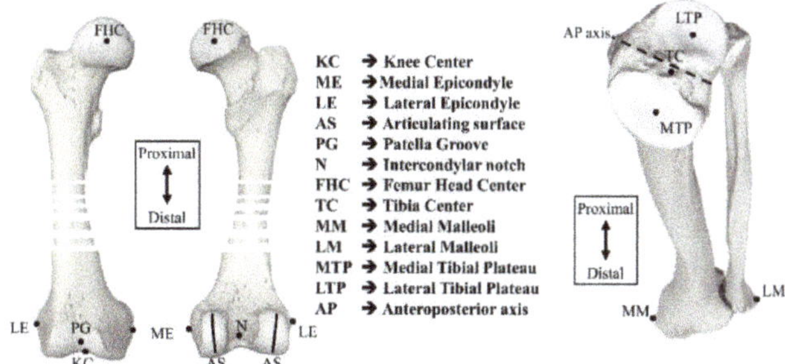

Figure 1. Anatomical points required for TKA [41]: femur and tibia.

2.1. Defining the Procedure for a Coordinate System

The algorithm to assign FCS and TCS is programmed in MATLAB. The anatomical points shown in Figure 1 are the required inputs for this algorithm. FCS and TCS are independent of each other.

2.1.1. Femoral Coordinate System

Assignment of FCS requires the identification of a knee center (KC), which becomes the origin of FCS. For this work, the femur mechanical axis is taken as the y-axis. The z-axis is marked between the lateral and medial sides. The x-axis is obtained by the cross product of the earlier marked axes and lies between the anterior and posterior of the body. Figure 2 shows a generalized four-step approach defined on an open-source [42] model.

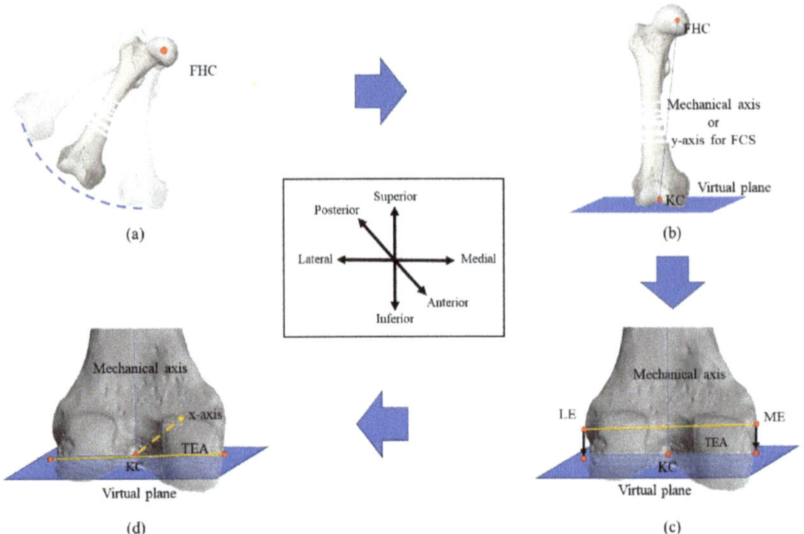

Figure 2. Generalized methodology to assign FCS using key anatomical points [41,42]. (**a**) FHC, (**b**) Mechanical axis and virtual plane, (**c**) Transverse axis, (**d**) Cross product.

Step 1: A fixed sensor is attached at the distal femur. Keeping the hip stationary, the circular motion of the femur allows the attached sensor to capture the surface of a sphere. A sphere-fitting algorithm fits an approximate sphere to the marked surface. The center of the sphere is the femur head center (FHC), as shown in Figure 2a. For the proposed work, a fast geometric fit algorithm for sphere is used because of its computational efficiency [43].

Step 2: During surgery, KC is marked by the surgeon. Currently, KC is taken above the center of the intramedullary canal hole, which is approximately 3 mm medial to the center of the femoral groove and 10 mm anterior to the posterior condylar ligament [44]. Figure 2b shows the process where FHC and KC define the femur mechanical axis and y-axis for FCS. A virtual plane parallel to a sagittal plane is created using Equation (1) for the projection of the transverse axis to be defined in step 3 [45]:

$$ax + by + cz + d = 0 \tag{1}$$

where, a, b, and c are the coefficients of the femur mechanical axis, and d is the distance between origin and plane.

Step 3: The transverse axis can be marked in several ways, for example by aTEA, Whiteside's line, and sphere fitting [9–11]. Once anatomical points on the distal femur are finalized by the surgeon, these points are projected onto the defined virtual plane, as shown in Figure 2c [46]. This projection ensures the orthogonality of the mechanical and transverse axes. Equations (2)–(4) represents the generalized equation for point projection:

$$x' = x_0 + at \tag{2}$$

$$y' = y_0 + bt \tag{3}$$

$$z' = z_0 + ct \tag{4}$$

where (x_0, y_0, z_0) are the coordinates of the point to be projected on the plane; (x', y', z') are the coordinates of a projected point; a, b, and c are the components of the normal vector of plane calculated in Equation (1); and *t* is the parameter such that the projected point is on the plane and line at the same time.

In the case that the anatomical points to accurately mark the transverse axis have deteriorated, the third axis is marked by Whiteside's line, since it is almost perpendicular to the epicondylar axis as calculated by Middleton and Palmer [15]; the average angle between Whiteside's line and the epicondylar axis is 91°.

Step 4: Once two axes are finalized, the cross product between the femur mechanical axis and the transverse axis or Whiteside's line completes the assignment of FCS, as shown in Figure 2d:

$$\vec{x} = \vec{y} \times \vec{z}$$ (5)

where '×' represents the cross product.

2.1.2. Tibial Coordinate System

The contribution of this work is related to the FCS only. The generalized method to assign TCS is explained because the algorithm used to define FCS can also define TCS, with change in only the first step. Figure 3 shows the steps carried out to define TCS [47]:

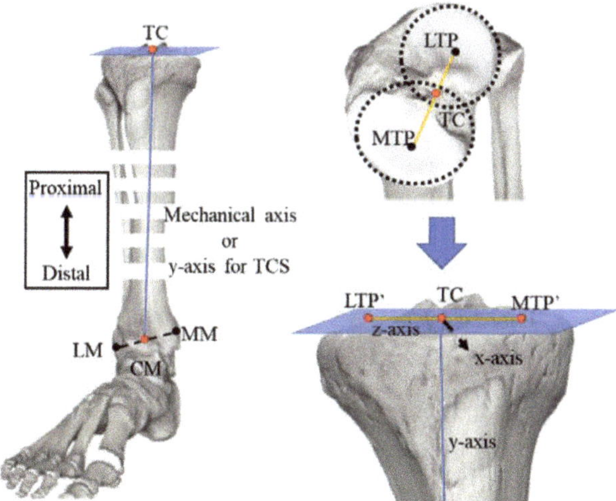

Figure 3. Generalized method to define TCS using anatomical points [41].

Step 1: The center of the malleoli/ankles is obtained by marking the lateral and medial malleoli. The midpoint of the axis defined by the coordinates of the lateral malleolus (LM) and medial malleolus (MM) is the center of malleoli (CM), which is not an exact average of LM and MM, but can be calculated by the following formula [48]:

$$(x, y, z)_{CM} = 0.57 \times (x, y, z)_{MM} + 0.46 \times (x, y, z)_{LM}$$ (6)

The tibia center (TC) is obtained by fitting a circle in the outer surfaces of the lateral tibial plateau (LTP) and medial tibial plateau (MTP); the midpoint of the line joining the centers of circles will define the TC. Then, TC and CM define the mechanical axis (y-axis) for TCS. For orthogonality of TCS, a plane is defined using TC and CM for the projection of MTP and LTP, using Equations (2)–(5).

Step 2: The centers of circles fitted in LTP and MTP define the z-axis. These centers are projected onto the virtual plane to create a z-axis orthogonal to the y-axis.

Step 3: The cross product between the z-axis and y-axis completes the definition of TCS.

2.2. Modified Whiteside's Line

Whiteside's line and aTEA are the most common transverse axes being used in commercial TKA devices because these anatomical axes are reported in clinical studies to be (almost) perpendicular to each other [15]. However, due to the small length of the Whiteside's line, a minute registration error leads to a large orientation error [21]. On the other hand, choosing an exact bony point to define an aTEA is nearly impossible, even under ideal conditions [30,31]. Utilizing the anatomical properties of the Whiteside's line and aTEA, a modified Whiteside's line method can be applied to define the transverse axis for FCS in a novel way. This method relies on the orthogonality of the Whiteside's line and aTEA, and the fact that the distance between the boniest points and a plane defined by the Whiteside's line will be the largest perpendicular distance among all points on the lateral and medial epicondyles.

The modified Whiteside's line requires marking multiple points on the lateral and medial sides of the epicondyle, referred to as a cloud of points. Figure 4a shows the position of clouds, N, and PG. This proposed algorithm is programmed in MATLAB. Within the algorithm, the center of N, KC, and FHC and PG, KC, and FHC defines two independent central planes C_N and C_{PG}, respectively, as shown in Figure 4b. The perpendicular distance between each point of the lateral cloud and C_N plane is calculated using Equation (7) [49].

$$\text{Distance} = \frac{|ax_o + by_o + cz_o + d|}{\sqrt{a^2 + b^2 + c^2}} \tag{7}$$

where (x_o, y_o, z_o) are the coordinates of the point and (a, b, c) are the components of the normal vector.

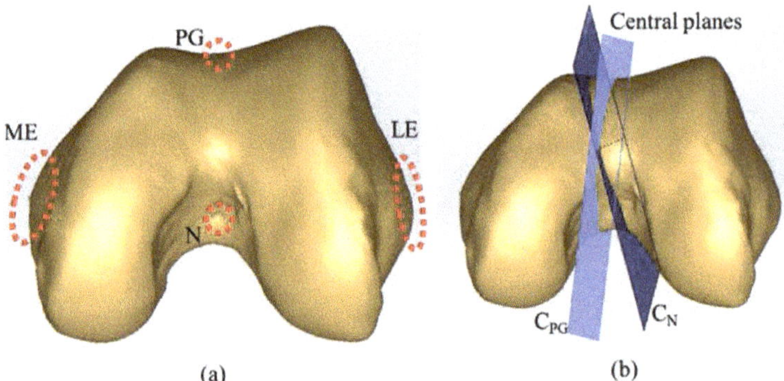

Figure 4. (**a**) N, PG, and multiple lateral and medial points for the modified Whiteside's line. (**b**) Central planes defined using (PG, KC, FHC) and (N, KC, FHC).

Similarly, the perpendicular distance between each medial point and C_{PG} is calculated. The points corresponding to the largest perpendicular distances on each side are selected to define the aTEA. The chosen points define a repeatable aTEA, because the highest point will always have the largest perpendicular distance. Algorithm 1 and Figure 5 describe the working principle of the MATLAB algorithm. The remaining procedure to define FCS is carried out as usual.

Algorithm 1: Modified Whiteside's line for FCS

Input: KC, FHC *// Chosen be probe sensor*
M_{Points}, L_{Points} *// A cloud of points on the lateral and medial sides*
Output: x_{axis}, y_{axis}, z_{axis} *// FCS*
y_{axis} := (FHC – KC)
// A plane normal to the y-axis and passing through KC
V_P := $\prod$ (y_{axis}, KC)
C_N := $\prod$ (N, KC, FHC)
C_{PG} := $\prod$ (PG, KC, FHC)
// Function 1: Selection of prominent point from M_{Points}
ME := **Function 1** (C_{PG}, M_{Points})
LE := **Function 1** (C_N, L_{Points})
// Function 2: Projection of prominent point on the virtual plane
$ME_{Projected}$:= **Function 2** (ME, V_P)
$LE_{Projected}$:= **Function 2** (LE, V_P)
z_{axis} := ($ME_{Projected}$ – $LE_{Projected}$)
x_{axis} := y_{axis} × z_{axis}
chosen point := **Function 1** (plane equation, cloud points)
for i := 1 to # cloud points
  i^{th} cloud point := (x, y, z)
  i^{th} cloud point, 4th column := | ax+by+cz+d | /sqrt($a^2 + b^2 + c^2$)
endfor
j := find indices of max(i^{th} cloud point, 4th column)
chosen point := j^{th} cloud point
endFunction 1
Projected point := **Function 2** (point, plane)
point := (x_o, y_o, z_o)
x := x_o + at; y := y_o + bt; z := z_o + ct;
substitute (x, y, z) in plane and solve for "t"
Projected point := substitute "t" in (x, y, z)
endFunction2

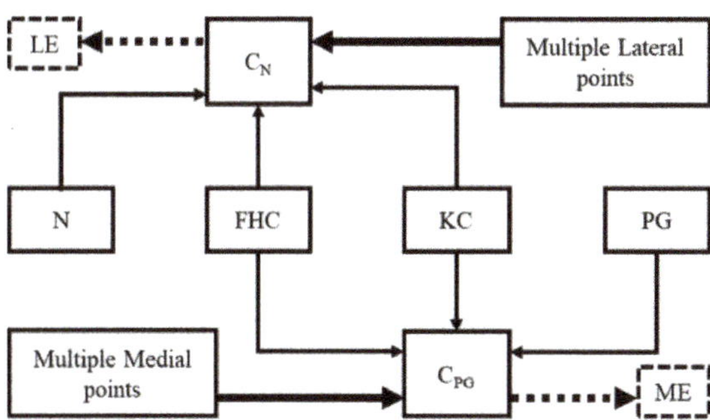

Figure 5. Working principle of the modified Whiteside's line.

3. Experimental Setup

3.1. CAD Model

During the actual surgery, all the anatomical points are recorded via either fixed or movable sensors. For this work, an open-source CT scan of a 44-year-old human bone is used. The weight and height of the patient are 85 kg and 185 cm, respectively [42]. All the anatomical points, including KC, Whiteside's line, and the highest points on the lateral and medial epicondyles (LE and ME) are marked in CAD. An alternative for this work, in contrast to real surgery, is that during surgery, FHC is obtained by the circular motion of the femur with a fixed sensor attached to the distal femur; however, for this experiment, FHC is obtained in commercial CAD software (Solidworks). Once all the points are marked, the bone model is printed, and anatomical points are captured, using a sensing system for experimental verification. During 3D printing, the bone model was linearly scaled down by 10% to print within the printing bed size of the 3D printer. Due to this downscaling, the size of the CAD model is larger than that of the printed bone. This change in size will not affect the results, as each axis of the FCS will be normalized. Figure 6 shows the marked CAD and 3D-printed models.

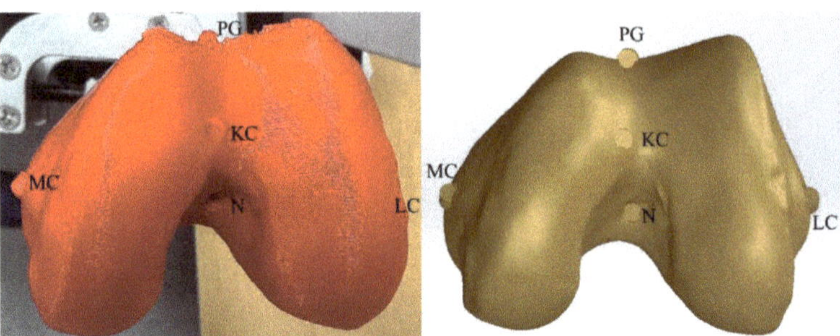

Figure 6. Printed bone model and CAD bone model.

3.2. Sensing System

The sensing system consists of a pair of OptiTrack Flex 3 cameras coupled with Motive 2.0 software (OptiTrack, Corvallis, OR, United States). This software requires a CS−200 calibration square and a probe sensor to capture the coordinates of different anatomical points. Figure 7 shows the experimental setup including infrared (IR) camera (OptiTrack Flex-3), fixed tracker (CS−200), 3D-printed bone model, bone fixture, and probe sensor. This passive system relies on the IR beam emitted by the OptiTrack camera. The spherical markers attached to the fixed and probe sensors reflect the IR beam. Motive 2.0 processes the reflected beam to provide the instantaneous position of the probe sensor with respect to the CS−200.

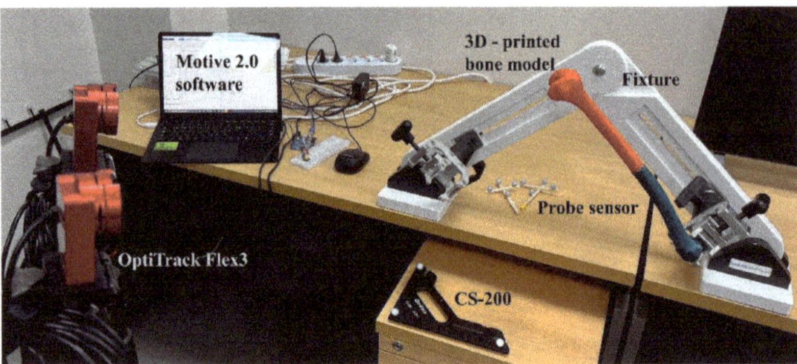

Figure 7. Experimental setup.

The CS−200 acts as a global coordinate system (GCS) for the probe sensor measurements. In this work, the data from the sensing system will be validated with CAD data. Since the GCS for the CAD model and sensing system are different, the CAD global coordinate system (CAD−GCS) is shifted to FCS in CAD, and the CAD-femoral coordinate system (CAD−FCS) acts as a common coordinate system (C−CS). To check the accuracy of the sensor measurement, FCS is defined on the 3D-printed bone, and a homogeneous transformation matrix (HTM) is created using Equation (8) [50].

$$^{A}T_{B} = \left[\begin{array}{c|c} \left(^{A}R_{B}\right) & \left(^{A}P_{B_{org}}\right) \\ \hline 0\,0\,0 & 1 \end{array} \right] \tag{8}$$

where "A" represents a base frame with respect to which frame "B" is defined. $^{A}R_{B}$ is a 3×3 rotation matrix to express the orientation of "B", and $^{A}P_{B_{org}}$ is a position vector between the origin of "A" and "B". Once the sensor-measured anatomical points are transformed into the sensor-femoral coordinate system (S−FCS) frames using Equation (9), the accuracy of the sensing system can be checked against CAD-measured anatomical points [51]:

$$^{B}P = {}^{B}T_{A} * P_{A} \tag{9}$$

where $^{B}T_{A}$ is the inverse HTM. P_{A} represents the position of any anatomical point in the sensor-global coordinate system (S−GCS) that is to be transformed into S−FCS. The coordinates of anatomical points in CAD−FCS can either be obtained directly in Solidworks or by repeating the procedure used to obtain the anatomical points in S−FCS by defining an HTM.

4. Results and Discussion

This section is divided into three parts. First, the results of the sensing system are compared with the CAD results to verify the authenticity of the experimental setup. The repeatability of current techniques to assign transverse axis is checked in Solidworks. In the end, the proposed technique is verified in the CAD and experimental environments.

4.1. Validation of the Sensing System with CAD Results

The points used to define the anatomical axes for FCS are marked on the CAD model before 3D printing, as shown in Figure 6. Table 1 gives each anatomical point in CAD−GCS. These points include KC, FHC, ME, and LE to define the HTM, and N and PG to define the two central planes C_N and C_{PG}, respectively. The unit vectors of CAD−FCS in CAD−GCS construct the rotation part, and the position of KC in GCS completes the HTM given in Equation (10):

$$^{CAD-GCS}T_{CAD-FCS} = \begin{bmatrix} -0.1546 & 0.2435 & 0.9575 & -44.64 \\ 0.0714 & -0.9639 & 0.2566 & 345.84 \\ 0.9854 & 0.1080 & 0.1317 & -40.47 \\ 0 & 0 & 0 & 1 \end{bmatrix} \tag{10}$$

Table 1. Anatomical points marked on CAD model in Solidworks.

Anatomical Point	CAD−GCS	CAD−GCS → CAD−FCS
KC	[−44.64, 345.84, −40.47]	[~0, ~0, ~0]
FHC	[49.95, −28.66, 1.51]	[0, 388.54, 0]
ME	[−9.53, 341.24, −45.05]	[−10.27, 12.49, 31.83]
LE	[−74.88, 324.47, −54.09]	[−10.27, 11.76, −36.23]
PG	[−44.82, 331.51, −24.65]	[14.60, 15.48, −1.77]
N	[−42.30, 332.39, −52.65]	[−13.32, 12.22, −2.81]

Equation (9) and $^{CAD-GCS}T_{CAD-FCS}$ transform CAD−GCS anatomical points into CAD−FCS. Table 1 gives the transformed anatomical points.

Similarly, an HTM from S−GCS to S−FCS is defined following the same procedure as for CAD−GCS to CAD−FCS. Table 2 shows the S−GCS coordinates of each anatomical points being used to define an HTM given in Equation (10). Additionally, N and PG will be used for defining C_N and C_{PG} planes, respectively. Furthermore, by using Equations (9) and (10), S−GCS anatomical points transformed into S−FCS are also shown in Table 2.

$$^{S-GCS}T_{S-FCS} = \begin{bmatrix} -0.5978 & -0.2584 & -0.7589 & 519.526 \\ 0.7186 & -0.5923 & -0.3644 & 271.440 \\ -0.3553 & -0.7631 & 0.5398 & 41.1153 \\ 0 & 0 & 0 & 1 \end{bmatrix} \tag{11}$$

Table 2. Anatomical points marked on the printed bone model with probe sensor.

Anatomical Point	S−GCS	S−GCS → S−FCS
KC	[519.53, 271.44, 41.12]	[~0, ~0, ~0]
FHC	[428.93, 63.77, −226.45]	[0, 350.61, 0]
ME	[500.89, 246.93, 51.80]	[−10.27, 11.19, 28.83]
LE	[547.84, 269.21, 17.85]	[−10.26, 11.75, −33.23]
PG	[522.99, 251.59, 35.60]	[14.37, 15.07, −1.63]
N	[506.83, 272.67, 28.73]	[−12.87, 12.00, −2.50]

4.2. Repeatability Challenge in Conventional Techniques

To verify the limitations of Whiteside's line, aTEA, and sphere fitting as reported in the literature, FCS is developed using each axis one at a time. Each anatomical axis is marked five times by viewing the distal femur from a different view. For example, LE and ME are marked by viewing the distal femur from top, bottom, anterior, posterior, and isometric views. In each view, the apparent highest points on the lateral and medial sides of the epicondyle are marked to define aTEA. Similarly, the sphere-fitting curve and Whiteside's line are marked to define the "test-FCSs". Figure 8 shows the orientation error between CAD−FCS and each test−FCS for five measurements of each method. It is observed that a small change in marking Whiteside's line led to large changes in orientation, and aTEA is close to CAD−FCS for two of the measurements. Similar results were observed by Victor et al. during a study of 12 cadaveric specimens [21]. Whiteside's was concluded to be the least consistent axis, with errors up to 11.67°. Similarly, the maximum error for aTEA was 6.16°.

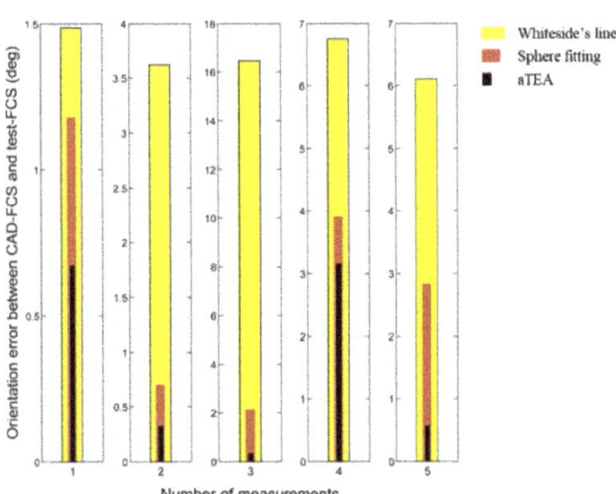

Figure 8. Repeatability of the current techniques to mark the transverse axis for FCS.

The results of the sphere-fitting line in between the aTEA and Whiteside's method are discussed. Sphere or cylinder fitting defines an accurate axis only when a fitting is done between (10 and 110)° of flexion [52,53]. For this experiment, the sphere is fitted between a full articulating surface, (0 to 110)°, (10 to 100)°, (0 to 100)°, and (10 to 110)° of flexion. Overall, it is observed that the conventional methods are not repeatable.

4.3. Modified Whiteside's Line—CAD Results

First, the proposed technique is tested using CAD data. Multiple points on the lateral and medial sides of the epicondyle are marked. The algorithm will choose the highest LE and ME points from the marked points. A total of 141 points were marked on the lateral side, and 145 points were marked on the medial point. Around the apparent highest lateral and medial points, the points that are also printed on the 3D bone model, dense points were marked with their surface-to-surface distance of no more than 0.25 mm, as shown in Figure 9. The spread of the cloud was 8.27 mm on the lateral side and 7.62 mm on the medial side. Similarly, 66 points in PG and 63 points in N with cloud spread of 4.15 mm and 5.35 mm, respectively, were marked for an algorithm to define C_{PG} and C_N, using one point at a time.

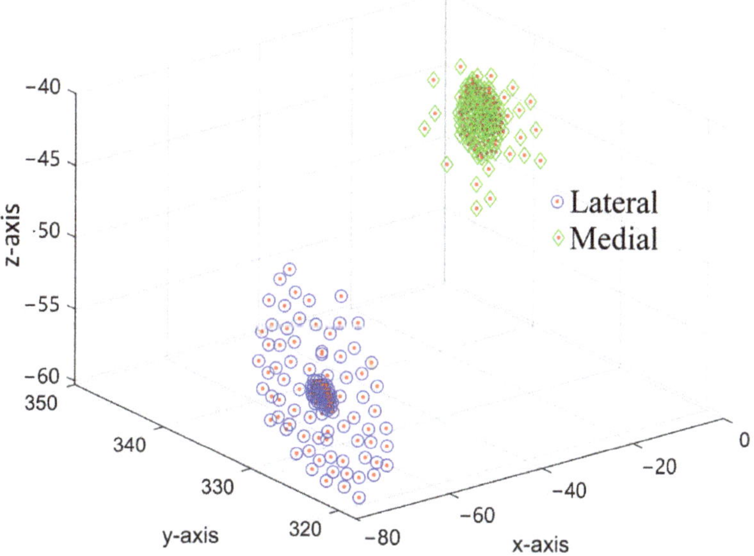

Figure 9. Clouds of lateral and medical points on epicondyle.

Based on the marked points, the algorithm defined FCS using each point, and compared its orientation with FCS defined by aTEA. For example, first, a constant LE marked on the bone is used while calculating ME by the proposed method. To get a new ME, each point marked in PG along with KC and FHC is used to define the C_{PG} plane. Based on the largest perpendicular distance, ME is chosen from the medial cloud. FCS is defined using the obtained points, and its orientation is compared with reference FCS defined using the marked aTEA, as described in Figure 6. Since the y-axis is the same in both FCSs, the angle between the z-axis is measured to obtain the orientation error. Once all points are used to define C_{PG}, ME is kept constant, and points marked in N are used to define C_N, to find the LE from the lateral cloud. Table 3 compares the orientation error between the FCSs obtained by Whiteside's line, aTEA, and the proposed modified Whiteside's line. The orientation error is measured by defining an anatomical point 3 mm away from its ideal position. To measure the error, first FCS is defined using an ideal anatomical point, then the anatomical point is marked with 3 mm of error in a lateral or medial direction to define a second FCS. Finally, the orientation between the first and second FCSs is measured to check the effect of 3 mm error on that anatomical axis.

Table 3. Comparison of orientation errors of each method depending on 3 mm registration error.

Method	Direction		Orientation Error (deg)
Whiteside's line	Anterior	Lateral	6.73
		Medial	5.77
	Posterior	Lateral	3.36
		Medial	5.50
aTEA	Lateral	Anterior	2.33
		Posterior	2.49
	Medial	Anterior	2.37
		Posterior	2.47
Modified Whiteside's line	Anterior	Lateral	0.72
		Medial	1.33

For the Whiteside's line, it is observed that 3 mm error in PG in the lateral direction caused a 6.73° deviation of FCS, while in the medial direction, 5.77° error was observed. For N in the lateral and medial directions, 3.36° and 5.50° error was observed, respectively. For an aTEA lateral side, 3 mm in anterior and posterior directions caused 2.33° and 2.49° deviation, respectively. Similarly, for the medial side, 2.37° and 2.47° deviations were observed. Before testing the modified Whiteside's line, it was made sure that the anatomical point coordinates used for testing aTEA are included in lateral and medial clouds; also, the C_N and C_{PG} were defined using the same points that were used to define the Whiteside's line. With the proposed method for 3 mm off of the marked PG position, 0.72° and 1.33° deviations were observed on the lateral and medial sides, respectively. For N, it was 0.64° and 2.06° on the lateral and medial sides, respectively. The maximum orientation error using the proposed method is 3.22° when PG is marked 19.80 mm away in a lateral direction and 5.54° when PG is 8.73 mm in the medial direction. Similarly, for N, maximum orientation error of 2.50° was observed at 19.20 mm in a lateral direction and 5.48° when N moved 12.75 mm in the medial direction. It is also observed that within the radius of 2.1 mm around the actual N, the mean orientation error was 0.4197°. The mean orientation error around 1.76 mm radius of PG was 0.3566°.

4.4. Modified Whiteside's Line—Experimental Results

For experimental validation, five points were marked in N and five points in PG using the probe sensor. The probe sensor captures 30 frames per second. For lateral and medial clouds, epicondyles were painted with a probe sensor for five seconds on each side to obtain 150 points in each cloud. The number of points added in the medial and lateral clouds are directly proportional to both the accuracy of the femoral implant and the surgery time. Figure 10 shows the sensor-marked PG and N points. While marking the N points, it was observed that marking the N point is easy compared to PG, as the N lies inside the notch, and is easy to identify. On the other hand, PG lies inside the curve with large curvature, so it needs to be marked carefully. However, during surgery, PG lies in front of the surgeon, so it can be accurately marked by an experienced surgeon. For experimental validation, the FCS is defined using the printed LE and ME points, as shown in Figure 6. Another FCS is defined by keeping LE the same as printed on the bone, and ME is defined using the modified Whiteside's line method. The error is a measure of the difference in orientation of both FCSs. Similarly, once all PG points are used to find ME, LE is chosen from the lateral cloud using all N points, one at a time.

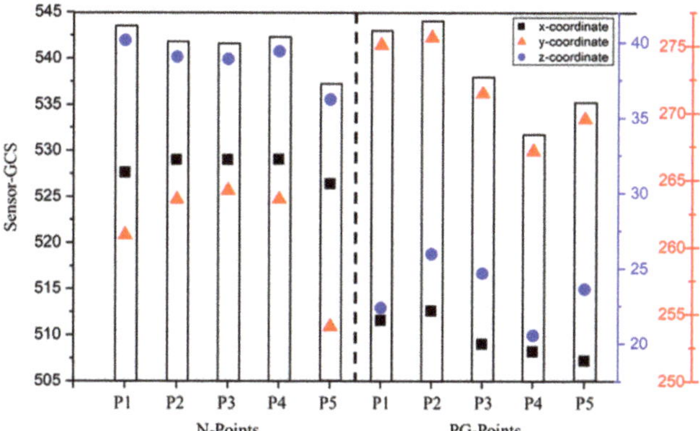

Figure 10. Coordinates of N and PG points in S–GCS.

Table 4 shows the orientation error for different points. PG and N points were marked around the marked PG and N points in all directions. With the modified Whiteside's line for the PG case, a minimum error of 0.3441° and a maximum error of 1.9197° were observed. The mean error and standard deviation for five measurements were 0.8481° and 0.6414°, respectively. For the N case, the minimum and maximum errors were 0.4586° and 0.9684°, respectively. The mean error and standard deviation for five measurements were 0.7566° and 0.1848°, respectively. These results show that the modified Whiteside's line produces near repeatable aTEA. Additionally, for conventional methods, as the error in the registration of anatomical points increases, the orientation error increases proportionally. On the other hand, the proposed method picks LE and ME from cloud points, so it defines a more reliable and repeatable TEA.

Table 4. The modified Whiteside's line experiment.

Modified Whiteside's Line		Orientation Error (deg)
$PG \rightarrow ME$ $LE \rightarrow Printed$	PG-P1	0.9268
	PG-P2	0.6428
	PG-P3	0.3441
	PG-P4	0.4069
	PG-P5	1.9197
$N \rightarrow LE$ $ME \rightarrow Printed$	N-P1	0.4586
	N-P2	0.7948
	N-P3	0.9684
	N-P4	0.7948
	N-P5	0.7663

5. Conclusions

The accuracy of imageless TKA relies on the anatomical points used to develop the FCS and TCS, as these coordinates are used to prepare the bone surfaces for an implant. Currently, imageless TKA devices use the femur mechanical axis as one of the axes of the FCS. The assignment of a transverse axis between lateral and medial sides is a critical task. The aTEA, Whiteside's line, and sphere-fitting line approach in articulating surfaces are three popular methods to assign the transverse axis; however, with a change in viewing angle, the repeatability of each of these anatomical axes is disturbed. In this work, a modified Whiteside's line approach is proposed that provides an adaptable technique to choose a near repeatable aTEA. This technique requires a cloud of points on the lateral and medial sides of epicondyles, and two central planes: C_N—defined using N, KC, and FHC; and C_{PG}—defined by PG, KC, and FHC. Based on the perpendicular distances between cloud points and C_N and C_{PG}, the highest points on the lateral and medial epicondyles are selected to define the aTEA. The proposed technique has been verified in Solidworks software, and experimentally by using the OptiTrack sensing system. Around 140 points were marked for each cloud, and 60 points were respectively marked for PG and N. FCS is defined using each PG and N point one at a time. The repeatability of the proposed method is verified by measuring the orientation of each developed FCS against a standard FCS.

Furthermore, the repeatability of the modified Whiteside's line, aTEA, and Whiteside's line are compared at 3 mm registration error. It is observed that the Whiteside's line is the most unrepeatable axis, with orientation error up to 6.73°; with a 3 mm error in an aTEA axis, the maximum orientation error was 2.49°. The modified Whiteside's line produced repeatable results with a maximum orientation error of 1.33° on PG and 2.06° on N sides. Additionally, with the conventional techniques, increase in registration error always increases the orientation error; however, with the proposed method, for PG with 19.80 mm and 8.73 mm errors in the lateral and medial directions caused only 3.22° and 5.54° orientation errors, respectively. For N cases, 2.50° and 5.48° of error were observed with registration errors of 19.20 mm and 12.75 mm in the lateral and medial directions, respectively. Additionally, within the radius of 2.01 mm around N and 1.76 mm around PG, the mean orientation error was 0.4197° and 0.3566°, respectively. Hence, the modified Whiteside's line has removed the sensitivity of the Whiteside's line and the uncertainty of aTEA by providing a repeatable transverse axis for FCS. In the future, a surface-fitting algorithm will be applied on the marked lateral and medial cloud points to avoid the malrotation error that could be caused in case the surgeon misses to register the optimal point to define the transepicondylar axis. Finally, the proposed model will be validated by a cadaveric study.

Author Contributions: Conceptualization, M.S. and J.L.; methodology, M.S. and J.L.; software, M.S., J.P. and J.Y.K.; writing—original draft preparation, M.S. and J.Y.K.; writing—review and editing, J.L. and H.S.K.; supervision, H.S.K.; project administration, H.S.K.; funding acquisition, H.S.K. All authors have read and agreed to the published version of the manuscript.

Funding: This work was supported by the Ministry of Trade, Industry, and Energy (MOTIE) and the Korea Institute for Advancement of Technology (KIAT), through the International Cooperative R&D program (Project No. P0016173).

Institutional Review Board Statement: Not applicable.

Informed Consent Statement: Not applicable.

Data Availability Statement: Not applicable.

Conflicts of Interest: The authors declare no conflict of interest.

Nomenclature

$^{A}T_{B}$	Transformation matrix from frame A to B
$^{A}P_{B_{org}}$	Distance between origins of frame B w.r.t frame A
$^{A}R_{B}$	Rotation matrix defining the orientation of frame B w.r.t frame A
AP axis	Anteroposterior axis
aTEA	anatomical Transepicondylar axis
CAD−GCS CAD	global coordinate system
CAD−FCS CAD	femoral coordinate system
C−CS	Common coordinate system: a common coordinate system between sensor and CAD system for validation of sensing system
CM	Center of malleoli
FCS	Femoral coordinate system

FHC	Femur head center
GCS	Global coordinate system
HTM	Homogeneous transformation matrix
KC	Knee center
LE	Lateral epicondylar
LM	Lateral malleoli
LTP	Lateral tibial plateau
C_N	Central plane defined by KC, FHC, and N
C_{PG}	Central plane defined by KC, FHC, and PG
Test-FCSs	Femoral coordinate systems defined using conventional methods to check their repeatability
Standard-FCS	Femoral coordinate system marked by anatomical points printed on the bone
ME	Medial epicondylar
MM	Medial malleoli
MTP	Medial tibial plateau
N	Intercondylar notch
PG	Patella groove
PCA	Posterior condylar axis
S−GCS	Sensor−global coordinate system
S−FCS	Sensor−femoral coordinate system
sTEA	surgical Transepicondylar axis
TC	Tibia center
TCS	Tibial coordinate system
TEA	Transepicondylar axis
TKA	Total knee arthroplasty

References

1. Richmond, J.C. Surgery for Osteoarthritis of the Knee. *Rheum. Dis. Clin. N. Am.* **2008**, *34*, 815–825. [CrossRef] [PubMed]
2. Carr, A.J.; Robertsson, O.; Graves, S.; Price, A.J.; Arden, N.K.; Judge, A.; Beard, D.J. Knee Replacement. *Lancet* **2012**, *379*, 1331–1340. [CrossRef]
3. Overhoff, H.M.; Lazovic, D.; Liebing, M.; Macher, C. Total Knee Arthroplasty: Coordinate System Definition and Planning Based on 3-D Ultrasound Image Volumes. *Int. Congr. Ser.* **2001**, *1230*, 292–299. [CrossRef]
4. Foley, K.A.; Muir, J.M. *Improving Accuracy in Total Knee Arthroplasty: A Cadaveric Comparison of a New Surgical Navigation Tool, Intellijoint KNEE, with Computed Tomography Imaging*; Intellijoint Surgical, Inc.: Waterloo, ON, USA, 2019.
5. Chauhan, S.K.; Scott, R.G.; Breidahl, W.; Beaver, R.J. Computer-Assisted Knee Arthroplasty versus a Conventional Jig-Based Technique. *J. Bone Jt. Surgery. Br. Vol.* **2004**, *86-B*, 372–377. [CrossRef] [PubMed]
6. Doro, L.C.; Hughes, R.E.; Miller, J.D.; Schultz, K.F.; Hallstrom, B.; Urquhart, A.G. The Reproducibility of a Kinematically-Derived Axis of the Knee versus Digitized Anatomical Landmarks Using a Knee Navigation System. *Open Biomed. Eng. J.* **2008**, *2*, 52–56. [CrossRef]
7. Stiehl, J.B. Computer Navigation in Primary Total Knee Arthroplasty. *J. Knee Surg.* **2007**, *20*, 158–164. [CrossRef] [PubMed]
8. van der Linden–van der Zwaag, H.M.J.; Valstar, E.R.; van der Molen, A.J.; Nelissen, R.G.H.H. Transepicondylar Axis Accuracy in Computer Assisted Knee Surgery: A Comparison of the CT-Based Measured Axis versus the CAS-Determined Axis. *Comput. Aided Surg.* **2008**, *13*, 200–206. [CrossRef] [PubMed]
9. Iwaki, H.; Pinskerova, V.; Freeman, M.A.R. Tibiofemoral Movement 1: The Shapes and Relative Movements of the Femur and Tibia in the Unloaded Cadaver Knee. *J. Bone Jt. Surgery. Br. Vol.* **2000**, *82-B*, 1189–1195. [CrossRef]
10. Renault, J.-B.; Aüllo-Rasser, G.; Donnez, M.; Parratte, S.; Chabrand, P. Articular-Surface-Based Automatic Anatomical Coordinate Systems for the Knee Bones. *J. Biomech.* **2018**, *80*, 171–178. [CrossRef] [PubMed]
11. Yoshioka, Y.; Siu, D.; Cooke, T.D. The Anatomy and Functional Axes of the Femur. *J. Bone Jt. Surg. Am.* **1987**, *69*, 873–880. [CrossRef]
12. Johal, P.; Williams, A.; Wragg, P.; Hunt, D.; Gedroyc, W. Tibio-Femoral Movement in the Living Knee. A Study of Weight Bearing and Non-Weight Bearing Knee Kinematics Using 'Interventional' MRI. *J. Biomech.* **2005**, *38*, 269–276. [CrossRef] [PubMed]
13. Nam, J.-H.; Koh, Y.-G.; Kang, K.; Park, J.-H.; Kang, K.-T. The Posterior Cortical Axis as an Alternative Reference for Femoral Component Placement in Total Knee Arthroplasty. *J. Orthop. Surg. Res.* **2020**, *15*, 603. [CrossRef] [PubMed]
14. Jabalameli, M.; Rahbar, M.; Bagheri, A.; Hadi, H.; Moradi, A.; Radi, M.; Mokhtari, T. Evaluation of Distal Femoral Rotational Alignment According to Transepicondylar Axis and Whiteside's Line: A Study in Iranian Population. *Shafa Orthop. J.* **2013**, *4*, 122–127.
15. Middleton, F.R.; Palmer, S.H. How Accurate Is Whiteside's Line as a Reference Axis in Total Knee Arthroplasty? *Knee* **2007**, *14*, 204–207. [CrossRef]

16. Whiteside, L.A.; Arima, J. The Anteroposterior Axis for Femoral Rotational Alignment in Valgus Total Knee Arthroplasty. *Clin. Orthop. Relat. Res.* **1995**, *321*, 168–172. [CrossRef]
17. Kobayashi, H.; Akamatsu, Y.; Kumagai, K.; Kusayama, Y.; Ishigatsubo, R.; Muramatsu, S.; Saito, T. The Surgical Epicondylar Axis Is a Consistent Reference of the Distal Femur in the Coronal and Axial Planes. *Knee Surg. Sports Traumatol. Arthrosc.* **2014**, *22*, 2947–2953. [CrossRef]
18. Valkering, K.P.; Tuinebreijer, W.E.; Sunnassee, Y.; van Geenen, R.C.I. Multiple Reference Axes Should Be Used to Improve Tibial Component Rotational Alignment: A Meta-Analysis. *J. ISAKOS* **2018**, *3*, 337–344. [CrossRef]
19. Tanavalee, A.; Yuktanandana, P.; Ngarmukos, C. Surgical Epicondylar Axis vs Anatomical Epicondylar Axis for Rotational Alignment of the Femoral Component in Total Knee Arthroplasty. *J. Med. Assoc. Thail. = Chotmaihet Thangphaet* **2001**, *84* (Suppl. 1), S401-8.
20. Yoshino, N.; Takai, S.; Ohtsuki, Y.; Hirasawa, Y. Computed Tomography Measurement of the Surgical and Clinical Transepicondylar Axis of the Distal Femur in Osteoarthritic Knees. *J. Arthroplast.* **2001**, *16*, 493–497. [CrossRef] [PubMed]
21. Victor, J.; Van Doninck, D.; Labey, L.; Van Glabbeek, F.; Parizel, P.; Bellemans, J. A Common Reference Frame for Describing Rotation of the Distal Femur: A Ct-Based Kinematic Study Using Cadavers. *J. Bone Jt. Surg. Br.* **2009**, *91*, 683–690. [CrossRef]
22. Kinzel, V.; Ledger, M.; Shakespeare, D. Can the Epicondylar Axis Be Defined Accurately in Total Knee Arthroplasty? *Knee* **2005**, *12*, 293–296. [CrossRef] [PubMed]
23. Pitocchi, J.; Wesseling, M.; van Lenthe, G.H.; Pérez, M.A. Finite Element Analysis of Custom Shoulder Implants Provides Accurate Prediction of Initial Stability. *Mathematics* **2020**, *8*, 1113. [CrossRef]
24. Witoolkollachit, P.; Seubchompoo, O. The Comparison of Femoral Component Rotational Alignment with Transepicondylar Axis in Mobile Bearing TKA, CT-Scan Study. *J. Med. Assoc. Thail. = Chotmaihet Thangphaet* **2008**, *91*, 1051–1058.
25. Stiehl, J.B.; Abbott, B.D. Morphology of the Transepicondylar Axis and Its Application in Primary and Revision Total Knee Arthroplasty. *J. Arthroplast.* **1995**, *10*, 785–789. [CrossRef]
26. Schnurr, C.; Nessler, J.; König, D.P. Is Referencing the Posterior Condyles Sufficient to Achieve a Rectangular Flexion Gap in Total Knee Arthroplasty? *Int. Orthop. (SICOT)* **2008**, *33*, 1561. [CrossRef] [PubMed]
27. Fehring, T.K.; Odum, S.; Griffin, W.L.; Mason, J.B.; Nadaud, M. Early Failures in Total Knee Arthroplasty. *Clin. Orthop. Relat. Res.* **2001**, *392*, 315–318. [CrossRef] [PubMed]
28. Geary, M.B.; Macknet, D.M.; Ransone, M.P.; Odum, S.D.; Springer, B.D. Why Do Revision Total Knee Arthroplasties Fail? A Single-Center Review of 1632 Revision Total Knees Comparing Historic and Modern Cohorts. *J Arthroplast.* **2020**, *35*, 2938–2943 [CrossRef] [PubMed]
29. Dalury, D.F.; Pomeroy, D.L.; Gorab, R.S.; Adams, M.J. Why Are Total Knee Arthroplasties Being Revised? *J. Arthroplast.* **2013**, *28*, 120–121. [CrossRef] [PubMed]
30. Stoeckl, B.; Nogler, M.; Krismer, M.; Beimel, C.; Moctezuma de la Barrera, J.-L.; Kessler, O. Reliability of the Transepicondylar Axis as an Anatomical Landmark in Total Knee Arthroplasty. *J. Arthroplast.* **2006**, *21*, 878–882. [CrossRef] [PubMed]
31. Jerosch, J.; Peuker, E.; Philipps, B.; Filler, T. Interindividual Reproducibility in Perioperative Rotational Alignment of Femoral Components in Knee Prosthetic Surgery Using the Transepicondylar Axis. *Knee Surg. Sports Traumatol. Arthrosc.* **2002**, *10*, 194–197. [CrossRef] [PubMed]
32. Davis, E.T.; Pagkalos, J.; Gallie, P.A.M.; Macgroarty, K.; Waddell, J.P.; Schemitsch, E.H. Defining the Errors in the Registration Process During Imageless Computer Navigation in Total Knee Arthroplasty: A Cadaveric Study. *J. Arthroplast.* **2014**, *29*, 698–701 [CrossRef] [PubMed]
33. Yau, W.P.; Leung, A.; Chiu, K.Y.; Tang, W.M.; Ng, T.P. Intraobserver Errors in Obtaining Visually Selected Anatomic Landmarks during Registration Process in Nonimage-Based Navigation-Assisted Total Knee Arthroplasty: A Cadaveric Experiment. *J. Arthroplast.* **2005**, *20*, 591–601. [CrossRef] [PubMed]
34. Yau, W.P.; Leung, A.; Liu, K.G.; Yan, C.H.; Wong, L.L.S.; Chiu, K.Y. Interobserver and Intra-Observer Errors in Obtaining Visually Selected Anatomical Landmarks during Registration Process in Non-Image-Based Navigation-Assisted Total Knee Arthroplasty. *J. Arthroplast.* **2007**, *22*, 1150–1161. [CrossRef] [PubMed]
35. Davis, E.T.; Pagkalos, J.; Gallie, P.A.M.; Macgroarty, K.; Waddell, J.P.; Schemitsch, E.H. A Comparison of Registration Errors with Imageless Computer Navigation during MIS Total Knee Arthroplasty versus Standard Incision Total Knee Arthroplasty: A Cadaveric Study. *Comput. Aided Surg.* **2015**, *20*, 7–13. [CrossRef] [PubMed]
36. Pagkalos, J.; Davis, E.; Gallie, P.; Macgroarty, K.; Waddell, J.; Schemitsch, E. The Effect of the Registration Process Error on Component Alignment during Imageless Computer Navigation for Knee Arthroplasty: A Cadaveric Study. *Orthop. Proc.* **2013**, *95-B*, 68. [CrossRef]
37. Liu, W.; Ding, H.; Zhu, Z.; Wang, G.; Zhou, Y. An Image-Free Surgical Navigation System for Total Knee Arthroplasty. In Proceedings of the 2011 4th International Conference on Biomedical Engineering and Informatics (BMEI), Shanghai, China, 15–17 October 2011; Volume 3, pp. 1353–1357.
38. Perrin, N.; Stindel, E.; Roux, C. BoneMorphing versus Freehand Localization of Anatomical Landmarks: Consequences for the Reproducibility of Implant Positioning in Total Knee Arthroplasty. *Comput. Aided Surg.* **2005**, *10*, 301–309. [CrossRef] [PubMed]
39. Bae, D.K.; Song, S.J. Computer Assisted Navigation in Knee Arthroplasty. *Clin. Orthop. Surg.* **2011**, *3*, 259–267. [CrossRef] [PubMed]

40. Stindel, E.; Briard, J.-L.; Lavallée, S.; Dubrana, F.; Plaweski, S.; Merloz, P.; Lefèvre, C.; Troccaz, J. Bone Morphing: 3D Reconstruction Without Pre- or Intraoperative Imaging—Concept and Applications. In *Navigation and Robotics in Total Joint and Spine Surgery*; Stiehl, J.B., Konermann, W.H., Haaker, R.G., Eds.; Springer: Berlin/Heidelberg, Germany, 2004; pp. 39–45. ISBN 978-3-642-59290-4.
41. Anatomy Standard Landing Page. Available online: https://www.anatomystandard.com/ (accessed on 8 August 2022).
42. Femur Bone | 3D CAD Model Library | GrabCAD. Available online: https://grabcad.com/library/femur-bone-3 (accessed on 8 August 2022).
43. YD, S. Fast Geometric Fit Algorithm for Sphere Using Exact Solution. *arXiv* **2015**, arXiv:1506.02776 [cs].
44. Diduch, D.R.; Iorio, R.; Long, W.J.; Scott, W.N. *Insall & Scott Surgery of the Knee*; Elsevier: Amsterdam, The Netherlands, 2018.
45. Schlatterer, B.; Linares, J.-M.; Chabrand, P.; Sprauel, J.-M.; Argenson, J.-N. Influence of the Optical System and Anatomic Points on Computer-Assisted Total Knee Arthroplasty. *Orthop. Traumatol. Surg. Res.* **2014**, *100*, 395–402. [CrossRef]
46. Walker, P.S.; Heller, Y.; Yildirim, G.; Immerman, I. Reference Axes for Comparing the Motion of Knee Replacements with the Anatomic Knee. *Knee* **2011**, *18*, 312–316. [CrossRef]
47. Lee, Y.S.; Park, S.J.; Shin, V.I.; Lee, J.H.; Kim, Y.H.; Song, E.K. Achievement of Targeted Posterior Slope in the Medial Opening Wedge High Tibial Osteotomy: A Mathematical Approach. *Ann. Biomed. Eng.* **2010**, *38*, 583–593. [CrossRef]
48. Yau, W.P.; Chiu, K.Y.; Tang, W.M. How Precise Is the Determination of Rotational Alignment of the Femoral Prosthesis in Total Knee Arthroplasty: An In Vivo Study. *J. Arthroplast.* **2007**, *22*, 1042–1048. [CrossRef] [PubMed]
49. Bundrick, C.M.; Sherry, D.L. Distance From a Point to a Line and a Point to a Plane Via Synthetic Methods. *Sch. Sci. Math.* **1978**, *78*, 304–306. [CrossRef]
50. Craig, J.J. *Introduction to Robotics: Mechanics and Control*; Pearson Educación: London, UK, 2005; ISBN 978-970-26-0772-4.
51. Sohail, M.; Butt, S.U.; Baqai, A.A. An Analytical Approach for Positioning Error and Mode Shape Analysis of n—Legged Parallel Manipulator. *IJCIM* **2018**, *31*, 677–691. [CrossRef]
52. Hancock, C.W.; Winston, M.J.; Bach, J.M.; Davidson, B.S.; Eckhoff, D.G. Cylindrical Axis, Not Epicondyles, Approximates Perpendicular to Knee Axes. *Clin. Orthop. Relat. Res.* **2013**, *471*, 2278–2283. [CrossRef] [PubMed]
53. Lozano, R.; Campanelli, V.; Howell, S.; Hull, M. Kinematic Alignment More Closely Restores the Groove Location and the Sulcus Angle of the Native Trochlea than Mechanical Alignment: Implications for Prosthetic Design. *Knee Surg. Sports Traumatol. Arthrosc.* **2019**, *27*, 1504–1513. [CrossRef]

 *mathematics*

Article

Atherosclerotic Plaque Segmentation Based on Strain Gradients: A Theoretical Framework

Álvaro T. Latorre [1,*], Miguel A. Martínez [1,2], Myriam Cilla [1,2], Jacques Ohayon [3,4] and Estefanía Peña [1,2,*]

1 Aragón Institute for Engineering Research (I3A), University of Zaragoza, 50018 Zaragoza, Spain; miguelam@unizar.es (M.A.M.); mcilla@unizar.es (M.C.)
2 CIBER de Bioingeniería, Biomateriales y Nanomedicina (CIBER-BBN) , 50018 Zaragoza, Spain
3 Laboratory TIMC-IMAG, CNRS UMR 5525, Grenoble-Alpes University, 38400 Grenoble, France; jacques.ohayon@univ-smb.fr
4 Mechanics and Material Department, Savoie Mont-Blanc University, Polytech Annecy, Le Bourget du Lac, 73000 Chambéry, France
* Correspondence: alatorr@unizar.es (Á.T.L.); fany@unizar.es (E.P.)

Abstract: *Background:* Atherosclerotic plaque detection is a clinical and technological problem that has been approached by different studies. Nowadays, intravascular ultrasound (IVUS) is the standard used to capture images of the coronary walls and to detect plaques. However, IVUS images are difficult to segment, which complicates obtaining geometric measurements of the plaque. *Objective:* IVUS, in combination with new techniques, allows estimation of strains in the coronary section. In this study, we have proposed the use of estimated strains to develop a methodology for plaque segmentation. *Methods:* The process is based on the representation of strain gradients and the combination of the Watershed and Gradient Vector Flow algorithms. Since it is a theoretical framework, the methodology was tested with idealized and real IVUS geometries. *Results:* We achieved measurements of the lipid area and fibrous cap thickness, which are essential clinical information, with promising results. The success of the segmentation depends on the plaque geometry and the strain gradient variable (SGV) that was selected. However, there are some SGV combinations that yield good results regardless of plaque geometry such as $|\nabla \varepsilon_{vMises}| + |\nabla \varepsilon_{r\theta}|$, $|\nabla \varepsilon_{yy}| + |\nabla \varepsilon_{rr}|$ or $|\nabla \varepsilon_{min}| + |\nabla \varepsilon_{Tresca}|$. These combinations of SGVs achieve good segmentations, with an accuracy between 97.10% and 94.39% in the best pairs. *Conclusions:* The new methodology provides fast segmentation from different strain variables, without an optimization step.

Keywords: atherosclerosis; fibrous cap thickness; finite element model; intravascular ultrasound; segmentation method; strain gradient

MSC: 74S05

Citation: Latorre, Á.T.; Martínez, M.A.; Cilla, M.; Ohayon, J.; Peña, E. Atherosclerotic Plaque Segmentation Based on Strain Gradients: A Theoretical Framework. *Mathematics* 2022, 10, 4020. https://doi.org/10.3390/math10214020

Academic Editor: Eva H. Dulf

Received: 26 September 2022
Accepted: 25 October 2022
Published: 29 October 2022

Publisher's Note: MDPI stays neutral with regard to jurisdictional claims in published maps and institutional affiliations.

1. Introduction

Cardiovascular diseases are the leading cause of death worldwide, with 17.9 million deaths per year, which represent 31% of the demises [1]. The majority of the coronary events are related to heart attack or cerebral strokes, which are commonly trigged by atherosclerotic plaque rupture [2]. The atherosclerotic plaque is the result of lipid deposition in the artery wall, which creates a lipid core surrounded by fibrotic tissue. The fibrotic tissue that separates the lipid core from the lumen is called the fibrous cap [3]. The rupture of the fibrous cap induces a thrombus in the artery that obstructs the blood flow, leading to an acute coronary event [4]. The vulnerability of the plaque is related to the risk of fibrous cap rupture and the thrombus formation. There are some geometrical parameters that are important for the vulnerability characterization. Some studies have suggested that atherosclerotic plaques with fibrous cap thicknesses (FCT) thinner than 65 µm and lipid cores with a large area are vulnerable and prone to rupture [3,5]. On the other hand, a large FCT usually indicates that the plaque is stable. However, the prediction of the plaque rupture is not only based on geometrical features, but also on the mechanical properties of the tissues [6,7]. Nowadays, intravascular ultrasound (IVUS) images are the gold standard for clinical diagnosis of atherosclerotic plaques in coronary arteries. IVUS images show a cross section of the artery wall in greyscale, and the segmentation usually depends on the cardiologist's experience. Each plaque tissue has different echo reflectivity characteristics, so its appearance within an IVUS image can be distinguished [8]. The segmentation can be performed manually; nevertheless, it requires clinical expertise, a high amount of time and, therefore, cost, and it depends on the image quality [9]. In order to solve this problem, new clinical techniques such as virtual histology intravascular ultrasound (VH-IVUS) have emerged. VH-IVUS is a clinical method for visualizing color-coded tissue maps, which provides an automated plaque characterization [10]. However, there are some limitations to this technique: first, it has a poor recognition of the FCT due to false detection of lipid core tissue and limitations in the plaque type classification (thin FCT, calcified plaque, or stable plaque) [11]. Second, only one computation can be performed per cardiac cycle, which reduces the number of IVUS frames used to characterize the plaque [12]. Third, the clinics have to be equipped with VH software.

That is why new techniques, mostly based on machine learning, have been developed to segment or characterize the atherosclerotic plaque tissues from IVUS images [9]. Methods based on Random Forest were used to classify IVUS image pixels into different tissues (dense calcium, necrotic, fibrotic tissue, and fibrofatty tissue) [12,13]. Although these strategies have achieved high classification accuracy (70–85%), the validation was performed with VH-IVUS and the results were unstable [13]. Other techniques, such as the the Neuro Fuzzy classifier, showed potential results in detecting fibrotic, lipidic and calcified tissues by classifying different pixels of the IVUS image [14]. Supporting vector machines have been used with IVUS and VH-IVUS images to classify the vulnerability of the plaques depending on the FCT (thin FCT vs. normal/stable FCT) [11,15] or to detect calcifications [16,17]. Recently, convolutional neural networks (CNN) have emerged strongly as a good classifier. CNN has also been used to classify plaque into thin or stable FCT [18], to detect calcifications in the IVUS frames [19,20], or to segment the lumen and outer contours [21,22]. Newer studies presented CNNs that detect different tissues of the atherosclerotic plaque with high accuracy [8,23]. However, the accuracy of the majority of the machine learning methods does not include the actual measure of the FCT, which plays a key role in the plaque vulnerability. Although some studies analyzed the FCT, they only used it as a classifier to characterize the plaque as thin or normal FCT [11,15,18]. The main limitation of these machine learning techniques arises from on the lack of a large public database to train and test the models [9]. This entails that, usually, each study proved the efficiency of their technique with less than 12 patients [8,11–13,15,20] and human plaque geometries vary greatly in each patient.

These machine learning methodologies give morphological information of the composition of the atherosclerotic plaque; however, the vulnerability also depends on the mechanical properties of the tissues. For this reason, another line of research focuses on segmentation by using mechanical properties such as strain or elasticity maps. For vulnerability characterizations, elastography is commonly used to obtain the elasticity map of the arterial wall [24–27]. Therefore, the main objective of many studies was to segment and characterize the mechanical properties of the different plaque tissues at the same time [25,28,29]. Different speckle estimators or optical flow methods can be used to track the pixels' motion or estimate the strains in IVUS images [27,30]. Then, the segmentation and mechanical property estimation procedures are linked and usually consist of an iterative optimization problem. Segmentation results depend on the number of inclusions evaluated at each iteration [28,29,31]. With this optimization process it is possible to estimate the mechanical properties of the arterial wall and, furthermore, the segmentation of the plaque geometry. This methodology allows to take measurements of the FCT, lipid area, and the stiffness of the tissues. These types of processes have been tested in silico with finite element (FE) models [25,28,32], in vitro with polyvinyl acetate (PVA) phantoms [24], and in vivo with IVUS images from patients. The main disadvantages of these techniques are the high computational cost and the fact that the result depends on the number of inclusions evaluated.

Our work continues the study of the state-of-the-art of atherosclerotic plaque vulnerability by separating the segmentation procedure from the estimation of the mechanical properties. The main contribution of this paper is the definition of a new intuitive segmentation tool to segment the atherosclerotic plaque tissues without iterative or optimization steps, thus reducing computational costs. In addition, the method allows segmentation based on the representation of a large number of variables. By knowing the exact number of tissues, this technique opens the opportunity to obtain mechanical properties in future studies. This is a theoretical framework to lay the groundwork for future research; therefore, the methodology was developed and validated with in silico data. We have simulated the estimated strains that could be obtained from IVUS images with speckle estimators, with FE models, and adding some noise to the strain fields. We have defined this process as simulated IVUS data, as we are recreating the type of data that could be extracted from IVUS images. Our segmentation process is based on the representation of the modulus of the strain gradients and Watershed and Gradient Vector Flow (W-GVF) algorithms. The results are mainly focused on the lipid core segmentation, because of the importance of measuring the FCT and the lipid area for plaque vulnerability. This methodology was studied by using different strain variables in the segmentation process with different geometries. We have modeled three idealized geometries to analyze the FCT influence on the segmentation and three real IVUS patient geometries. In all of the analyzed cases, the proposed method was able to segment the lipid core and to measure the lipid area and FCT with enough accuracy.

2. Materials and Methods

The structure of the methodology was divided into five steps and it is schematized in Figure 1. The first step was to simulate IVUS data by computing different FE models, and then the FE results were analyzed mimicking two consecutive pictures taken by an IVUS. In the second step, some noise was added to the FE strains to mimic the intrinsic noise of the IVUS images. After that, the different strain gradient variables (SGVs) were computed in order to use them for the lipid segmentation process. Finally, after the segmentation we analyzed the performance of the results.

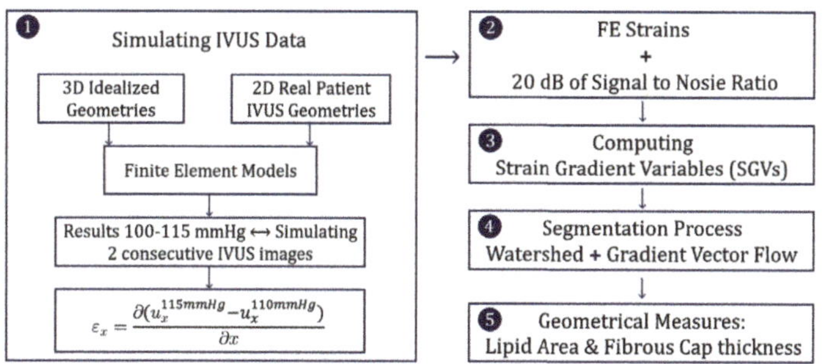

$$\varepsilon_x = \frac{\partial(u_x^{115mmHg} - u_x^{110mmHg})}{\partial x}$$

Figure 1. Scheme of the five steps that define the methodology.

2.1. Simulating IVUS Data

2.1.1. Geometries

As this was a theoretical study, the IVUS data were simulated by using FE models with idealized and real patient geometries. The idealized geometries consisted of a 3D geometry with a 13 mm long atheroma plaque and with a lipid core length of 6.5 mm and different FCTs [33]. We analyzed three different FCTs, trying to cover the different geometric possibilities. A FCT of 65 μm was considered to represent a vulnerable case [5],

300 μm represented a stable plaque, and finally, 150 μm was an intermediate value between the two extremes. The geometry was reconstructed following the Glagov results [34] and the Finet law [6]. The model was constructed with symmetric conditions, so only a quarter of the geometry is shown in Figure 2a. The model had different tissues: adventitia, healthy media and intima, fibrotic tissue, and lipid core. As one of the aims of this study was to check the influence of the FCT on the segmentation process, three thicknesses, 65, 150, and 300 μm, were considered, and they are represented in Figure 2b–d. Despite the use of a 3D model, we only analyzed the section of maximum stenosis with the plane strain assumption in order to reproduce the IVUS technique. On the other hand, the real patient geometries were obtained from three IVUS images of human coronary plaques that were manually segmented by an expert cardiologist in a previous study [35]. Both IVUS images and the cardiologist segmentation of the three plaques are shown in Figure 3. When there was lack of information in the axial direction, in these cases the FE models were 2D. In two IVUS geometries, only the fibrotic tissue and the lipid core were considered, and on the third plaque a calcification was also included.

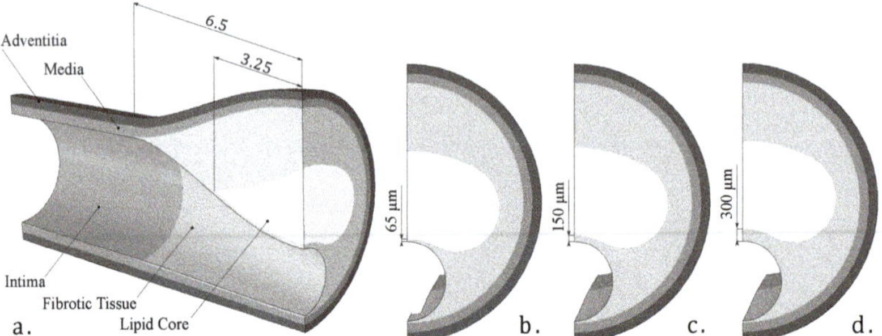

Figure 2. (**a**) 3D Idealized geometry; (**b**) fibrous cap thickness of 65 microns; (**c**) fibrous cap thickness of 150 microns; (**d**) fibrous cap thickness of 300 microns.

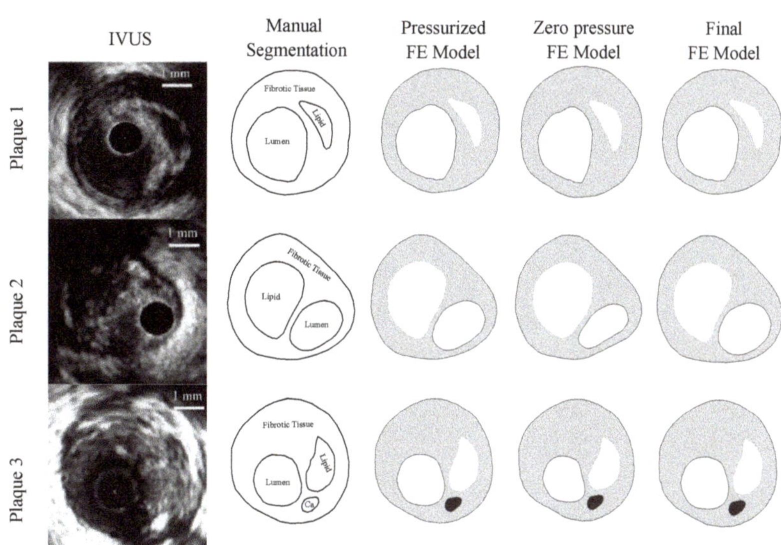

Figure 3. The first column presents IVUS images [35] of the three different plaques; the second column is the manual segmentation performed by a cardiologist [35]; the third column is the IVUS reconstruction in Abaqus of the pressurized geometry; the fourth column shows the plaque models with zero-pressure geometry in Abaqus. These geometries were used to initiate the FE simulations. The fifth column is the final FE model after applying an internal pressure of 115 mmHg to the previous geometry.

2.1.2. Modeling of Tissue Behavior

All tissues were modeled as hyperelastic, non-lineal, and incompressible with the constitutive model proposed by Gasser et al. [36]. The healthy tissues (adventitia, media, and intima) were modeled as anisotropic with two families of fibers. Conversely, the unhealthy tissues (fibrotic tissue and lipid core) were considered as an isotropic behavior model by the use of parameter $\kappa = 1/3$ in the *Gasser* model. All tissue was assumed to be fully incompressible (D = 0) in the idealized geometries and quasi-incompressible (D = 0.49) in the real IVUS geometries. The material parameters of the equation in (1) were fitted from experimental curves obtained from the bibliography [37,38] using the software Hyperfit [39]. All the parameters of (1) are reflected in Table 1, where α is the angle of the fibers with respect to the circumferential direction. The calcification of the third IVUS plaque was modeled with an isotropic neo-Hookean material model [35].

$$\Psi = \frac{1}{D} \cdot [J-1]^2 + \mu[I_1 - 3] + \frac{k_1}{2k_2} \sum_{i=4,6} \left(exp\left(k_2[\kappa[I_1 - 3] + [1 - 3\kappa][I_i - 1]]^2\right) - 1\right), \quad (1)$$

Table 1. Fitted parameters for the hyperelastic model.

Tissue	μ [kPa]	k_1 [kPa]	k_2 [-]	κ[-]	α[°]
Adventitia	4.22	547.67	568.01	0.26	±61.80
Media	0.7	206.16	58.55	0.29	±28.35
Intima	3.41	109.10	101.04	0.21	±52.72
Fibrotic	4.79	17,654.91	0.51	1/3	-
Lipid Core	0.025	956.76	70	1/3	-
Calcification	1875	-	-	-	-

2.1.3. FE Models

The FE models were created in the commercial software Abaqus [40], where the boundary conditions and loads were imposed. In the 3D idealized FE models, not only were the symmetrical conditions imposed, but also the contact with the heart was mimicked by avoiding displacement in a contour line on the outside of the adventitia [33]. The blood pressure imposed inside the artery was 115 mmHg, which is the average pressure in patients with high normal pressure and grade 1 hypertension [41]. On the other hand, the FE models of real IVUS geometries were 2D, so they were solved following the plain strain assumption and the rigid body motion was constrained by two contour points with zero displacements. Furthermore, the IVUS geometry was previously reconstructed from pressurized images (third column in Figure 3). Therefore, it was necessary to obtain the zero-pressure geometry to be used as the initial geometry. For this purpose, we assumed that IVUS images were taken with an internal blood pressure of 110 mmHg, and the zero-pressure geometry was recovered using the pull back algorithm defined by Raghavan et al. [42]. After applying the pull back method, the resulted geometry was extracted as the initial geometry (fourth column in Figure 3). Finally, it was possible to impose the pressure of 115 mmHg in the lumen and achieve the final pressurized geometry (fifth column in Figure 3). In all of the FE models (idealized and real IVUS geometries), the origin of the coordinate system was located in the center of the lumen in order to simulate the position of the IVUS catheter.

A sensitivity mesh analysis was performed to assure good precision and low computational cost. In the 3D idealized geometries, the maximum stenosis section, which was the most important part of the study, was meshed with small-sized elements. The fibrous cap between the lipid and the lumen had a smaller size depending on the thickness (e.g., 0.02 mm in the thickness of 65 microns that is on the order of IVUS technique precision). The rest of the 3D model had larger-sized elements, because of the lack of importance in the segmentation process. The element type selected for 3D was the hybrid quadratic tetrahedral elements with hybrid formulation to avoid numerical problems due to incompressibility (C3D10H), with at least three elements in the FCT in each case. In the 2D IVUS models, the element type was the plain strain hybrid three-node linear element (CPE3H) with additionaly, at least, three elements between the lumen and the lipid core.

2.1.4. Strain Variables

After simulating all of the FE models, the post-processing of the data was performed in Matlab 2021b [43]. The nodal coordinates (X and Y) and displacements (u_x and u_y) of the steps at 110 mmHg and 115 mmHg were collected. Then, the relative node displacements between both pressure steps were computed, as we can see in the example of Equation (2). This data processing attempted to simulate the data obtained by speckle estimators on two consecutive pictures taken by an IVUS in a 5 mmHg pressure increment [24,25,28,31]. Afterwards, the strains were calculated under the infinitesimal strain theory. Despite having the displacement field through the entire idealized 3D FE models, we analyzed the maximum stenosis section with the hypothesis of plane strain to obtain results that are closer to what happens in the IVUS. Different strain variables were computed: strains referring to Cartesian coordinates ($\varepsilon_{xx}, \varepsilon_{yy}$ and ε_{xy}); strains in cylindrical coordinates ($\varepsilon_{rr}, \varepsilon_{\theta\theta}$ and $\varepsilon_{r\theta}$); principal strains (ε_{max} and ε_{min}), and equivalent strains of the von Mises, Tresca, and Anisotropic index (FA) [44], defined in Equation (3). Some variables, such as principal strains [45] or equivalent strains, do not depend on the coordinate system. Thus, their value will be the same regardless of the positions of the IVUS catheter.

$$u_x = u_x^{115mmHg} - u_x^{110mmHg},\tag{2}$$

$$FA = \frac{\sqrt{3}}{\sqrt{2}} \cdot \frac{\sqrt{(\lambda_1 - \lambda)^2 + (\lambda_2 - \lambda)^2 + (\lambda_3 - \lambda)^2}}{\sqrt{\lambda_1^2 + \lambda_2^2 + \lambda_3^2}}, \tag{3}$$

$$\lambda = \frac{\lambda_1 + \lambda_2 + \lambda_3}{3}, \qquad \lambda_i = (\varepsilon_{max} \text{ or } \varepsilon_{med} \text{ or } \varepsilon_{min}) + 1,$$

Sumi et al. [46] developed a method to obtain a relationship between the vector gradient of the Young modulus and the strain tensor components for the plane stress approach. This criterion was adapted by Le Floc'h et al. [28] under the plane strain assumption, and they developed the elastic gradient of the material ($dW_{simplified}$) by neglecting the shear strains and computing only with the radial strains; Equation (4). This variable was selected due to the good results when marking the lipid core contour shown in different studies [28,29].

$$dW_{simplified} = -\frac{1}{\varepsilon_{rr}} \left(\frac{\partial \varepsilon_{rr}}{\partial r} + \frac{2\varepsilon_{rr}}{r} \right) dr - \frac{1}{\varepsilon_{rr}} \frac{\partial \varepsilon_{rr}}{\partial \theta} d\theta, \tag{4}$$

In this work, the parameter dW was also calculated without any type of simplification, and it is developed in Equations (5)–(7). Furthermore, the absolute value of this parameter was computed for segmentation purposes, and it was represented as $|dW|$.

$$H_r = \frac{-1}{\varepsilon_{rr}^2 + \varepsilon_{r\theta}^2} \cdot \left[\varepsilon_{rr} \cdot \left(\frac{\partial \varepsilon_{rr}}{\partial r} + \frac{1}{r} \cdot \frac{\partial \varepsilon_{r\theta}}{\partial \theta} + \frac{2 \cdot \varepsilon_{rr}}{r} \right) + \varepsilon_{r\theta} \cdot \left(\frac{\partial \varepsilon_{r\theta}}{\partial r} - \frac{1}{r} \cdot \frac{\partial \varepsilon_r}{\partial \theta} + \frac{2 \cdot \varepsilon_{r\theta}}{r} \right) \right], \tag{5}$$

$$H_\theta = \frac{-1}{\varepsilon_{rr}^2 + \varepsilon_{r\theta}^2} \cdot \left[\varepsilon_{r\theta} \cdot \left(\frac{\partial \varepsilon_{rr}}{\partial r} + \frac{1}{r} \cdot \frac{\partial \varepsilon_{r\theta}}{\partial \theta} + \frac{2 \cdot \varepsilon_{rr}}{r} \right) - \varepsilon_{rr} \cdot \left(\frac{\partial \varepsilon_{r\theta}}{\partial rr} - \frac{1}{r} \cdot \frac{\partial \varepsilon_{rr}}{\partial \theta} + \frac{2 \cdot \varepsilon_{r\theta}}{r} \right) \right], \tag{6}$$

$$dW = \vec{H} \cdot \vec{dx} = [H_r \; H_\theta] \cdot \begin{bmatrix} dr \\ r \cdot d\theta \end{bmatrix}, \tag{7}$$

2.2. Adding Noise

The strain information was obtained from the FE models (clean strains). Nevertheless, the in vivo IVUS images have some noise and the speckle estimated strains will also contain that noise. Therefore, to reproduce more realistic strains we added white Gaussian noise to the different FE strain fields. We used an SNR of 20 dB in the FE strains [24] . However, the segmentation procedure was studied in all geometries with and without noise so as to analyze the robustness of the process.

2.3. Computing SGVs

Once all of the strain variables were obtained (with and without noise), the next step was to obtain the modulus of their gradient. For instance, $|\nabla \varepsilon_{rr}|$ represents the modulus of the gradient of the radial strains. These SGVs allowed to highlight the contours of the different tissues of the plaque. Each SGV marked different parts of the tissue contours; this is why the segmentation procedure could use one or two combined SGVs to extract the entire lipid contour. We computed the modulus of the gradient of all strain variables, except for dW. By definition, dW showed the contours of the areas with different stiffness. The use of this single variable marked the tissue contours. At the end, there were 14 single SGVs and 91 possible combinations of two SGVs.

2.4. Segmentation Process

The methodology was based on the representation of two combined SGVs or a single SGV and image segmentation algorithms. The W-GVF processes were imposed on the SGVs representation to extract the lipid core. The watershed process used the contour and the grayscale representation to treat a set of pixels as a topography separating the lipid core. The W-GVF algorithm allowed to segment different tissues as the lipid core or the calcifications. In this work, only the results of the lipid core are presented due to their relevance to FCT measurements and plaque vulnerability. After the segmentation, the lipid was smoothed in order to reduce the sharp areas of the segmentation. The method was tested in all geometries (three idealized and three real geometries) with all of the 105 SGVs. A sensitivity analysis of different relevant variables in the segmentation process was performed. For this analysis only the idealized geometry with 150 μm of FCT was considered. These variables were related to the plaque morphology or related to the IVUS technology:

- Plaque-related variables: We analyzed the influence of considering the fibrotic tissue as fully incompressible or with different degrees of quasi-incompressibility. We have also considered four different fibrotic tissues (default, stiff, medium, and soft tissues). Furthermore, some inclusions were added to the FE model, mimicking the presence of micro calcifications. These inclusions were simplified as spheres with calcification properties presented in Table 1, and four diameters were studied (10, 50, 150, and 300 μm).
- IVUS-related variables: The influence of the catheter position was studied by changing the origin and orientation of the coordinate system in the FE models. It was also important to check if the segmentation methodology was affected by the blood pressure. In addition, the pressure increment between both steps was also studied.

Although the methodology was mainly focused on the lipid core segmentation, different areas were segmented as well. Lumen was segmented using the W-GVF technique in each geometry in order to measure the FCT. Large calcifications, such as the one in the third real IVUS geometry, were segmented by using the same segmentation process. On the other hand, fibrotic tissue could be easily segmented as the difference of the whole plaque minus the segmented lipid and lumen. Finally, adventitia and media could be segmented with the W-GVF technique; however, this segmentation has no clinical application due to the fact that IVUS images provide little information on the outermost tissues.

2.5. Geometrical Measures

After the lipid and the lumen were segmented, it was possible to assess the FCT as the minimum distance between them. The area of the lipid core was also computed. Both measurements are closely related to the risk of plaque rupture [5]. The indices I_1, I_2, and I_3 were defined in order to quantify the accuracy of the segmentation for each SGV or combination of two SGVs. The first index (I_1) in Equation (8) is the relative error between the real and the measured FCT (t_{real} and $t_{measure}$). The second index (I_2) in Equation (9) defines the percentage of the lipid area that was correctly segmented (true positive area). This index could be represented as the white area in Figure 4. The third index (I_3) in Equation (10) corresponds to the extra lipid area that was segmented (false positive area). It could be represented as the green area in Figure 4. The second and third indices were defined to quantify not only the lipid area value, but also the correct segmentation of its shape. In order to quantify the segmentation using only one index, we defined the Segmentation Index (SI) as a linear combination of the previous indices. The final SI parameter was defined in Equation (11) and its value was directly related to the performance of the segmentation. SGV combinations with an $SI \geq 90\%$ provided measurements of the lipid area and the fibrous cap thickness with high precision. An SI between 90–85% meant that the segmentation had trouble with one measure, normally the fibrous cap thickness. SI values in the range of 85–75% indicated a poor lipid segmentation. Finally, values of $SI \leq 75\%$ were for those SGVs with a high measurement error or those that could not segment the lipid.

$$I_1 = \left| \frac{t_{measure} - t_{real}}{t_{real}} \right| \cdot 100, \tag{8}$$

$$I_2 = \frac{Area_{RealSegmented}}{Area_{Real}} \cdot 100, \tag{9}$$

$$I_3 = \frac{Area_{FalseSegmented}}{Area_{Real}} \cdot 100, \tag{10}$$

$$SI = \frac{(100 - I_1) + I_2 + (100 - I_3)}{3}, \tag{11}$$

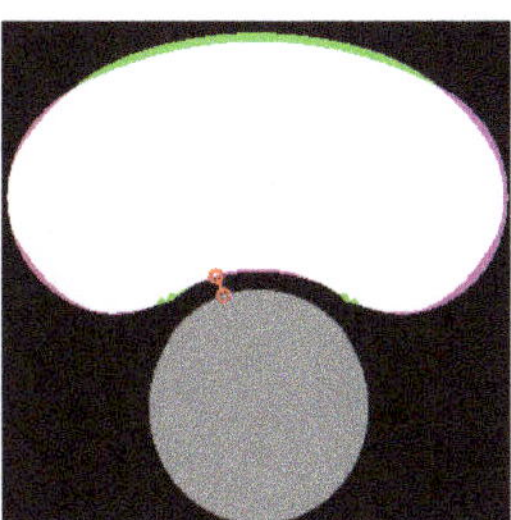

Figure 4. Comparison between the real and the segmented lipid core. The lumen is represented in gray color, the true positive area in white, the false negative area in green, the actual area that was not segmented in purple, and the measure of the FCT is the red line.

3. Results

This section presents the results of the lipid segmentation. It is divided into the results in idealized geometries, results in real IVUS geometries, and an analysis of the best SGVs after considering all of the geometries.The results were analyzed with the FE strains and the strains with 20 dB of SNR.

3.1. Idealized Geometries

The process is summarized in Figure 5, where the idealized geometries with 65, 150, and 300 μm of fibrous cap thickness are represented. A combination of the SGVs $|\nabla \varepsilon_{min}|$ and $|\nabla FA|$ was chosen due to its high segmentation performance. Both SGV variables are shown in each idealized geometry in Figure 5a,b, respectively. The combination of both is represented in Figure 5c and it was used as an input for the W-GVF process. The final result is shown in Figure 5d, where every segmented part has a different color. The segmented lipid core is represented before and after the smooth treatment in Figure 5e, and it is shown as an overlap between the actual and segmented lipid core. In this representation, the white area displays the well-segmented area, the purple one is the actual lipid that is not segmented, and inversely, the green area is the extra area wrongly segmented by the procedure.

In the idealized geometries, the lipid area was similar in all scenarios, so the number of successful SGV combinations depended only on the fibrous cap thickness. The box-plot in Figure 6 represents the SI distribution of the 105 SGV combinations in each geometry. The geometries were segmented using the FE strains (clean strains) and the strains with 20 dB of noise. By analyzing only the influence of the FCT, the results show that the mean (represented by an asterisk) and the median of the SI value increased with the FCT. In addition, the interquartile range decreased with greater thicknesses. This fact can be observed in Figure 6 for the segmentation with or without noise. On the other hand, the noise addition led to an increase of the outliers and the interquartile range. The mean and median decreased after considering the noise. Despite having different SI results in each geometry, there were some combinations of SGVs with proper results for all thicknesses. This was the case of $|\nabla \varepsilon_{yy}| + |\nabla \varepsilon_{\theta\theta}|$, $|\nabla \varepsilon_{vMises}| + |\nabla \varepsilon_{r\theta}|$, $|\nabla FA| + |\nabla \varepsilon_{\theta\theta}|$ or $|\nabla \varepsilon_{max}| + |\nabla \varepsilon_{rr}|$. It was also possible to yield good results with only one variable such as $|\nabla \varepsilon_{yy}|$, $|\nabla FA|$, dW, $|\nabla \varepsilon_{Tresca}|$ and $|\nabla \varepsilon_{xy}|$. However, there were variables with low SI values for all cases, such as $dW_{simplified}$, $|\nabla \varepsilon_{r\theta}|$, so they were discarded.

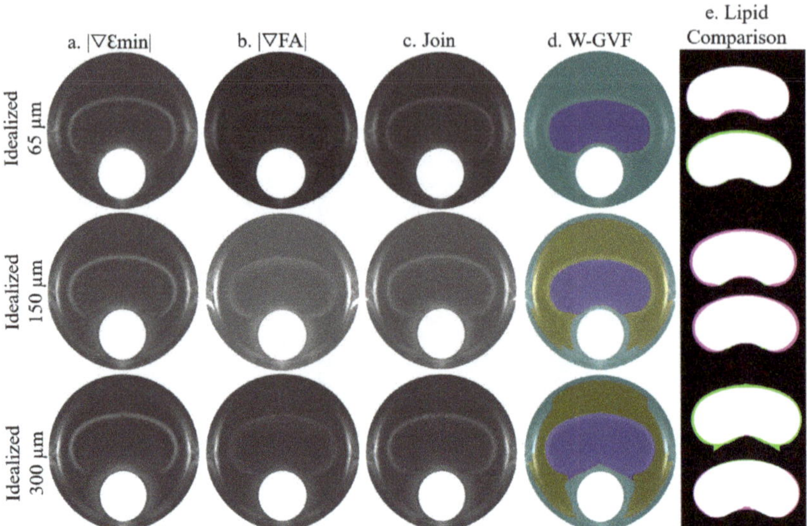

Figure 5. Influence of the FCT on the segmentation procedure analyzed with clean strains. The rows represent the segmentation process with the geometries of 65, 150, and 300 μm of FCT. The segmentation process consists of the combination of two SGVs, in these cases $|\nabla\varepsilon_{min}|$ and $|\nabla FA|$ in (**a**,**b**), respectively. The combination of both is represented in (**c**). This representation is the input for the W-GVF and its results are represented in (**d**); finally, the overlap between the actual and the segmented lipid core before and after the smooth treatment is represented in (**e**), where the true positive area is in white, the false negative area in green and the actual area that is not segmented in purple.

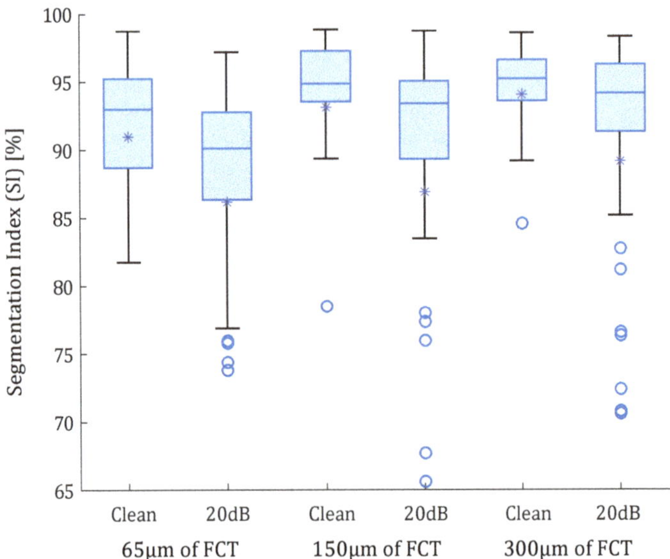

Figure 6. Box-plots of the SI values of the idealized geometries. From left to right: 65, 150, and 300 μm of fibrous cap thickness. Each geometry was analyzed with clean strains and with 20 dB of SNR. The median values were represented with a horizontal line. The median values were 93% and 90.14% for the idealized geometry with 65 μm FCT with and without noise; 94.88% and 93.42% for geometry of 150 μm FCT; and 95.23% and 94.17% for 300 μm. Mean values are represented with asterisks. Outliers are represented with circles. Some outliers were below 65% but are not shown.

3.2. Real IVUS Geometries

The proposed methodology was tested with the three real IVUS plaque geometries. The lipid cores had different shapes, areas (1.65, 5.53, and 1.93 mm^2), and FCTs (330, 175, and 209 µm). Figure 7 is an example of lipid segmentation with the SGV combination of the invariants $|\nabla\varepsilon_{min}|$ and $|\nabla FA|$ for the case of clean strains. The performance of the segmentation is represented in the box-plot shown in Figure 8, where the segmentation was not only affected by the FCT, but also by the lipid core area. In all cases, the noise addition increased the interquartile range and decreased the mean and the median (except in the first plaque, where the median increased after the 20 dB). A single SGV such as $|\nabla\varepsilon_{min}|$, $|dW|$, $|\nabla\varepsilon_{rr}|$, dW, $|\nabla\varepsilon_{Tresca}|$ or combinations such as $|\nabla\varepsilon_{yy}| + |\nabla\varepsilon_{rr}|$, $|\nabla\varepsilon_{yy}| + |\nabla\varepsilon_{min}|$, or $|\nabla\varepsilon_{min}| + |\nabla\varepsilon_{Mises}|$ still had promising results for these geometries. As what happened in the idealized cases, the SGV $|\nabla\varepsilon_{r\theta}|$ did not show any adequate SI for any plaque.

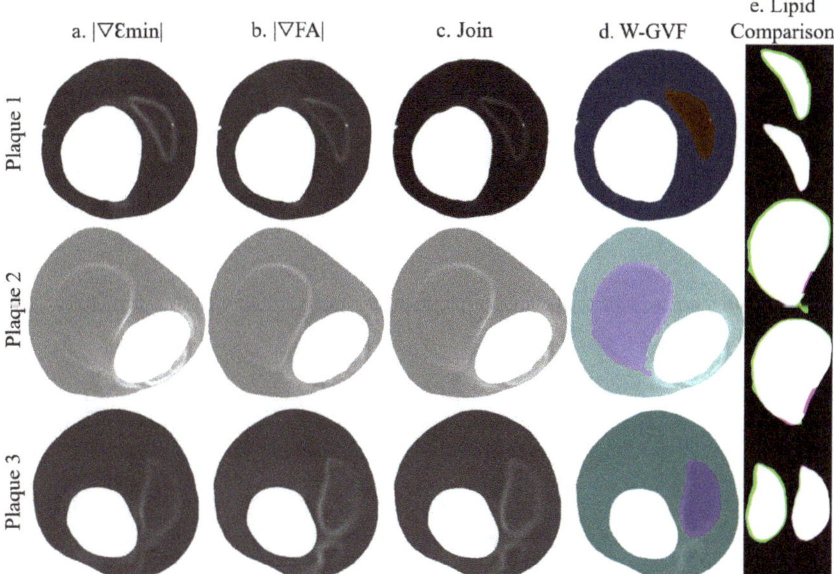

Figure 7. Segmentation procedure in the IVUS geometries with clean strains. The rows represent the segmentation process with the plaques 1, 2, and 3; (**a**) representation of $|\nabla\varepsilon_{min}|$; (**b**) representation of $|\nabla FA|$; (**c**) the combination of $|\nabla\varepsilon_{min}| + |\nabla FA|$; (**d**) W-GVF results; and (**e**) segmented lipid before and after the smooth treatment, where the true positive area is in white, the false negative area in green and the actual area that is not segmented in purple.

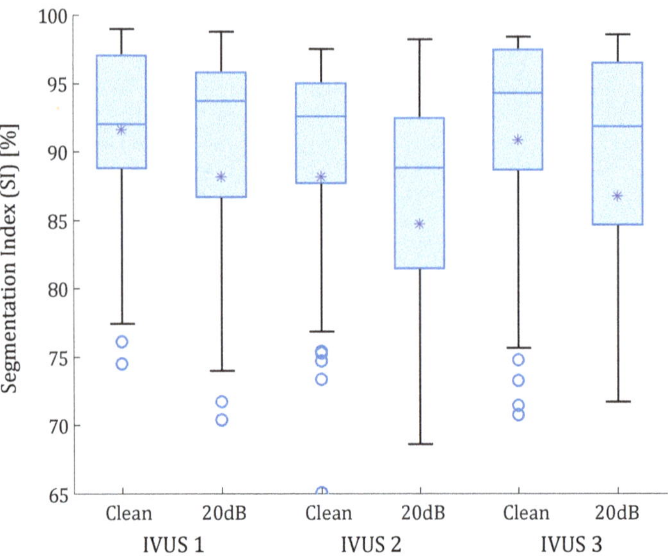

Figure 8. Box-plots of the SI values of the real IVUS geometries. The three plaques were analyzed with clean strains and with 20 dB of SNR. The median values are represented with a horizontal line. The median values were 92.02% and 93.74% for the first IVUS geometry, with and without noise; 92.60% and 88.82% for the second IVUS geometry; and 94.30% 91.84% for third IVUS geometry, respectively. Mean values are represented with asterisks. Outliers are represented with circles. Some outliers were below 65% but are not shown.

3.3. SGV Candidates

Finally, the mean SI value ($\bar{SI}$) was computed by considering all of the geometries together. All of the SGVs from FE strains, alone or in combination with others, were analyzed, and the best five single SGVs and fifteen SGV combinations were collected in Table 2. This table shows the SGVs or SGV combinations with better performance in the segmentation process, allowing to have good accuracy not only in the area, but also in the FCT. The mean SI value of those combinations with 20 dB of noise was included ($\bar{SI}_{noise}$) to visualize the noise influence. Single SGVs, such as dW or $|\nabla\varepsilon_{rr}|$, presented good segmentation results. Furthermore, $|\nabla\varepsilon_{vMises}|$ or $|\nabla\varepsilon_{min}|$ had good performance as well, and in these cases they had the advantage of being invariants, so they will be not affected by the catheter position or coordinate system. However, the invariants have the disadvantage of needing the entire strain tensor. On the other hand, there were a great number of SGV combinations with high segmentation accuracy regardless of the analyzed case, such as $|\nabla\varepsilon_{vMises}|+|\nabla\varepsilon_{r\theta}|$, or $|\nabla\varepsilon_{yy}|+|\nabla\varepsilon_{rr}|$. Additionally, combination of invariants appeared to have good results, such as $|\nabla\varepsilon_{min}|+|\nabla\varepsilon_{Tresca}|$. Table 2 shows that SGV combinations achieved higher SI values than single SGVs.

Table 2. Summary of the results for the best five single SGVs and fifteen SGV combinations for the lipid segmentation of all of the 105 possible combinations based on the results of the idealized and IVUS geometries with FE strains. The SGVs with the SI value in dark green mean a perfect segmentation; light green stands for good segmentations; yellow for SGV indicates some problems in segmenting the fibrous cap or the lipid area. $\bar{SI}$ and $\bar{SI}_{noise}$ represent the mean SI value without and with noise, respectively.

		Segmentation Index (SI)							
		Idealized Geometry			Real IVUS Geometry			$\bar{SI}$	$\bar{SI}_{noise}$
		65 μm	150 μm	300 μm	Plaque 1	Plaque 2	Plaque 3		
One SGV	dW	95.20	97.13	94.31	96.98	92.06	92.50	94.70	90.62
	$|dW|$	92.33	94.55	96.02	93.99	94.66	96.23	94.63	86.04
	$|\nabla\varepsilon_{vMises}|$	93.49	93.60	97.65	90.74	94.99	97.08	94.59	94.47
	$|\nabla\varepsilon_{rr}|$	97.65	94.40	93.07	98.48	86.78	96.32	94.45	92.27
	$|\nabla\varepsilon_{min}|$	86.07	93.77	97.86	97.73	93.43	97.46	94.39	93.29
Combination of two SGVs	$|\nabla\varepsilon_{vMises}| + |\nabla\varepsilon_{r\theta}|$	95.87	97.63	97.09	98.61	96.14	97.28	97.10	95.22
	$|\nabla\varepsilon_{yy}| + |\nabla\varepsilon_{rr}|$	95.74	97.93	94.79	98.53	97.51	96.76	96.88	95.68
	$|\nabla\varepsilon_{yy}| + |\nabla\varepsilon_{min}|$	95.74	98.21	94.23	98.53	95.55	98.28	96.75	94.28
	$|\nabla\varepsilon_{min}| + |\nabla\varepsilon_{Tresca}|$	96.97	96.32	96.09	97.36	95.98	97.43	96.69	93.67
	$|\nabla\varepsilon_{max}| + |\nabla\varepsilon_{rr}|$	97.73	94.08	98.58	97.76	92.33	98.03	96.42	92.88
	$|\nabla\varepsilon_{min}| + |\nabla FA|$	97.85	93.01	96.17	97.46	95.24	97.47	96.20	94.76
	$|\nabla\varepsilon_{min}| + |\nabla\varepsilon_{vMises}|$	95.17	93.28	96.87	97.43	95.98	98.37	96.18	95.10
	$|\nabla\varepsilon_{rr}| + |\nabla\varepsilon_{r\theta}|$	92.81	97.49	97.95	98.94	93.13	96.43	96.13	88.68
	$|\nabla\varepsilon_{Tresca}| + |\nabla\varepsilon_{rr}|$	93.30	96.25	95.81	97.34	96.12	97.77	96.10	93.67
	$|\nabla\varepsilon_{yy}| + |\nabla\varepsilon_{Tresca}|$	93.06	97.64	95.17	98.41	93.55	97.94	95.96	93.09
	$|\nabla FA| + |\nabla\varepsilon_{rr}|$	92.87	94.88	96.53	97.51	95.96	97.49	95.87	92.88
	$|\nabla\varepsilon_{max}| + |\nabla\varepsilon_{min}|$	97.53	94.46	92.74	96.34	95.40	97.96	95.74	93.65
	$|\nabla FA| + |\nabla\varepsilon_{r\theta}|$	93.68	97.40	95.04	98.56	90.99	97.53	95.53	92.69
	$|\nabla\varepsilon_{xx}| + |\nabla FA|$	95.93	95.79	95.51	92.44	95.29	98.23	95.53	92.03
	$|\nabla\varepsilon_{xy}| + |\nabla FA|$	94.32	94.21	94.84	96.49	95.12	97.16	95.36	93.04

The elastic gradient of the material (dW) was calculated in a simplified way, following Le Floc'h et al. [29] and with the whole 2D strain tensor. In order to have the lipid contour marked and to achieve a better combination with other SGVs, the absolute value $|dW|$ was computed. These variables are shown in Figure 9. The variable $dW_{Simplified}$ had a mean SI value of 32.13% in the six geometries without noise and 28.48% with noise, whereas dW without simplifications achieved a mean SI value of 94.70% without noise and 90.62% with 20 dB of noise and $|dW|$ 94.63% and 86.04% without and with noise, respectively.

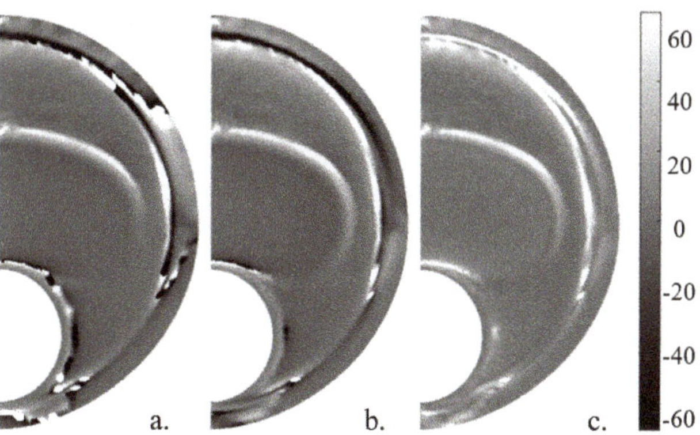

Figure 9. Idealized geometry with 150 μm thickness with the different dW represented without noise. (**a**) $dW_{simplified}$, (**b**) dW computed with the 2D strain tensor, and (**c**) $|dW|$.

3.4. Sensitivity Analysis

Once the SGV candidates were analyzed, we selected the best four SGV combinations and the best single SGV of Table 2 to evaluate the robustness of the proposed segmentation methodology. For this purpose, we studied the influence of plaque and IVUS variables in the segmentation results only within the idealized geometry with 150 μm of FCT. Figure 10 shows all of the results of the sensitivity study. Each SGV had seven variables to analyze (incompressibility, different material behavior of fibrotic tissues, addition of inclusions, different catheter positions, total pressures and pressure increments, and noise addition), and all are color-coded in Figure 10. The SI values obtained with the FE strains were taken as a reference. In this reference case, the model was analyzed as fully incompressible, with a stiffer tissue, the catheter placed in the center of the lumen, with a pressure analysis of 110–115 mmHg and without noise. The SI values achieved with that model are represented in Figure 10 with a horizontal dotted line.

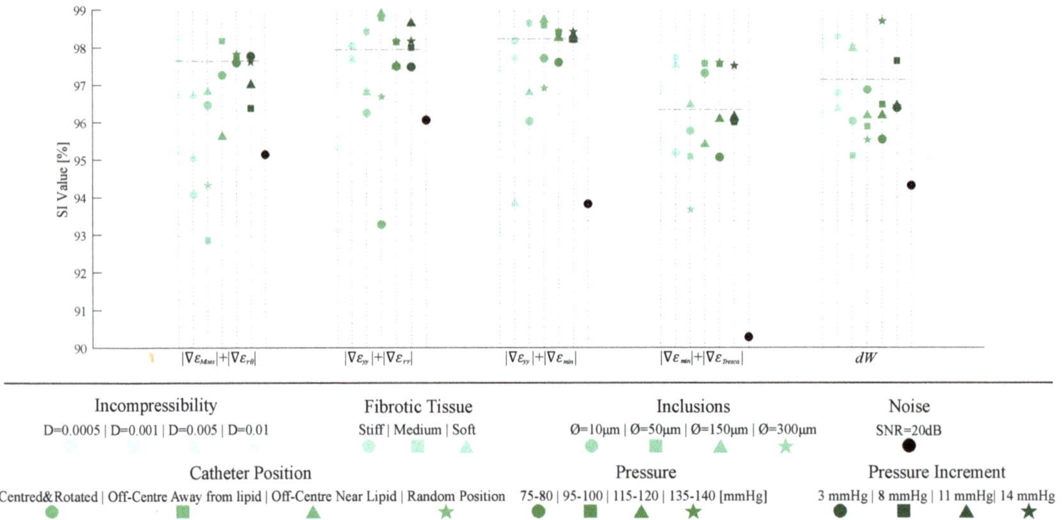

Figure 10. Graphic summary of the influence of incompressibility, changing fibrotic tissues, addition of inclusions, different catheter positions, different pressures and pressure increments, and noise addition on the segmentation process. The considered SGVs are $|\nabla\varepsilon_{Mises}|+|\nabla\varepsilon_{r\theta}|$, $|\nabla\varepsilon_{yy}|+|\nabla\varepsilon_{rr}|$, $|\nabla\varepsilon_{yy}|+|\nabla\varepsilon_{min}|$, $|\nabla\varepsilon_{min}|+|\nabla\varepsilon_{Tresca}|$, and dW. Each SGV has different variables to analyze, divided by colors. Each plaque- or IVUS-related variable had different cases that were differentiated by shape markers.

4. Discussion

The clinical detection and segmentation of the atherosclerotic plaque is still a challenging step for an early diagnosis. Machine learning techniques are mainly focused either on the segmentation of the lumen and outer contours [21] or on the detection of calcifications [16,17] or vulnerable FCT [15]. However, it is difficult to find a quantitative study of FCT or the lipid area. Other techniques based on IVUS allow to estimate the 2D strain field in the arterial wall [27,30]. From the strain map, with iterative optimization tools, it is possible to estimate the mechanical properties and the segmentation of the plaque [28,29,31]. However, these cases had high computational cost and depended on the inclusions evaluated. The most difficult issue was to obtain an accurate segmentation of the plaque that could provide good estimations of the FCT. In this paper, we have presented the theoretical basis to segment the tissues of the atherosclerotic plaque from the representation of different strain variables without any iteration or the need for a large database to train the methodology. The segmentation procedure was based on the Watershed and Gradient Vector Flow algorithms that extract the tissues. In this article, we focused on the lipid core segmentation due to its role in plaque vulnerability [5]. The accuracy of the results depends on the represented SGVs. This approach allowed us to obtain measurements of the lipid area and FCT and it achieved promising results with many different SGV combinations. The methodology was developed and validated with computational models designed from idealized and real IVUS geometries. The aim was to check if the strain map method was able to segment the lipid and also to know which are the best SGVs to achieve it. Furthermore, the process was performed with different morphological and IVUS technical variations to prove the versatility of the proposed method.

4.1. Segmentation Analysis

4.1.1. Idealized Geometries

Using idealized geometries, the results show that as the FCT was smaller, there were less SGVs available to obtain a proper segmentation. The box-plot of Figure 6 shows that the variability of SI values decreases with larger thicknesses, while the segmentation performance increases. The explanation is that the amount of data available on the fibrous cap is lower and the SGVs in low thicknesses; it marked the lipid contour close to the lumen and it was more difficult for the segmentation process to track the lipid. Therefore, the measure of the FCT presents higher errors. Consequently, the precision of the IVUS technology will delimit the amount of strain information in the fibrous cap and thus the segmentation performance. After analyzing the 105 possible SGV combinations in each geometry with clean FE strains, the median SI value was always above 93% in all cases. That highlights the strong segmentation capacity of the proposed methodology. Previous studies also showed a decrease in the FCT segmentation at lower thicknesses, increasing the relative error of the segmented FCT [28,31]. However, this proposed methodology opened the possibility of using a large number of SGVs for segmentation. Some combinations of SGV, such as dW, $|\nabla \varepsilon_{vMises}| + |\nabla \varepsilon_{r\theta}|$ or $|\nabla \varepsilon_{yy}| + |\nabla \varepsilon_{rr}|$, had errors close to zero and with negligible variation between thicknesses. After adding a 20 dB SNR to the FE strains, the SGVs were not as smooth as in the previous cases; however, the median SI of each group decreased by most 3%, but all of them were above 90%. Hence, the segmentation method proved that there were large amounts of SGVs that allowed to extract the lipid core regardless of the FCT or the noise addition. Furthermore, the areas of the lumen and the complete plaque were automatically segmented with the proposed method. This provided results similar to those obtained with machine learning [21,22] but without the need to train a neural network.

4.1.2. IVUS Geometries

The IVUS geometries allowed testing the segmentation procedure on real human atherosclerotic coronary plaque geometries with distinct lipid core areas and FCT. The results showed that the W-GVF process relied not only on the fibrous cap thickness, but also on the lipid area and stenosis degree, showing a strong dependency on the real IVUS plaques. The segmentation performance varied greatly in each geometry due to their differences. In all cases the median of the SI value was above 88% (with or without noise). The noise addition slightly reduced the segmentation except for the first IVUS plaque, where noisy strains improved the median SI value by 1.87%. The noise addition also increases the variability of the performance. The third IVUS plaque had a calcification that was segmented using the W-GVF technique with a relative error smaller than a 5%. Nevertheless, the study showed that the segmentation procedure was able to track lipids with different areas and morphologies. Since these geometries were analyzed in other studies [28,32], it was possible to compare the segmentation results with those of previous studies. Previous work reached a relative FCT error of between 1.4 and 15.5% depending on the geometry [28]. Other methodologies obtained a mean error above 16% by measuring the FCT on other real geometries [31,47]. In this work, the FCT measure depended on the chosen SGV. For the SGVs analyzed in Table 2, the mean relative error was 3.11% in the IVUS geometries.

4.2. SGV Candidates

Different studies rate the segmentation performance with a qualitative index, such as thin vs. stable FCT [11,15,18], or by detecting the calcified plaques [16,17]. Those studies that used quantitative values only considered the error of the segmented FCT and lipid area [28,31,47]. Our SI included values to consider the relative error of the FCT, the value of the segmented lipid and its shape, which facilitates the comparison of segmentations between geometries. Studies that used a strain variable for segmenting only used a single strain variable (commonly the modified Sumi's transformation ($dW_{Simplified}$) [28,46,47] or von Mises strains [31]). On the contrary, this methodology made it possible to use 105 different strain variables (single SGVs or combinations of SGVs).

The combination of idealized and real IVUS results showed that the segmentation results depended mainly on the FCT and the lipid core area. Nevertheless, there were some variables with a high SI by itself regardless of the geometry case. That was the case of SGVs such as dW, $|\nabla\varepsilon_{min}|$ or $|\nabla\varepsilon_{rr}|$, which obtained an overall SI value of more than 90% with or without noise. By the combination of different SGV, some variables with low accuracy increased the results. That was the case of $|\nabla\varepsilon_{r\theta}|$, which had a mean SI value of 67.24% and in combination with $|\nabla\varepsilon_{vMises}|$ was the best pair, achieving a mean SI of 97.1% (95.22% with noise). SGV combinations reached higher precision than single SGVs. Some examples were $|\nabla\varepsilon_{yy}|+|\nabla\varepsilon_{rr}|$ or $|\nabla\varepsilon_{min}|+|\nabla\varepsilon_{Tresca}|$ with SI mean values of 96.88–95.68% and 96.69–93.67%, respectively (with clean strains–with 20 dB of SNR, repsectively). All of the SGV combinations presented in Table 2 allowed for a very precise segmentation of the lipid in all scenarios with different SGV options.

On the other hand, after analyzing the elastic gradient of the material (dW), all dW SGVs had a similar representation; however, the $dW_{simplified}$ proposed by Le Floc'h et al. [28] had more trouble marking the shoulder of the lipid core and obtained worse segmentation results than dW and $|dW|$. The variable dW marked the whole lipid contour, and alone proved to be the best SGV for the lipid segmentation. Its absolute value, $|dW|$, was computed in order to obtain better results in combination with the others SGVs; however, it showed similar SI values to dW and it was the second-best single SGV option.

4.3. Sensitivity analysis

The best single SGVs and SGV combinations were analyzed in a sensitivity analysis to study the influence of different variables (plaque- and IVUS-related variables). We concluded that the incompressibility of the materials did not affect the segmentation performance in the five selected SGVs. By changing the material behavior of the fibrotic tissue, the SI values remained above 92%, except in one case for dW. This parameter was related to the different stiffness between tissues, and it would fail if the lipid and fibrotic tissue had similar stiffness. The addition of inclusions could affect the lipid segmentation if they were placed close to the lipid contour. Nonetheless, the methodology appeared to obtain similar results. Principal strains are widely used because of their non-dependence on the coordinate system [45]; however, the catheter position seemed to have no influence on the process, since the worst SI value obtained was 93.23%. Finally, the pressure and pressure increments did not affect the method due to the fact that the tool is based on gradients of the strains, and they had a similar representation regardless of the modification of pressures. Overall, the results suggest that the proposed methodology had no dependency on the analyzed cases, showing a strong robustness.

4.4. Relevance for Clinical Applications

The new developed technology is an intuitive segmentation tool that could provide morphological information on the atherosclerotic plaque. It could help to reduce segmentation human errors and it could assist clinicians with a new accurate diagnosis support.

From IVUS images and the use of a strain estimator, we can use the representation of different moduli of the gradient of variables to detect the lipid or inclusions contours in a fast way. After that, with the W-GVF segmentation procedure it is possible to extract the lipid core or other tissues and take measurements, which are directly involved with plaque vulnerability. The results show that the performance of certain SGV combinations depended on plaque morphology; however, the SGV combination between $|\nabla\varepsilon_{min}|$ and $|\nabla\varepsilon_{vMises}|$ presented good segmentation performances for the lipid core, regardless of the plaque geometry. Additionally, other combinations showed in Table 2 appear to be accurate enough for good clinical diagnosis. Single SGVs such as $|\nabla\varepsilon_{rr}|$ or dW provided accurate segmentations. In addition, ε_{rr} is the strain variable that can be extracted from IVUS with the highest accuracy, so it will be one of the main candidates for clinical application.

The highlights of this work are presented in the following list in order to summarize the results of the research.

i. A segmentation process based on strain representation was presented to extract the different tissues of an atherosclerotic plaque. This methodology achieved high accuracy in measuring FCT and the lipid core area. These measurements play a key role in the vulnerability of the plaque.

ii. Unlike other segmentation processes, this method does not require a database to be trained or an optimization process, as it relies on image processing rather than machine learning or analysis of the mechanical properties of the tissues. In addition, it could be performed with many different strain variables instead of a single one [27,28,31,47]. Thus, there are different possibilities to obtain the segmentation using only one variable or combining different SGVs.

iii. The results show that the performance of the segmentation was linked to the plaque geometry and the selected SGVs. However, there were some SGVs with good results regardless of the geometry. The method also showed good robustness in sensitivity analysis, providing accurate results with different catheter positions, pressures, and noise addition.

4.5. Limitations

This study has two main limitations that have to be mentioned:

- Since this work was a theoretical framework, the methodology was only tested with computational models of in silico data. Therefore, the next step would be to prove the segmentation methodology with in vitro and in vivo IVUS data from patients with coronary atherosclerotic plaques. After analyzing the methodology with noise, which simulates the intrinsic noise of IVUS data, the results for segmentation are expected to be valid on real IVUS data.
- In the finite element analysis we only have considered the load of the blood pressure. We have disregarded the residual stress and the influence of heart motion. As the methodology is based on gradients and not on absolute strain/stress values, we could expect a minimum influence of the residual stress on this segmentation methodology.

5. Conclusions

In this paper we have proposed a new method to segment the atheroma plaque based on strain measurements with low computational and time costs. This work is a theoretical framework and has been developed and tested with FE models with idealized and real IVUS geometries. The representation of the SGVs opens the possibility of segmenting the lipid core with different strain variables. This representation is used in the W-GVF segmentation to extract the tissues and measure the FCT and lipid area for the clinical diagnosis. We made an extensive study of the strain variables used for the W-GVF segmentation and selected only those who detect the lipid core and other tissues. The method had good results in all scenarios, showing an SI value higher than 94% (with noise). There are strain variables such as $|\nabla\varepsilon_{vMises}|$ in combination with $|\nabla\varepsilon_{r\theta}|$ or $|\nabla\varepsilon_{yy}| + |\nabla\varepsilon_{rr}|$ that achieved accurate results regardless of the geometry, morphological changes, or noise addition. Furthermore, single SGVs such as dW or $|\nabla\varepsilon_{min}|$ could provide the lipid segmentation. Although the methodology has to be tested with in vivo data, it has promising preliminary results.

Author Contributions: Conceptualization of the study: J.O. and E.P. Investigation and methodology: A.T.L. Supervision: M.A.M., M.C. and E.P. Writing: A.T.L. Review and editing: A.T.L., M.C., M.A.M., J.O. and E.P. Funding acquisition, M.A.M. and E.P. All authors have read and agreed to the published version of the manuscript.

Funding: This work was supported by the Spanish Ministry of Science and Technology through research project PID2019-107517RB-I00, the regional Government of Aragón support for the funding of the research project T24-20R, and grant CUS/581/2020. CIBER Actions are financed by the Instituto de Salud Carlos III with assistance from the European Regional Development Fund. Jacques Ohayon was supported by the SIMR project (2019–2023) operated by the French National Research Agency.

Data Availability Statement: Not applicable.

Acknowledgments: We would like to thank Santiago Cruz (Department of Electronic Engineering and Communications of University of Zaragoza) for the helpful discussion on the segmentation procedure.

Conflicts of Interest: The authors declare no conflict of interest.

Abbreviations

The following abbreviations are used in this manuscript:

IVUS	Intravascular Ultrasound
FCT	Fibrous Cap Thickness
VH	Virtual Histology
FE	Finite Element
CNN	Convolutional Neural Networks
SGV	Strain Gradient Variable
SNR	Signal-to-Noise Ratio
SI	Segmentation Index
W-GVF	Watershed-(Gradient Vector Flow)

References

1. Roth, G.A.; Abate, D.; Abate, K.H.; Abay, S.M.; Abbafati, C.; Abbasi, N.; Abbastabar, H.; Abd-Allah, F.; Abdela, J.; Abdelalim, A.; et al. Global, regional, and national age-sex-specific mortality for 282 causes of death in 195 countries and territories, 1980–2017: A systematic analysis for the Global Burden of Disease Study 2017. *Lancet* **2018**, *392*, 1736–1788. [CrossRef]
2. Casscells, W.; Naghavi, M.; Willerson, J.T. Vulnerable atherosclerotic plaque: A multifocal disease. *Circulation* **2003**, *107*, 2072–2075. [CrossRef]
3. Jebari-Benslaiman, S.; Galicia-García, U.; Larrea-Sebal, A.; Olaetxea, J.R.; Alloza, I.; Vandenbroeck, K.; Benito-Vicente, A.; Martín, C. Pathophysiology of Atherosclerosis. *Int. J. Mol. Sci.* **2022**, *23*, 3346. [CrossRef]
4. Libby, P.; Theroux, P. Pathophysiology of coronary artery disease. *Circulation* **2005**, *111*, 3481–3488. [CrossRef]
5. Cilla, M.; Peña, E.; Martínez, M.A. 3D computational parametric analysis of eccentric atheroma plaque influence of axial and circumferential residual stresses. *Biomech. Model. Mechanobiol.* **2012**, *11*, 1001–1013. [CrossRef] [PubMed]
6. Finet, G.; Ohayon, J.; Rioufol, G. Biomechanical interaction between cap thickness, lipid core composition and blood pressure in vulnerable coronary plaque: Impact on stability or instability. *Coron. Artery Dis.* **2004**, *15*, 13–20. [CrossRef]
7. Guo, X.; Zhu, J.; Maehara, A.; Monoly, D.; Samady, H.; Wang, L.; Billiar, K.L.; Zheng, J.; Yang, C.; Mintz, G.S.; et al. Quantify patient-specific coronary material property and its impact on stress/strain calculations using in vivo IVUS data and 3D FSI models: A pilot study. *Biomech. Model. Mechanobiol.* **2017**, *16*, 333–344. [CrossRef] [PubMed]
8. Olender, M.L.; Athanasiou, L.S.; Michalis, L.K.; Fotiadis, D.I.; Edelman, E.R. A Domain Enriched Deep Learning Approach to Classify Atherosclerosis Using Intravascular Ultrasound Imaging. *IEEE J. Sel. Top. Signal Process.* **2020**, *14*, 1210–1220. [CrossRef]
9. Gudigar, A.; Nayak, S.; Samanth, J.; Raghavendra, U.; Ashwal, A.J.; Barua, P.D.; Hasan, M.N.; Ciaccio, E.J.; Tan, R.S.; Rajendra Acharya, U. Recent Trends in Artificial Intelligence-Assisted Coronary Atherosclerotic Plaque Characterization. *Int. J. Environ. Res. Public Health* **2021**, *18*, 3. [CrossRef] [PubMed]
10. Kubo, T.; Nakamura, N.; Matsuo, Y.; Okumoto, Y.; Wu, X.; Choi, S.Y.; Komukai, K.; Tanimoto, T.; Ino, Y.; Kitabata, H.; et al Virtual Histology Intravascular Ultrasound Compared With Optical Coherence Tomography for Identification of Thin-Cap Fibroatheroma. *Int. Heart J.* **2011**, *52*, 175–179. [CrossRef] [PubMed]

11. Rezaei, Z.; Selamat, A.; Taki, A.; Mohd Rahim, M.S.; Abdul Kadir, M.R. Automatic plaque segmentation based on hybrid fuzzy clustering and k nearest neighborhood using virtual histology intravascular ultrasound images. *Appl. Soft Comput.* **2017**, *53*, 380–395. [CrossRef]

12. Athanasiou, L.S.; Karvelis, P.S.; Tsakanikas, V.D.; Naka, K.K.; Michalis, L.K.; Bourantas, C.V.; Fotiadis, D.I. A Novel Semiautomated Atherosclerotic Plaque Characterization Method Using Grayscale Intravascular Ultrasound Images: Comparison with Virtual Histology. *IEEE Trans. Inf. Technol. Biomed.* **2012**, *16*, 391–400. [CrossRef] [PubMed]

13. Kim, G.; Lee, J.; Hwang, Y. A novel intensity-based multi-level classification approach for coronary plaque characterization in intravascular ultrasound images. *Biomed. Eng. Online* **2018**, *17*, 1–15. [CrossRef]

14. Selvathi, D.; Emimal, N.; Selvaraj, H. Automated Characterization of Atheromatous Plaque in Intravascular Ultrasound Images Using Neuro Fuzzy Classifier. *Int. J. Electron. Telecommun.* **2012**, *58*, 425–431. [CrossRef]

15. Rezaei, Z.; Selamat, A.; Taki, A.; Mohd Rahim, M.S.; Abdul Kadir, M.R.; Penhaker, M.; Krejcar, O.; Kuca, K.; Herrera-Viedma, E.; Fujita, H. Thin Cap Fibroatheroma Detection in Virtual Histology Images Using Geometric and Texture Features. *Appl. Sci.* **2018**, *8*, 1632. [CrossRef]

16. Sofian, H.; Than, J.C.M.; Mohammad, S.; Mohd Noor, N. Calcification Detection of Coronary Artery Disease in Intravascular Ultrasound Image: Deep Feature Learning Approach. *Int. J. Integr. Eng.* **2018**, *10*, 43–57. [CrossRef]

17. Liu, S.; Neleman, T.; Hartman, E.M.; Ligthart, J.M.; Witberg, K.T.; van der Steen, A.F.; Wentzel, J.J.; Daemen, J.; van Soest, G. Automated Quantitative Assessment of Coronary Calcification Using Intravascular Ultrasound. *Ultrasound Med. Biol.* **2020**, *46*, 2801–2809. [CrossRef] [PubMed]

18. Jun, T.; Kang, S.J.; Lee, J. Automated detection of vulnerable plaque in intravascular ultrasound images. *Med. Biol. Eng. Comput.* **2019**, *57*, 863–876. [CrossRef]

19. Balocco, S.; Gonzalez, M.; Nanculef, R.; Radeva, P.; Thomas, G. Calcified Plaque Detection in IVUS Sequences: Preliminary Results Using Convolutional Nets. In *International Workshop on Artificial Intelligence and Pattern Recognition*; Springer: Berlin/Heidelberg, Germany, 2018; pp. 34–42. [CrossRef]

20. Sofian, H.; Ming, J.; Muhammad, S.; Noor, N. Calcification detection using convolutional neural network architectures in intravascular ultrasound images. *Indones. J. Electr. Eng. Comput. Sci.* **2019**, *17*, 1313–1321. [CrossRef]

21. Du, H.; Ling, L.; Yu, W.; Wu, P.; Yang, Y.; Chu, M.; Yang, J.; Yang, W.; Tu, S. Convolutional networks for the segmentation of intravascular ultrasound images: Evaluation on a multicenter dataset. *Comput. Methods Programs Biomed.* **2022**, *215*, 1065–1099. [CrossRef]

22. Balakrishna, C.; Dadashzadeh, S.; Soltaninejad, S. Automatic detection of lumen and media in the IVUS images using U-Net with VGG16 Encoder. *arXiv* **2018**, arXiv:1806.07554.

23. Li, Y.C.; Shen, T.Y.; Chen, C.C.; Chang, W.T.; Lee, P.Y.; Huang, C.C.J. Automatic Detection of Atherosclerotic Plaque and Calcification From Intravascular Ultrasound Images by Using Deep Convolutional Neural Networks. *IEEE Trans. Ultrason. Ferroelectr. Freq. Control* **2021**, *68*, 1762–1772. [CrossRef] [PubMed]

24. Porée, J.; Garcia, D.; Chayer, B.; Ohayon, J.; Cloutier, G. Noninvasive vascular elastography with plane strain incompressibility assumption using ultrafast coherent compound plane wave imaging. *IEEE Trans. Med. Imaging* **2015**, *34*, 2618–2631. [CrossRef] [PubMed]

25. Gómez, A.; Tacheau, A.; Finet, G.; Lagache, M.; Martiel, J.L.; Le Floc'h, S.; Yazdani, S.K.; Elias-Zuñiga, A.; Pettigrew, R.I.; Cloutier, G.; et al. Intraluminal ultrasonic palpation imaging technique revisited for anisotropic characterization of healthy and atherosclerotic coronary arteries: A feasibility study. *Ultrasound Med. Biol.* **2019**, *45*, 35–49. [CrossRef] [PubMed]

26. Li, H.; Porée, J.; Roy Cardinal, M.H.; Cloutier, G. Two-dimensional affine model-based estimators for principal strain vascular ultrasound elastography with compound plane wave and transverse oscillation beamforming. *Ultrasonics* **2019**, *91*, 77–91. [CrossRef] [PubMed]

27. Li, H.; Porée, J.; Chayer, B.; Cardinal, M.H.R.; Cloutier, G. Parameterized Strain Estimation for Vascular Ultrasound Elastography With Sparse Representation. *IEEE Trans. Med. Imaging* **2020**, *39*, 3788–3800. [CrossRef]

28. Le Floc'h, S.; Ohayon, J.; Tracqui, P.; Finet, G.; Gharib, A.M.; Maurice, R.L.; Cloutier, G.; Pettigrew, R.I. Vulnerable atherosclerotic plaque elasticity reconstruction based on a segmentation-driven optimization procedure using strain measurements: Theoretical framework. *IEEE Trans. Med. Imaging* **2009**, *28*, 1126–1137. [CrossRef]

29. Le Floc'h, S.; Cloutier, G.; Saijo, Y.; Finet, G.; Yazdani, S.K.; Deleaval, F.; Rioufol, G.; Pettigrew, R.I.; Ohayon, J. A four-criterion selection procedure for atherosclerotic plaque elasticity reconstruction based on in vivo coronary intravascular ultrasound radial strain sequences. *Ultrasound Med. Biol.* **2012**, *38*, 2084–2097. [CrossRef]

30. Maurice, R.L.; Ohayon, J.; Finet, G.; Cloutier, G. Adapting the Lagrangian speckle model estimator for endovascular elastography: Theory and validation with simulated radio-frequency data. *J. Acoust. Soc. Am.* **2004**, *116*, 1276–1286. [CrossRef] [PubMed]

31. Porée, J.; Chayer, B.; Soulez, G.; Ohayon, J.; Cloutier, G. Noninvasive vascular modulography method for imaging the local elasticity of atherosclerotic plaques: Simulation and in vitro vessel phantom study. *IEEE Trans. Ultrason. Ferroelectr. Freq. Control* **2017**, *64*, 1805–1817. [CrossRef]

32. Bouvier, A.; Deleaval, F.; Doyley, M.; Yazdani, S.; Finet, G.; Le Floc'h, S.; Cloutier, G.; Pettigrew, R.; Ohayon, J. A direct vulnerable atherosclerotic plaque elasticity reconstruction method based on an original material-finite element formulation: Theoretical framework. *Phys. Med. Biol.* **2013**, *58*, 8457–8476. [CrossRef]

33. Peña, E.; Cilla, M.; Latorre, Á.T.; Martínez, M.A.; Gómez, A.; Pettigrew, R.I.; Finet, G.; Ohayon, J. Emergent biomechanical factors predicting vulnerable coronary atherosclerotic plaque rupture. In *Biomechanics of Coronary Atherosclerotic Plaque*; Academic Press: Cambridge, MA, USA, 2021; pp. 361–380. [CrossRef]

34. Glagov, S.; Weisenberg, E.; Zarins, C.; Stankunavicius, R.; Kolettis, G. Compensatory enlargement of human atherosclerotic coronary arteries. *N. Engl. J. Med.* **1987**, *316*, 1371–1375. [CrossRef]

35. Le Floc'h, S. Modulographie Vasculaire: Application à l'Identification In-Vivo du Module de Young local des Plaques d'Athérosclérose. Ph.D. Thesis, Université Joseph-Fourier-Grenoble I, Saint-Martin-de-Re, France, 2009.

36. Gasser, T.C.; Ogden, R.W.; Holzapfel, G.A. Hyperelastic modelling of arterial layers with distributed collagen fibre orientations. *J. R. Soc. Interface* **2006**, *3*, 15–35. [CrossRef] [PubMed]

37. Versluis, A.; Bank, A.J.; Douglas, W.H. Fatigue and plaque rupture in myocardial infarction. *J. Biomech.* **2006**, *39*, 339–347. [CrossRef]

38. Holzapfel, G.A.; Sommer, G.; Gasser, C.T.; Regitnig, P. Determination of layer-specific mechanical properties of human coronary arteries with nonatherosclerotic intimal thickening and related constitutive modeling. *Am. J. Physiol.-Heart Circ. Physiol.* **2005**, *289*, H2048–H2058. [CrossRef] [PubMed]

39. Skacel, P. Hyperfit: Sotfware for Fitting of Hyperelastic Constitutive Models. 2018. Available online: http://www.hyperfit.wz.cz (accessed on 16 September 2020).

40. Dassault Systèmes Simulia Corp. ABAQUS/Standard User's Manual, Version 6.14. 2014.

41. Ramzy, D. Definition of hypertension and pressure goals during treatment (ESC-ESH Guidelines 2018). *Eur. Soc. Cardiol. J.* **2019**, *17*.

42. Raghavan, M.; Ma, B.; Fillinger, M. Non-invasive determination of zero-pressure geometry of arterial aneurysms. *Ann. Biomed. Eng.* **2006**, *34*, 1414–1419. [CrossRef] [PubMed]

43. MATLAB. *R2021b*; The MathWorks Inc.: Natick, MA, USA, 2021.

44. Soleimanifard, S.; Abd-Elmoniem, K.Z.; Agarwal, H.K.; Tomas, M.S.; Sasano, T.; Vonken, E.; Youssef, A.; Abraham, M.R.; Abraham, T.P.; Prince, J.L. Identification of myocardial infarction using three-dimensional strain tensor fractional anisotropy. In Proceedings of the 2010 IEEE International Symposium on Biomedical Imaging: From Nano to Macro, Rotterdam, The Netherlands, 14–17 April 2010; pp. 468–471. [CrossRef]

45. Wang, D.; Chayer, B.; Destrempes, F.; Gesnik, M.; Tournoux, F.; Cloutier, G. Deformability of ascending thoracic aorta aneurysms assessed using ultrafast ultrasound and a principal strain estimator: In vitro evaluation and in vivo feasibility. *Med. Phys.* **2022**, *49*, 1759–1775. [CrossRef]

46. Sumi, C.; Nakayama, K. A robust numerical solution to reconstruct a globally relative shear modulus distribution from strain measurements. *IEEE Trans. Med. Imaging* **1998**, *17*, 419–428. [CrossRef]

47. Tacheau, A.; Le Floc'h, S.; Finet, G.; Doyley, M.M.; Pettigrew, R.I.; Cloutier, G.; Ohayon, J. The imaging modulography technique revisited for high-definition intravascular ultrasound: Theoretical framework. *Ultrasound Med. Biol.* **2016**, *42*, 727–741. [CrossRef]

 *mathematics*

Communication

Computational Study of the Microsphere Concentration in Blood during Radioembolization

Unai Lertxundi [1], Jorge Aramburu [1,2,*], Macarena Rodríguez-Fraile [3,4], Bruno Sangro [3,5] and Raúl Antón [1,2,3]

1   Thermal and Fluids Engineering Division, Department of Mechanical Engineering and Materials, TECNUN Escuela de Ingeniería, Universidad de Navarra, 20018 Donostia-San Sebastian, Spain
2   Centro de Ingeniería Biomédica (CBIO), Universidad de Navarra, 31009 Pamplona, Spain
3   IdiSNA (Instituto de Investigación Sanitaria de Navarra), 31008 Pamplona, Spain
4   Department of Nuclear Medicine, Clínica Universidad de Navarra, 31008 Pamplona, Spain
5   Liver Unit and CIBEREHD, Clínica Universidad de Navarra, 31008 Pamplona, Spain
*   Correspondence: jaramburu@tecnun.es

Abstract: Computational fluid dynamics techniques are increasingly used to computer simulate radioembolization, a transcatheter intraarterial treatment for patients with inoperable tumors, and analyze the influence of treatment parameters on the microsphere distribution. Ongoing clinical research studies are exploring the influence of the microsphere density in tumors on the treatment outcome. In this preliminary study, we computationally analyzed the influence of the microsphere concentration in the vial on the microsphere concentration in the blood. A patient-specific case was used to simulate the blood flow and the microsphere transport during three radioembolization procedures in which the only parameter varied was the concentration of microspheres in the vial and the span of injection, resulting in three simulations with the same number of microspheres injected. Results showed that a time-varying microsphere concentration in the blood at the outlets of the computational domain can be analyzed using CFD, and also showed that there was a direct relationship between the variation of microsphere concentration in the vial and the variation of microsphere concentration in the blood. Future research will focus on elucidating the relationship between the microsphere concentration in the vial, the microsphere concentration in the blood, and the final microsphere distribution in the tissue.

Keywords: computational fluid dynamics; hemodynamics; liver cancer; dosimetry; drug delivery; tumor targeting; patient-specific; treatment planning

MSC: 76Z05

Citation: Lertxundi, U.; Aramburu, J.; Rodríguez-Fraile, M.; Sangro, B.; Antón, R. Computational Study of the Microsphere Concentration in Blood during Radioembolization. *Mathematics* **2022**, *10*, 4280. https://doi.org/10.3390/math10224280

Academic Editor: Mauro Malvè

Received: 29 September 2022
Accepted: 14 November 2022
Published: 16 November 2022

Publisher's Note: MDPI stays neutral with regard to jurisdictional claims in published maps and institutional affiliations.

1. Introduction

Radioembolization is a treatment for patients with inoperable malignant liver tumors, one of the deadliest types of cancer worldwide [1]. The treatment is carried out using a microcatheter placed in the hepatic artery to infuse radiolabeled microspheres into the hepatic arterial bloodstream, which carries the microspheres to the tumoral bed, where they get lodged and irradiate tumoricidal doses of radiation [2]. The treatment outcome depends on the dose absorbed by tumors, meaning that it depends on the final distribution of microspheres in tumors. The absorbed tumoricidal dose to be achieved depends on the types of microspheres, which can be resin or glass microspheres loaded with isotope yttrium-90 or with isotope holmium-166 [3].

In the last decade, many computational fluid dynamics (CFD) studies have analyzed the microsphere distribution in hepatic arterial trees during radioembolization by analyzing the hepatic artery hemodynamics and microsphere transport, and assessed the influence of various parameters (e.g., injection velocity [4,5], microcatheter location [6–9], catheter type [10], etc.) on the microsphere distribution at the outlets of the trees. However, to date, no study has focused on the analysis of the microsphere concentration in the blood (i.e., the number of microspheres per unit blood volume) as these microspheres exit the hepatic arteries under study. The analysis of this parameter could be included when developing an integrated software package to provide clinicians with practical recommendations about the optimal treatment parameters based on CFD simulations [11].

We hypothesize that the microsphere concentration in the blood could play a role in the distal penetration of microspheres and tumor microsphere coverage, and therefore in the final microsphere deposition in tumors. Indeed, recent publications have stressed the importance of glass microsphere density in tumors (i.e., the number of microspheres per unit tumor volume) [12,13], the number of microspheres injected based on the vascularity of tumors [14], and the method of administration [15] on the treatment outcome. Additionally, the microsphere volume fraction, which is directly related to the microsphere concentration in the blood, could play a role in the penetration of microspheres because the probability of clogging distal arterioles with vessel-to-microsphere diameter ratios below 3 depends on the microsphere volume fraction [16]. Such clogging may occur when microspheres would group together before reaching distal vessels with diameters below the microsphere diameter.

In addition to studying the microsphere distribution at the outlets of the arterial trees, CFD simulations allow for analyzing the microsphere concentration in the blood at such outlets. In order to explore this parameter, the aim of the present study was to analyze the influence of injection conditions on the microsphere concentration in the blood at the outlets of a patient-specific hepatic artery tree using CFD.

2. Materials and Methods

2.1. Patient Data

This study was based on a patient-specific case. To conduct the present study, the protocol 186/2018 was approved by the ethics committee of the University of Navarra and informed consent was signed by the patient. This patient was a 69-year-old male with multinodular hepatocellular carcinoma (HCC) involving liver segments S2, S3, S6, S7, and S8.

For CFD simulations, the geometries of the hepatic artery and microcatheter, and boundary conditions, were needed. The hepatic artery geometry and boundary conditions were extracted or derived from the information provided by MeVis (MeVis Medical Solutions AG, Bremen, Germany), that is, the three-dimensional hepatic artery, the healthy and tumor tissue volumes per liver segment, and information regarding the specific artery branches that feed each liver segment. In this case, the geometry of the hepatic artery starts at a 4.5 mm diameter proper hepatic artery (PHA) level and bifurcates until 43 outlets are obtained. Figure 1a shows the liver geometry, with the locations of tumors and the hepatic artery tree, and Figure 1b shows a model of the hepatic artery with the microcatheter, the location of which is indicated with an arrowhead. The hemodynamic conditions were derived based on volumetry analysis (volumes of normal and tumor tissue per segment) and perfusion-CT analysis (arterial perfusion of normal and tumor tissue), which were used to calculate the blood flow rate at the inlet of the PHA and the flow distribution in the 43 outlets [17]. This patient also participated in a previous study by the authors to validate the simulation methodology by comparing the microsphere distributions measured in vivo with the microsphere distributions calculated based on the results of CFD simulations that reproduced the actual treatments [18]. The segment-to-segment tissue volumes and arterial blood flow rates are tabulated in Table 1. Even though, previously, three microsphere injections were administered to the patient, in this study, only one injection was simulated, with the microcatheter placed in the PHA, 20 mm away from the bifurcation. The microcatheter was modeled as a standard end-hole microcatheter with an inner diameter of 0.65 mm and outer diameter of 0.9 mm.

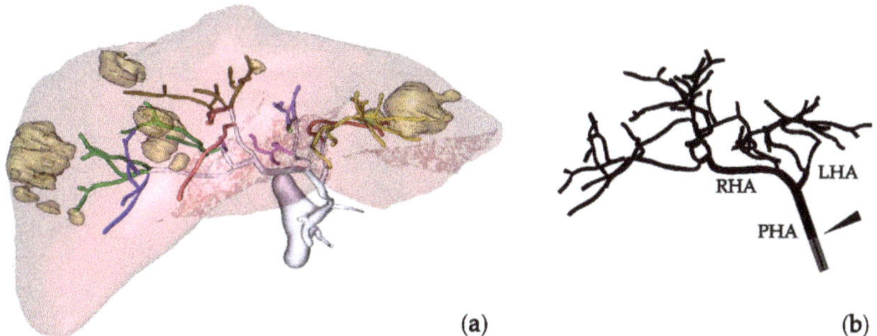

Figure 1. (a) MeVis study showing the liver, hepatic arterial tree, and tumor nodules. (b) Hepatic artery tree model with the microcatheter tip indicated by an arrowhead. PHA: proper hepatic artery; LHA: left hepatic artery; RHA: right hepatic artery.

Table 1. Patient's liver mass volumes and blood flow rates per segment.

Segment	Healthy Tissue Volume (mL)	Tumor Volume (mL)	Volumetric Flow Rate (mL/min)
S1	62.0		3.2
S2	127.2	0.8	6.8
S3	170.2	10.8	15.3
S4a	73.0		3.8
S4b	11.0		0.6
S5	124.0		6.4
S6	169.0		8.9
S7	349.6	23.4	32.3
S8	200.0	4.0	11.8

2.2. Simulation of Radioembolization

The geometry of the hepatic artery was imported into the software package SpaceClaim (Ansys Inc., Canonsburg, PA, USA), where the microcatheter geometry was also added. The geometries were discretized in Fluent Meshing 2021R1 (Ansys Inc.) using poly-hexcore elements, yielding a sufficiently fine mesh of 2.5 million elements.

Regarding the governing equations and boundary conditions, the model used by Aramburu et al. [6] was used: blood flow was simulated using the conservation of mass and the conservation of linear momentum in a laminar regime for an incompressible (1050 kg/m^3) and non-Newtonian fluid with the viscosity modeled using a modified Quemada model. Microspheres were modeled as 32 µm, 1600 kg/m^3 spheres and their dynamics were modeled with Newton's second law of motion. Four forces were considered to be acting on the microspheres: virtual mass force, gravitational force, pressure gradient force, and drag force. The interaction of blood flow with microspheres was modeled as bidirectional, but the interaction between microspheres was neglected. As for the boundary conditions, a periodic pulsatile fully-developed velocity profile with a period of one second representing one cardiac cycle was prescribed at the inlet, flow fractions at the outlets, and the no-slip condition at the walls. Microspheres were injected through the inlet of the microcatheter and elastic collisions were assumed at the walls.

The equations of the model were solved numerically using Fluent 2021R1 (Ansys Inc.). The SIMPLE scheme (i.e., Semi-Implicit Method for Pressure Linked Equations) was used for the pressure and velocity coupling, the least squares cell-based algorithm for the computation of gradients, and second-order schemes to interpolate pressure and momentum. The convergence criterion was to attain scaled residuals of 10^{-5}. The time step was fixed at a value of 2 milliseconds with a maximum of 80 iterations per time step. A total of seven cardiac cycles were simulated: the first cycle, i.e., one second, was for flow convergence (from $t = -1$ s to $t = 0$ s), during the second cycle the microspheres were injected (from $t = 0$

s to $t = 1$ s), and the remaining five cycles ensured that the majority of the injected microspheres exited the domain (from $t = 1$ s to $t = 6$ s). Simulation results were analyzed from $t = 0$ s onward. In this study, depending on the simulation, the injection of microspheres sometimes lasted more than one cycle, in which cases the remaining cycles for microsphere exiting were fewer than five cycles.

2.3. Study Design and Postprocessing

The objective of this study was to analyze the influence of injection conditions on the microsphere concentration in the blood at the outlets. The number of injected microspheres can be calculated using Equation (1):

$$N = C_{\mathrm{vial}} U_{\mathrm{inj}} A_c T, \tag{1}$$

where N (microspheres, hereafter MS) is the total number of microspheres injected in the simulation, C_{vial} (MS/mL) is the number of microspheres per unit milliliter in the vial, U_{inj} (cm/s) is the injection velocity, A_c (cm^2) is the cross-sectional area of the microcatheter, and T is the injection span (s).

In this study, three simulations were run. In these simulations, N was kept constant, meaning that the same number of microspheres was injected, and U_{inj} was also kept constant, so C_{vial} and T were modified, as shown in Table 2.

Table 2. Characteristics of the simulations of the study. In all simulations, the number of injected microspheres and injection velocity were kept constant; in Simulation #2, the injection span and the concentration of microspheres in the vial were twice and half of the injection span and the concentration of microspheres in the vial in Simulation #1, respectively; in Simulation #3, the injection span and the concentration of microspheres in the vial were three times and one-third of the injection span and the concentration of microspheres in the vial in Simulation #1, respectively.

Case	N (MS)	C_{vial} (MS/mL)	U_{inj} (cm/s)	T (s)
Simulation #1	4.45×10^5	6.9×10^6	19	1
Simulation #2	4.45×10^5	3.45×10^6	19	2
Simulation #3	4.45×10^5	2.3×10^6	19	3

Regarding the postprocessing of simulation results, the overall outlet-to-outlet and segment-to-segment microsphere distributions were extracted, as well as the concentration of microspheres in the blood in the PHA and at the outlets over time. To calculate the concentration of microspheres in the blood in the PHA, Equation (2) was used:

$$C_{\mathrm{PHA}} = \frac{N_{\mathrm{PHA}}}{V_{\mathrm{PHA}}}, \tag{2}$$

where C_{PHA} (MS/mL) is the concentration of microspheres in the blood in the PHA calculated for a given span, N_{PHA} (MS) is the number of microspheres that were infused into the PHA during the same span, and V_{PHA} (mL) is the volume of blood and the volume of injection fluid that flowed through the PHA during the same span. To calculate the concentration of microspheres in blood at the outlets Equation (3) was used:

$$C_i = \frac{N_i}{V_{\mathrm{blood},i}}, \tag{3}$$

where C_i (MS/mL) is the concentration of microspheres in the blood for outlet i (with i from 1 to 43), calculated for a given span, N_i (MS) is the number of microspheres that exited through outlet i during the same span, and $V_{\mathrm{blood},i}$ (mL) is the volume of blood that flowed through outlet i during the same span. In this study, a span of 20 milliseconds was selected.

3. Results

First, the segment-to-segment microsphere distribution was analyzed. The total number of microspheres injected was the same in all simulations (4.45×10^5 MS), so the results were normalized with respect to the total number of microspheres injected. For example, a value of 50% for a given segment means that half of the injected microspheres reached that segment. When analyzing the differences between simulations, the absolute differences were computed, i.e., the differences in percent (%). These results are shown in Figure 2a, where the blood flow distribution and segment-to-segment microsphere distribution are shown. Despite the disease being bilobar (present in segments S2, S3, S6, S7, and S8), almost no microspheres reached the left lobe, and those which did reach the left lobe reached segments S1 and S4. Therefore, the injection point used in this study did not produce an effective microsphere distribution. In fact, three injections were given to this patient. It can be noted that for the segments of the right lobe (S5, S6, S7, and S8), the microsphere distribution followed a trend where the greater the blood flow rate to a given segment, the greater the number of microspheres exiting toward that segment. Moreover, differences can be seen between simulations. For example, microspheres flowed toward segments S1 and S4 in simulations #2 and #3, while no microsphere flowed toward the left lobe in simulation #1. In all simulations, the segment receiving the most microspheres was segment S7, which is the segment with the biggest nodule (a 23 mL nodule). It can also be noted that even though segment S3 received 17% of the total blood flow, no microspheres reached that segment. Additionally, the microsphere distributions were different in the simulations, with an absolute difference of 16 percent between simulations #1 and #3 for segment S7. The average absolute difference between simulations in all segments was 5 percent.

Second, the concentration of microspheres in the blood over time in the PHA and at each outlet was studied. In this study, the concentration of microspheres in the vial was of the order of 10^6 MS/mL, which was reduced to a concentration of microspheres in the blood of the order of 10^5 MS/mL in the PHA, and was further reduced to a concentration of microspheres in the blood of the order of 10^4 MS/mL (and below that value) in segmental arteries, meaning that the concentration decreased as microspheres flowed toward distal vessels. Figure 2b shows the microsphere concentration in blood at the PHA level over time. A constant injection flow and microsphere injection rate resulted in a microsphere concentration in the blood with a minimum value during systole and a maximum value during diastole. The geometry had 43 outlets, but only the outlets with a flow fraction greater than 1% and a concentration of microspheres in the blood greater than 40,000 MS/mL are reported, resulting in an analysis of eight outlets: outlet 21 feeding segment S6, outlets 26, 27, 29, 32, and 33 feeding segment S7, and outlets 34 and 38 feeding segment S8. These results are shown in Figure 2c–j. In each panel, the time-dependent microsphere concentration in the blood is plotted for simulations #1, #2, and #3. Additionally, the shape of the blood flow rate is plotted in dotted lines, and the percentage of blood flow feeding that outlet and the percentage of microspheres exiting that outlet are indicated. Figure A1 in Appendix A shows the same information as in Figure 2 for the outlets that had a flow of microspheres but did not meet the criteria of having a flow fraction greater than 1% and a concentration of microspheres in the blood greater than 40,000 MS/mL.

Regarding the percentage of microspheres exiting the eight outlets of Figure 2, this value was, in general, different from the blood flow percentage, but it also differed slightly among simulations. The mean value of the difference in microsphere distributions in these outlets was 3 percent, with the greatest difference being 8 percent for outlet 29 between simulations #1 and #3 (see Figure 2f).

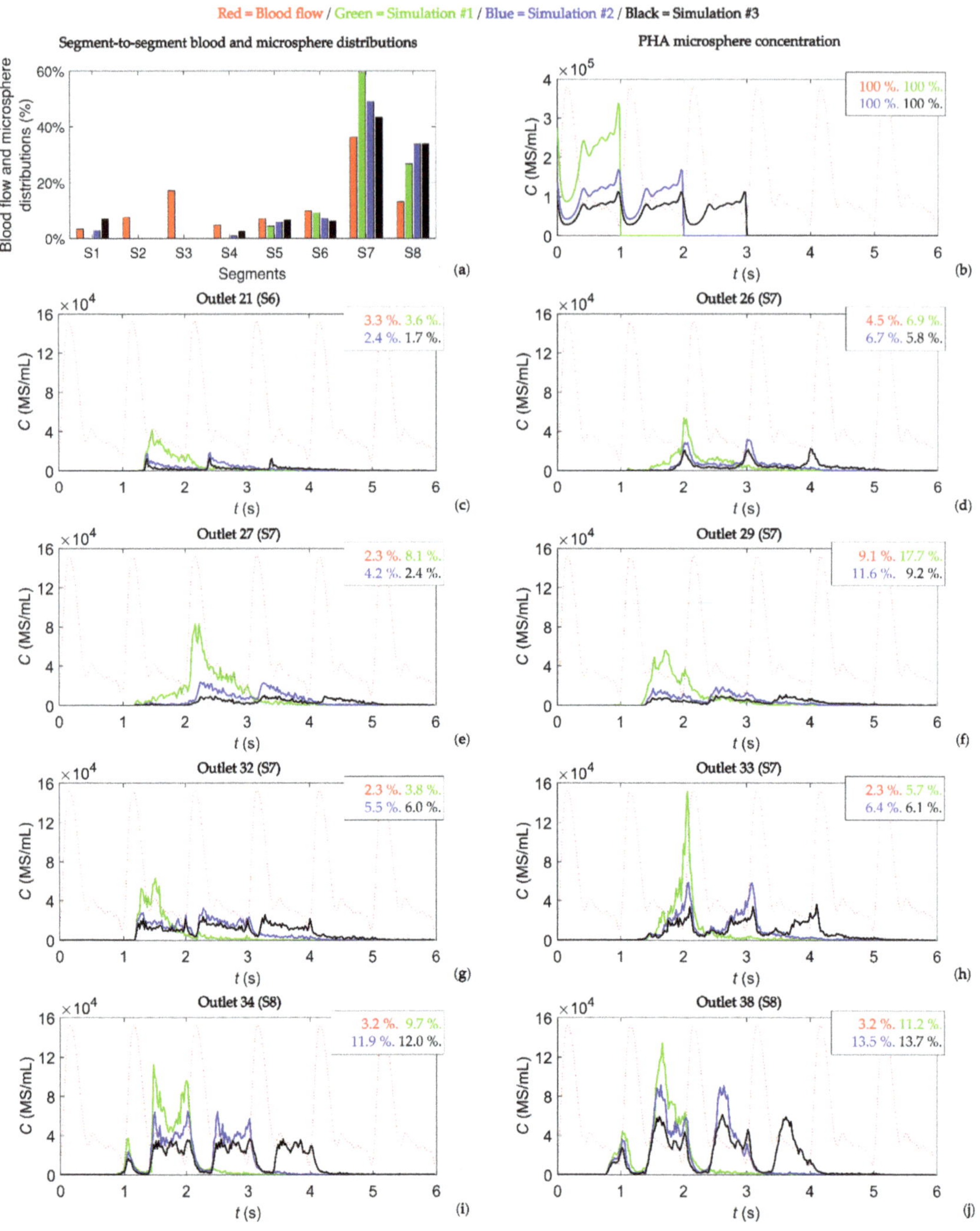

Figure 2. (**a**) Segment-to-segment blood flow and microsphere distributions. (**b**) Concentration of microspheres in the blood at the PHA level. (**c–j**) Concentration of microspheres in the blood reaching outlets 21, 26, 27, 29, 32, 33, 34, and 38 over time. These outlets feed tumor-bearing segments S6, S7, and S8. In each panel (**b–j**), the shape of the blood flow is in a dotted red line and the percentages indicate the percentage of blood, and percentage of microspheres exiting through those outlets.

As for the microsphere concentration in the blood over time, the following trend was seen for the outlets in Figure 2c–j for a periodic flow and a constant microsphere infusion: (i) the greater the concentration of microspheres in the vial, the greater the microsphere concentration in the blood at the outlets, and (ii) the concentration of microspheres in the blood was periodic-like, meaning that the same shape was repeated over the cardiac cycles with a period similar to that of the cardiac cycle. In this case, for simulations #1, #2, and #3 (with microspheres injected during one, two, and three cardiac cycles, respectively), one-cycle, two-cycle, and three-cycle outlet-specific patterns were observed. This trend was not seen at the outlets shown in Figure A1.

It is also important to note that the peaks of the periodic-like patterns did not coincide with any specific moment of the cardiac cycle (e.g., the systole or diastole) (see Figure 2c–j). Microspheres were injected at a constant rate, and their velocity matched the blood flow velocity shortly after they were incorporated into the bloodstream. If the microspheres had traveled through straight streamlines, the concentration of microspheres in the blood at the outlet should have had the same pattern as that in the PHA (Figure 2b). However, the tortuosity of arteries made the blood flow and microsphere trajectories intricate, and this fact made the concentration of microspheres inconstant. Likewise, there were some non-periodic peaks for outlets 34 and 38 at the beginning of microsphere crossing (see Figure 2i,j), which could be due to transient effects. When analyzing the peak values, if the outlets of Figure 2 were considered, on average, the concentration values were reduced by 52% from simulation #1 to simulation #2, and by 69% from simulation #1 to simulation #3. These values were similar to the decrease in the concentration of microspheres in the vial between simulations #1 and #2 and simulations #1 and #3 (50% and 67%, respectively). If all the outlets were considered, these average peak values decreased to 27% for simulation #2 and 45% for simulation #3.

4. Discussion

The objective of this study was to analyze the influence of the microsphere concentration in the vial on the microsphere distribution in the blood using CFD techniques. The concentration of microspheres in the blood depends on the injection conditions (e.g., the microsphere concentration in the vial and injection velocity) and the hemodynamic characteristics of patients. This concentration could play a role in the final distribution of microspheres within tumors because of potential microsphere aggregation, distal vessel clogging, and hemodynamic redistributions in the microvasculature. Indeed, the study of the microsphere density in tumors is an active area of ongoing research. Recent studies have shown that the resin microsphere distribution in tumors determines the treatment outcome [12,13], and that the number of microspheres injected and the method of administration play a role in the microsphere distribution [14,15].

Regarding segment-to-segment microsphere distributions, our results showed that the microsphere dynamics' and hemodynamics' interaction could play a role in microsphere transport and therefore should be considered. Indeed, a difference of 16 percent was obtained between simulations #1 and #3 in segment S7, and no microspheres targeted the left lobe in simulation #1, with microspheres targeting the left lobe only in simulations #2 and #3; the only difference between simulations was the concentration of microspheres in the vial, and therefore the rate at which microspheres were injected.

With regard to the microsphere concentration in the blood, periodic patterns were observed. As for the specific patterns, no relationship was observed between the concentration of microspheres in the blood in the PHA and the microsphere concentration in the blood at the outlets, nor between the shape of the microsphere distribution in the blood and the shape of the blood flow. The microsphere concentration decreased as the flow reached distal vessels of the hepatic artery tree studied, from 10^5 MS/mL in the PHA to 10^4 MS/mL or lower values. However, these results showed a relationship between the microsphere concentration in the vial and the peak value of the microsphere concentration in the blood. In fact, the concentration of microspheres in the vial was reduced by 50%

between simulations #1 and #2, and on average the peak of microsphere concentrations in the blood was reduced by 52%. Between simulations #1 and #3, the concentration of microspheres in the vial was reduced by 67%, and on average the peak of microsphere concentrations in the blood was reduced by 69%. This relationship could be of interest for future research. Additionally, the influence of the microsphere concentration in the blood in distal arterioles and capillaries needs to be assessed. As the ratio of the vessel diameter to the microsphere diameter approaches a value of 3, the probability of clogging depends on the microsphere volume fraction [16], which is similar to the microsphere concentration in the blood. Microspheres could become clogged before reaching the distal arterioles and capillaries with diameters similar to that of the microsphere, influencing the final microsphere distribution in tissues.

5. Conclusions

This study showed the potential impact of the microsphere concentration in the vial on the microsphere concentration in the blood during radioembolization. A relationship between the concentration of microspheres in the vial and the peak of the microsphere concentration in the blood was observed. Further research on the influence of injection characteristics (i.e., injection velocity and concentration of microspheres in the vial) is needed to confirm the relationship observed in this study, in addition to studying the effects of the microsphere concentration in the blood on the microsphere distribution in tumors. The results of this and future studies could be of use for planning optimal patient-specific radioembolization procedures.

Author Contributions: Conceptualization, J.A. and R.A.; methodology, J.A. and R.A.; formal analysis, U.L.; investigation, U.L. and J.A.; writing—original draft preparation, U.L. and J.A.; writing—review and editing, M.R.-F., B.S. and R.A.; visualization, U.L.; supervision, J.A. and R.A.; project administration, M.R.-F. and J.A.; funding acquisition, M.R.-F., J.A. and R.A. All authors have read and agreed to the published version of the manuscript.

Funding: This research was funded by Diputación Foral de Gipuzkoa (the Provincial Council of Gipuzkoa), grant number 2021-CIEN-000076-04-01, and by the PI18/00692 project, integrated in the 2013–2016 National R&D Plan and co-financed by the ISCIII General Division for Research Evaluation and Promotion and the European Regional Development Fund. The APC was funded by the PI18/00692 project, integrated in the 2013–2016 National R&D Plan and co-financed by the ISCIII General Division for Research Evaluation and Promotion and the European Regional Development Fund.

Data Availability Statement: Data and codes are available on request from the authors.

Acknowledgments: U.L. gratefully acknowledges the financial support of Eusko Jaurlaritzako Hezkuntza Saila (Basque Government Department of Education) through the Non-Doctor Research Personnel Predoctoral Training Program.

Conflicts of Interest: J.A. and R.A. have received speaker honoraria from Sirtex Medical and research fees from Terumo and Sirtex Medical. M.R.-F. has received consultation fees and speaker honoraria from Sirtex Medical. B.S. has received consulting fees from BTG, Sirtex Medical and Terumo; speaker fees from Sirtex Medical and Terumo; and institutional research funding from Sirtex Medical. The funders had no role in the design of the study; in the collection, analyses, or interpretation of data; in the writing of the manuscript; or in the decision to publish the results.

Appendix A

This appendix consists of Figure A1, which complements Figure 2 in showing the results of the study.

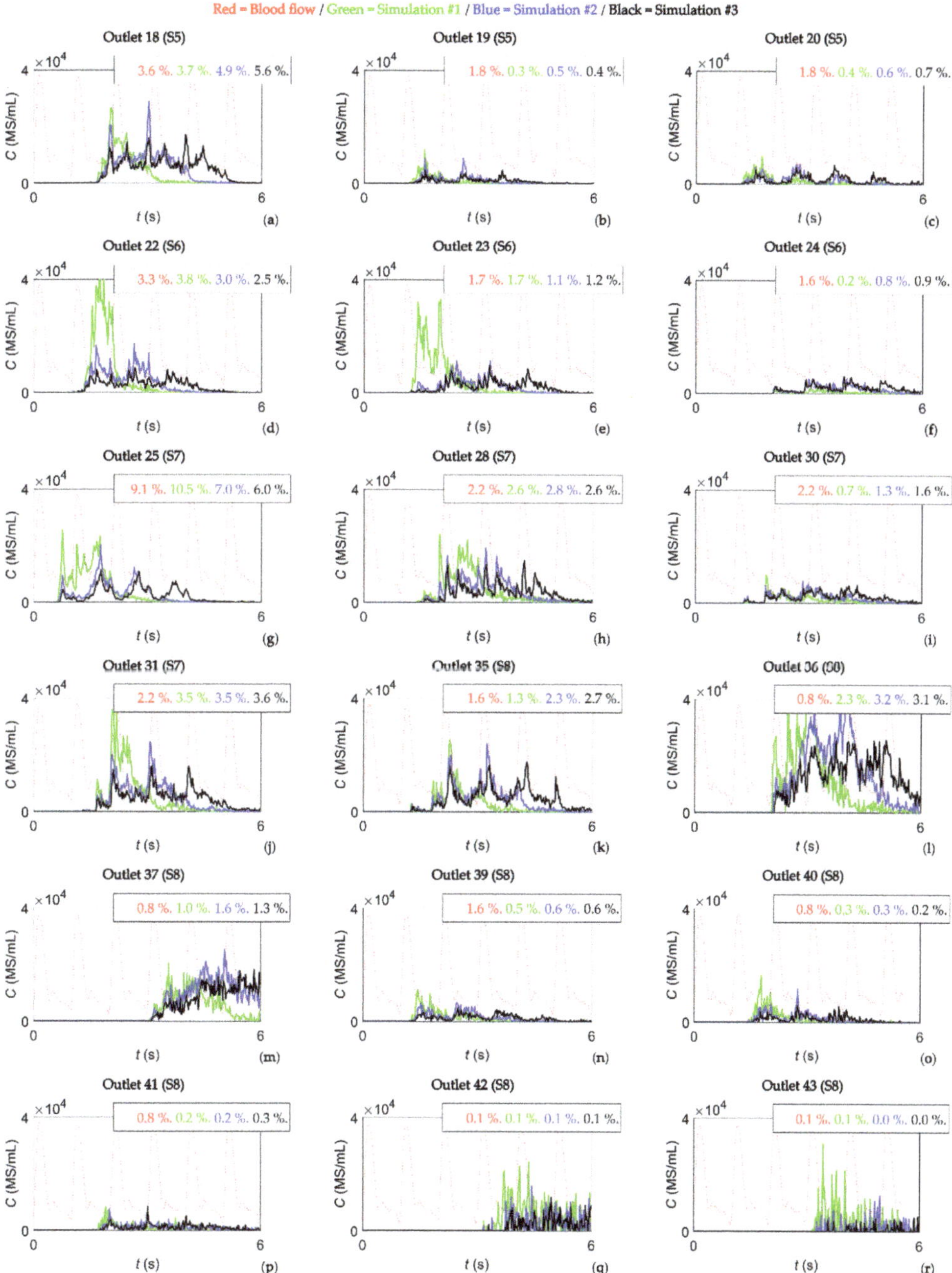

Figure A1. (**a–r**) Concentration of microspheres in the blood reaching outlets 18, 19, 20, 22, 23, 24, 25, 28, 30, 31, 35, 36, 37, 39, 40, 41, 42, and 43 over time. In each panel, the shape of the blood flow is a dotted red line and the percentages indicate the percentage of blood and percentage of microspheres exiting through those outlets.

References

1. Padia, S.A.; Lewandowski, R.J.; Johnson, G.E.; Sze, D.Y.; Ward, T.J.; Gaba, R.C.; Baerlocher, M.O.; Gates, V.L.; Riaz, A.; Brown, D.B.; et al. Radioembolization of Hepatic Malignancies: Background, Quality Improvement Guidelines, and Future Directions. *J. Vasc. Interv. Radiol.* **2017**, *28*, 1–15. [CrossRef] [PubMed]
2. Sangro, B.; Iñarrairaegui, M.; Bilbao, J.I. Radioembolization for Hepatocellular Carcinoma. *J. Hepatol.* **2012**, *56*, 464–473. [CrossRef] [PubMed]
3. Garin, E.; Guiu, B.; Edeline, J.; Rolland, Y.; Palard, X. Trans-Arterial Radioembolization Dosimetry in 2022. *Cardiovasc. Interv. Radiol.* **2022**, *45*, 1608–1621. [CrossRef] [PubMed]
4. Basciano, C.A.; Kleinstreuer, C.; Kennedy, A.S. Computational Fluid Dynamics Modeling of 90Y Microspheres in Human Hepatic Tumors. *J. Nucl. Med. Radiat. Ther.* **2011**, *2*, 2. [CrossRef]
5. Kleinstreuer, C.; Basciano, C.A.; Childress, E.M.; Kennedy, A.S. A New Catheter for Tumor Targeting With Radioactive Microspheres in Representative Hepatic Artery Systems. Part I: Impact of Catheter Presence on Local Blood Flow and Microsphere Delivery. *J. Biomech. Eng.* **2012**, *134*, 051004. [CrossRef] [PubMed]
6. Aramburu, J.; Antón, R.; Rivas, A.; Ramos, J.C.; Sangro, B.; Bilbao, J.I. Computational Particle-Haemodynamics Analysis of Liver Radioembolization Pretreatment as an Actual Treatment Surrogate. *Int. J. Numer. Methods Biomed. Eng.* **2017**, *33*, e02791. [CrossRef] [PubMed]
7. Basciano, C.A.; Kleinstreuer, C.; Kennedy, A.S.; Dezarn, W.A.; Childress, E. Computer Modeling of Controlled Microsphere Release and Targeting in a Representative Hepatic Artery System. *Ann. Biomed. Eng.* **2010**, *38*, 1862–1879. [CrossRef] [PubMed]
8. Bomberna, T.; Koudehi, G.A.; Claerebout, C.; Verslype, C.; Maleux, G.; Debbaut, C. Transarterial Drug Delivery for Liver Cancer: Numerical Simulations and Experimental Validation of Particle Distribution in Patient-Specific Livers. *Expert Opin. Drug Deliv.* **2020**, *18*, 409–422. [CrossRef] [PubMed]
9. Bomberna, T.; Vermijs, S.; Lejoly, M.; Verslype, C.; Bonne, L.; Maleux, G.; Debbaut, C. A Hybrid Particle-Flow CFD Modeling Approach in Truncated Hepatic Arterial Trees for Liver Radioembolization: A Patient-Specific Case Study. *Front. Bioeng. Biotechnol.* **2022**, *10*, 914979. [CrossRef] [PubMed]
10. Aramburu, J.; Antón, R.; Rivas, A.; Ramos, J.C.; Sangro, B.; Bilbao, J.I. Computational Assessment of the Effects of the Catheter Type on Particle–Hemodynamics during Liver Radioembolization. *J. Biomech.* **2016**, *49*, 3705–3713. [CrossRef] [PubMed]
11. Aramburu, J.; Antón, R.; Rodríguez-Fraile, M.; Sangro, B.; Bilbao, J.I. Computational Fluid Dynamics Modeling of Liver Radioembolization: A Review. *Cardiovasc. Interv. Radiol.* **2021**, *45*, 12–20. [CrossRef] [PubMed]
12. Chiesa, C.; Mazzaglia, S.; Maccauro, M. Spatial Density and Tumor Dosimetry Are Important in Radiation Segmentectomy with 90Y Glass Microspheres. *Eur. J. Nucl. Med. Mol. Imaging* **2022**, *49*, 3607–3609. [CrossRef] [PubMed]
13. Pasciak, A.S.; Abiola, G.; Liddell, R.P.; Crookston, N.; Besharati, S.; Donahue, D.; Thompson, R.E.; Frey, E.; Anders, R.A.; Dreher, M.R.; et al. The Number of Microspheres in Y90 Radioembolization Directly Affects Normal Tissue Radiation Exposure. *Eur. J. Nucl. Med. Mol. Imaging* **2020**, *47*, 816–827. [CrossRef] [PubMed]
14. Maxwell, A.W.P.; Mendoza, H.G.; Sellitti, M.J.; Camacho, J.C.; Deipolyi, A.R.; Ziv, E.; Sofocleous, C.T.; Yarmohammadi, H.; Maybody, M.; Humm, J.L.; et al. Optimizing 90 Y Particle Density Improves Outcomes after Radioembolization. *Cardiovasc. Interv. Radiol.* **2022**, *45*, 958–969. [CrossRef] [PubMed]
15. Miller, S.R.; Jernigan, S.R.; Abraham, R.J.; Buckner, G.D. Comparison of Bolus and Dual Syringe Administration on Glass Yttrium-90 Microsphere Deposition in an In Vitro Microvascular Hepatic Tumor Model. *J. Vasc. Interv. Radiol.* 2022; *in press*. [CrossRef] [PubMed]
16. Marin, A.; Lhuissier, H.; Rossi, M.; Kähler, C.J. Clogging in Constricted Suspension Flows. *Phys. Rev. E* **2018**, *97*, 021102. [CrossRef] [PubMed]
17. Aramburu, J.; Antón, R.; Rivas, A.; Ramos, J.C.; Sangro, B.; Bilbao, J.I. Liver Cancer Arterial Perfusion Modelling and CFD Boundary Conditions Methodology: A Case Study of the Haemodynamics of a Patient-Specific Hepatic Artery in Literature-Based Healthy and Tumour-Bearing Liver Scenarios. *Int. J. Numer. Methods Biomed. Eng.* **2016**, *32*, e02764. [CrossRef] [PubMed]
18. Antón, R.; Antoñana, J.; Aramburu, J.; Ezponda, A.; Prieto, E.; Andonegui, A.; Ortega, J.; Vivas, I.; Sancho, L.; Sangro, B.; et al. A Proof-of-Concept Study of the in-Vivo Validation of a Computational Fluid Dynamics Model of Personalized Radioembolization. *Sci. Rep.* **2021**, *11*, 3895. [CrossRef] [PubMed]

Article

An Image-Based Framework for the Analysis of the Murine Microvasculature: From Tissue Clarification to Computational Hemodynamics

Santiago Mañosas [1], Aritz Sanz [1], Cristina Ederra [2], Ainhoa Urbiola [2], Elvira Rojas-de-Miguel [1,3,4], Ainhoa Ostiz [1,3,4], Iván Cortés-Domínguez [2,3], Natalia Ramírez [1,3,4], Carlos Ortíz-de-Solórzano [2,3,5], Arantxa Villanueva [1,3] and Mauro Malvè [1,6,*]

[1] Department of Engineering, Campus Arrosadía s/n, Universidad Pública de Navarra (UPNA), 31006 Pamplona, Spain
[2] Imaging Platform, Center for Applied Medical Research (CIMA), University of Navarra, 31008 Pamplona, Spain
[3] Instituto de Investigación Sanitaria de Navarra (IdiSNA), Irunlarrea 3, 31008 Pamplona, Spain
[4] Oncohematology Research Group, Navarrabiomed, Hospital Universitario de Navarra (HUN), Irunlarrea 3, 31008 Pamplona, Spain
[5] CIBERONC, Centro de Investigación Biomédica en Red-Cáncer, Instituto de Salud Carlos III, C/Monforte de Lemos 3-5, Pabellón 11, Planta 0, 28029 Madrid, Spain
[6] CIBER-BBN, Centro de Investigación Biomédica en Red-Bioingeniería, Biomateriales y Nanomedicina, C/Poeta Mariano Esquillor s/n, 50018 Zaragoza, Spain
* Correspondence: mauro.malve@unavarra.es

Citation: Mañosas, S.; Sanz, A.; Ederra, C.; Urbiola, A.; Rojas-de-Miguel, E.; Ostiz, A.; Cortés-Domínguez, I.; Ramírez, N.; Ortíz-de-Solórzano, C.; Villanueva, A.; et al. An Image-Based Framework for the Analysis of the Murine Microvasculature: From Tissue Clarification to Computational Hemodynamics. *Mathematics* **2022**, *10*, 4593. https://doi.org/10.3390/math10234593

Academic Editor: Fernando Simoes

Received: 4 November 2022
Accepted: 28 November 2022
Published: 4 December 2022

Publisher's Note: MDPI stays neutral with regard to jurisdictional claims in published maps and institutional affiliations.

Abstract: The blood–brain barrier is a unique physiological structure acting as a filter for every molecule reaching the brain through the blood. For this reason, an effective pharmacologic treatment supplied to a patient by systemic circulation should first be capable of crossing the barrier. Standard cell cultures (or those based on microfluidic devices) and animal models have been used to study the human blood–brain barrier. Unfortunately, these tools have not yet reached a state of maturity because of the complexity of this physiological process aggravated by a high heterogeneity that is not easily recapitulated experimentally. In fact, the extensive research that has been performed and the preclinical trials carried out provided sometimes contradictory results, and the functionality of the barrier function is still not fully understood. In this study, we have combined tissue clarification, advanced microscopy and image analysis to develop a one-dimensional computational model of the microvasculature hemodynamics inside the mouse brain. This model can provide information about the flow regime, the pressure field and the wall shear stress among other fluid dynamics variables inside the barrier. Although it is a simplified model of the cerebral microvasculature, it allows a first insight on into the blood–brain barrier hemodynamics and offers several additional possibilities to systematically study the barrier microcirculatory processes.

Keywords: blood–brain barrier microvasculature; cortical capillary network; tissue clarification; imaging technique; numerical model; microvascular hemodynamics

MSC: 76-10; 92C55; 68U10; 92C10

1. Introduction

The increase of performances of the medical imaging technique in the last decades has allowed non-invasive information of geometries and associated morphologies of large cerebral arteries. In biomedical engineering, this information can be further used for generating 1D to 3D computational models to shed light on cerebrovascular hemodynamics [1,2]. Unfortunately, this process becomes more and more complicated once the vascular scale is becoming smaller, reaching the blood–brain barrier capillaries (BBB), whose number is huge (more than 10 billion) [3].

In recent years, extensive research has been oriented toward the microcirculatory flow proposing complex mathematical models based on cerebrovascular images [3–8]. However, image segmentation has inherently several challenges. First of all, patient-specific human images are difficult to be obtained. The acquisition of the images with the necessary resolution in vivo is not feasible at the micro-scale. Furthermore, the use of cadavers for obtaining useful images also affects as the capillaries tend to collapse once the blood flow and pressure reduce after death [9]. For this reason, murine and rodent images have been utilized as baseline geometry for studying BBB microcirculation [10–16]. Other works have presented imaging-driven modeling for hemodynamics in zebrafish microvasculature and mammalian hearts [17–20]. Models oriented to the vascular topology and transport efficiency have been presented by Katifori and coworkers [21,22].

Nowadays, a wide range of imaging techniques are available. In the literature, corrosion casting [23], confocal microscopy [4], computerized tomography angiography and quantitative magnetic resonance angiography [2], two-photon imaging [8,24,25] and synchrotron radiation-based X-ray tomographic microscopy [26–28] have been mostly adopted depending on the specific necessities of the researchers. A combination of some of these methods can thus be of advantage for limiting the weakness of each method and it is applied in the reconstruction protocols of the microvasculature. In this sense, a recent study by Waelchli et al. [29] provides a detailed visualization and quantification of the 3D brain vasculature using resin-based vascular corrosion casting, scanning electron microscopy, synchrotron radiation and desktop microcomputed tomography imaging. These imaging modalities can provide a large field of view of the vascular network but at the same time low resolution. On the contrary, high resolutions are associated with a smaller field [9,30]. Micro-computerized tomography (micro-CT) is a powerful tool for visualizing large vessels but, as aforementioned, it is not capable of imaging properly the microvasculature due to the lack of resolution [31]. For this reason, its use for the microvasculature needs to be modified using additional techniques. With the aim of improving micro-CT performances, Hlushchuk et al. [32] for instance presented an innovative high-resolution micro-CT imaging of animal brain vasculature. Ghanavati et al. [33] proposed a surgical protocol for improving the surgical perfusion of cerebral blood vessels throughout the murine brain and thus obtaining more consistent cerebrovascular images by X-ray micro-CT.

An additional issue is that the cerebral tissue is opaque so that conventional light microscopy is inefficient due to the light scattering provoked by lipids [34,35]. To solve this problem and allow microscopic light to penetrate the brain tissue, several optical techniques have recently been developed. All of them are based on clearing the tissue using chemical procedures. Dodt and coworkers [36] used a mixture of benzyl alcohol and benzyl benzoate to match the refractive index of fixed tissue. However, this protocol only permitted a partial tissue clarification as the clearing solutions led to the rapid loss of fluorescent signals. In a later study, using a so-called 3D Imaging of Solvent Cleared Organs approach (3DISCO), they found that a fast optical clearing can be obtained [37]. Further studies [38,39], using different chemical approaches also achieved rapid tissue clearing encountering similar instability issues.

Tissue-clearing techniques emerged in the last decade to allow high-resolution 3D imaging of biological tissues. Numerous tissue clearing methods are currently available such as DISCO (iDISCO, uDISCO and 3DISCO) [37,40,41], CLARITY [34,42–47], and seeDB [39], among others. Most of these protocols reduce the light scattering provoked by the presence of the lipid and homogenize the RI, obtaining more transparent tissues [48]. Susaki and coworkers [35] developed a whole-brain clearing and imaging method called CUBIC (Clear, Unobstructed Brain Imaging Cocktails and Computational analysis). CUBIC is a comprehensive experimental method involving the immersion of brain samples in chemical cocktails containing aminoalcohols, which enables rapid whole-brain imaging with single-photon excitation microscopy. In parallel, they also improved their methodology developing the so-called Advanced CUBIC clearing method. This improvement was based on hydration and extended the clearing process to several organs of a mouse, allowing high-resolution 3D imaging [49,50]. Advanced CUBIC was time consuming, had a limited efficiency for clearing organs with high pigment content and adopted the same time for different samples. For these reasons, Res et al. [48] introduced a new ultrasound processing to reduce the clearing time and proposed a new decolorization cocktail to remove pigments. With this optimization method, also called CUBIC-Plus, they enable a considerable shortening of the time acquisition of high-resolution 3D images of the lung. Pinheiro et al. [51] developed an improved clearing protocol, called CUBIC-f, for optimizing fragile samples. Hasegawa et al. [52] introduced CUBIC-kidney for kidney research applications. Based on the CUBIC methodology, Murakami et al. [53] proposed a fluorescent-protein-compatible clearing and homogeneous expansion protocol based on an aqueous chemical solution (CUBIC-X). The expansion of the brain sample allowed the construction of a point-based mouse brain atlas that allows the analysis of numerous samples providing a platform for different organs in the biomedical research, the so-called CUBIC-Atlas [54].

Notwithstanding that imaging techniques are continuously progressing, there are no specific techniques that alone are capable of providing the entire cerebral blood vessels for further reconstruction of comprehensive 3D models [9]. In general, the data obtained after the medical imaging techniques require considerable additional work before these can be treated by computer-aided design programs and computational software. An important pre-processing is for instance necessary for closing all the gaps of the acquired data, simplifying and smoothing the segments of the network that represent the vessels, avoiding noise and generating surfaces that form the limits of the computational domain [3,5,8,9,11,12]. For this reason, in the literature, idealized synthetic computational models based on mathematical algorithms are considered a valuable way to study the cerebral microvasculature. They avoid some of the limitations affecting images-based methods due to the considered microscale [9]. It is about the binary branching trees or networks that mimic the vascular bed morphology. However, the cerebral microvasculature presents for instance loops and anastomoses that cannot be taken into account using simple fractal networks [9]. Hence, simplified binary fractal trees [55–57] have been progressively more and more replaced by complex networks. These models are useful tools for different purposes. In the literature, computer methods have been often based on brain animal images due to the impossibility of invasive experimentation in humans [4,6]. Some studies have been used for interpreting optical measurements acquired in rodents [58,59]. Others were oriented to the analysis of intracellular transport phenomena on length scales not accessible to imaging methods [60,61]. Sherwin et al. [62] introduced a 1D model of a vascular network in space-time variables. Boas et al. [63] presented a symmetric binary vascular network composed by 190 segments to investigate the steady-state or transient response to specific diameter variations of the arteriolar region. Reichold et al. [28] proposed a computational methodology based on anatomical data obtained by synchrotron radiation X-Ray Tomography for simulating rat cerebral blood flow. They presented qualitative results of a fully three-dimensional intra-cortical vasculature structure modeled as a vascular graph. Lorthois and coworkers [3–6,64,65] have provided a large quantitative data focused on the microcirculation of the human cerebral cortex. Recently, they have introduced an analytical model capable of describing the coupling between arteriolar and venular trees, which were modeled using a vascular network approach and the capillary tree, modeled as a continuum porous medium. The research group of Linninger, Hartung and coworkers [2,9,13,14,16,24,66–68] has extensively worked in the microvascular architecture hemodynamics obtaining detailed information on the cerebral microcirculation inside the cerebral cortex by means of vascular networks and numerical algorithms. They analyzed the tissue metabolism coupled to micro-hemodynamics [66], and latterly, they introduced an alternative method to the binary tree for obtaining a more realistic microcirculatory network. This model was generated using Voronoi tessellation first and was later improved by introducing novel closed networks. It finally includes an arterial and a venous tree with capillary connection synthesized with a single algorithm that allows reducing the computational costs [9].

To the best of our knowledge, there are no studies in the literature that combine tissue clearing, advanced microscopy with images treatment, geometrical reconstruction and numerical simulations. In this study, we aimed to introduce a novel protocol that combines all these techniques for obtaining detailed information about cerebrovascular cortical microcirculation. Concretely, we propose the combination of tissue clearing and advanced microscopy techniques with image treatment, geometrical reconstruction and numerical simulations. With the proposed protocol, we provided a 1D image-based computational model of the cerebral murine microvasculature that allows solving instantaneously the associated hemodynamics. Blood flow features and a quantitative evaluation of the microvascular morphology at the brain cortical territory can be predicted. In particular, different regions and depths of the BBB were considered and investigated with the aim of helping understand its microvascular functionalities and characteristics.

2. Materials and Methods

As described in the previous section, the cerebral tissue is opaque due to the presence of lipids, so conventional light microscopy is inefficient [34,35]. Hence, the lipids need firstly to be removed from this tissue for allowing the light passage without scattering or absorption, matching the refraction index (RI) between the tissue and medium. The tissue clarification is a chemical process of delipidation, decoloring and RI matching.
In this work, the mouse brain samples were treated as follows:

1. Fixation: Samples were fixed using paraformaldehyde (PFA) after transcardiacally mice perfusion and dissection before post-fixation with PFA.
2. Sectioning: Here, 500 μm thick brain slices were sectioned using a vibratome.
3. Clearing: Sections were cleared using the CUBIC protocol.
4. Staining: Delipidated sections were stained with FITC-Lectin and an arteriole-specific dye Alexa Fluor 633 hydrazide.
5. Imaging: Here, 500 μm slices were analyzed using an advanced two-photon microscopy.

Figure 1 shows the step-by-step the process followed. In the next subsections, each step is described.

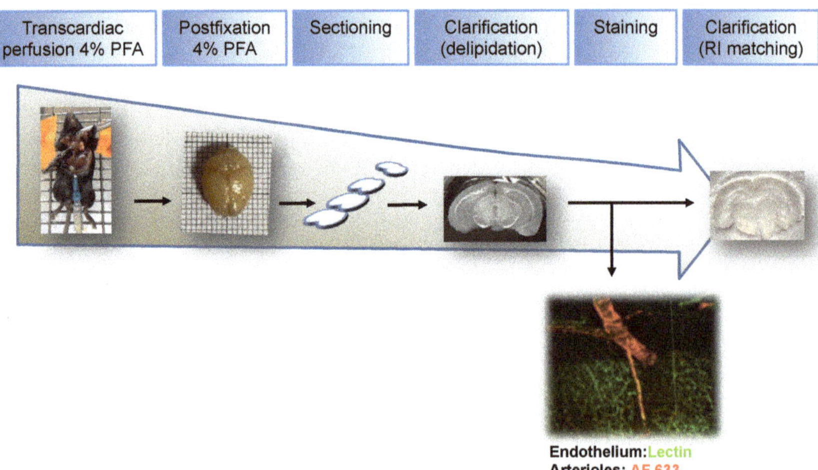

Figure 1. From mouse transcardiac perfusion to staining and cleared tissue.

2.1. Fixation, Sectioning and Tissue Optical Clearing

Specifically, all animal procedures of this study followed European and Spanish legal regulations and were performed under an ethical protocol approved by the University of Navarra Committee for Ethical Use of Laboratory Animal (076-19). Concretely, *C57B6* mice were euthanized and transcardiacally perfused with PBS (pH = 7.4) and 10 mL of 4% paraformaldehyde in PBS. Mice's brains were dissected, post-fixed overnight in 4% PFA and washed with PBS. Then, 500 μm thick brain slices were sectioned using a vibratome (VT1000S, Leica, Leica Biosystems Technologies, Danaher Corporation, Washington, DC, USA) and kept in PBS solution at 4 °C. Sections were cleared following the CUBIC protocol [50]. The method consists of two phases: delipidation with ScaleCUBIC Reagent-1 (urea 25 wt %, Quadrol 25 wt %, Triton X-100 15 wt % and dH$_2$O) and refractive index matching with ScaleCUBIC Reagent 2 (urea 25 wt %, sucrose 50 wt %, triethanolamine 10 wt % and dH$_2$O). The brain slices were first incubated with ScaleCUBIC Reagent-1 for 4 days at 37 °C while gentle shaking, which was followed by PBS washing for 16 h at room temperature. After that, the tissue staining was performed, and the slices were immersed in ScaleCUBIC Reagent 2 until their visualization in the microscope.

2.2. Tissue Staining

Staining of the endothelium in the brain tissue sections was performed between the first and second phases of the clearing process. To this end, delipidated sections immersed in PBS were blocked with BSA 4% and incubated with a 50 μg/mL solution of FITC-Lectin (Sigma-Aldrich, Merck KGaA, Darmstadt, Germany, USA) during 24 h at room temperature. Finally, the sections were incubated with arteriole-specific dye Alexa Fluor 633 hydrazide solution (Thermofisher Scientific, Waltham, MA, USA) (2 μm) for 1 h. Following the arterioles down to the capillaries in the acquired 3D volumes, we determined the areas of connection between the arterial and venous system. This allowed establishing the inlets and outlets of the computational model and hence obtaining later the correct flow direction in the simulations.

2.3. Two-Photon Excitation Microscopy

Image stacks (1 mm × 1 mm × 0.5 mm in size) were collected using a Zeiss LSM 880 (Carl Zeiss, Jena, Germany) equipped with a two-photon femtosecond pulsed laser (MaiTai DeepSee, Spectra-Physics, Milpitas, CA, USA), tuned to a central wavelength of 800 nm, using a 25×/1.8 objective (LD LCI Plan-Apochromat 25×/0.8, Carl Zeiss). Tiles of z-stack scan from 500 μm sections were acquired in the non-descanned mode after spectral separation and emission re-filtering using 500–550 nm and 645–685 nm BP filters for Lectin and Alexa 633 signals, respectively. In Figure 2, a lectin-stained vessels region is shown with a close-up view to a cubic sample.

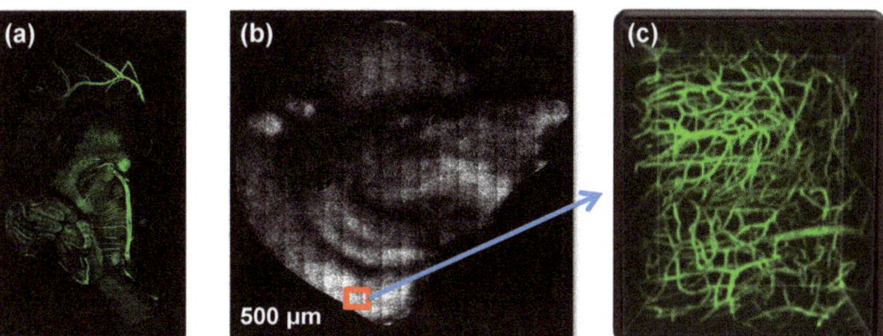

Figure 2. Tissue staining: The tissue was divided in cubic samples of 500 μm: (**a**) endothelial labeling with lectin; (**b**) cubic samples; (**c**) close-up view of the single cubic sample box in red in (**b**).

From the cubic samples represented in Figure 2b (a single region is highlighted in red), five cortical regions were selected and further used for images acquisition, geometrical reconstruction and computational simulation, as it will be shown in the next sections. The reason why these specific regions were chosen is related to the aim of the study. The cortical regions are areas of the brain located in the cerebral cortex where the BBB is localized and thus are the target of our study because the associated microvasculature is the main filter to pharmacologic drugs. Hence, these cortical regions are of particular interest versus other brain internal areas where there is hardly any capillary network (for example, white matter situated in subcortical regions).

2.4. Image Analysis

The following procedure was applied to each vessel region obtained from the previous steps. All the image treatment was performed in MaTLaB (The MathWorks, Natick, MA, USA), using an appropriate in-house code.

Each volume was first filtered with a 3D Gaussian filter (size: $3 \times 3 \times 3$ voxels, standard deviation: $\sigma = 0.5$). Even this could lead to a possible loss of information, and the images of the present study are mainly dominated by Poisson noise; thus, we cannot neglect the presence of Gaussian noise that needs to be filtered. Then, a non-local means filter was applied to each slice [69] in the volume to reduce the Poisson noise [70] resulting from the acquisition procedure in the microscope. The vessel-like patterns were enhanced by using morphological filters with linear structuring elements, L_i, as it was performed in [71], as an adaptation of the top-hat method. In a first stage, an opening is carried out using the previously filtered volume, named as S_f, with structuring elements of varying orientation, L_i. In this way, each L_i was composed by a length of 51 voxels and a width of 1 voxel. Eight different orientations were defined in the xy-plane, i.e., horizontal plane (angular variation was $\pi/8$, i.e., $22.5°$) to which eight additional angular variations were added in the z-axis direction, resulting in 64 possible structuring elements, i.e., L_i with $i = 1, \ldots, 64$.

For each one of the L_i, a volume was obtained as result of an opening operation. A new volume, named S_0, was constructed assigning to each voxel the maximum value voxel from the 64 opened volumes previously calculated, as shown in Equation (1):

$$S_0 = max_{i=1,\ldots,64}\left\{\gamma_{L_i}\left(S_f\right)\right\} \tag{1}$$

where γ is the opening operator. This step was finished by means of a geodesic reconstruction using S_f as a mask, obtaining the opened volume S_{op}, as shown in Equation (2):

$$S_{op} = \Gamma_{S_f}^{rec}(S_0) \tag{2}$$

where Γ is the geodesic reconstruction (opening) operator. The next step was to open S_f with the different L_i and subtract to S_{op}. Then, we added every volume calculated.

$$S_{sum} = \sum_{i=1}^{64}\left(S_{op} - \gamma_{L_i}\left(S_f\right)\right) \tag{3}$$

Once the vascular structure was enhanced, the binarizing of the volume was performed by using an adaptive threshold [72] of size 71 and a Gaussian statistic (Figure 3a,b). Finally, the biggest connected region was selected and cleaned with a morphological closing ($3 \times 3 \times 3$) and a 3D hole-filling approach [73] (Figure 3c).

Vessel Measurements

Once the volumes were binarized (Figure 3), the main geometrical features were extracted from each region. As first step, the vessel structures were skeletonized using the corresponding morphological operator by obtaining the medial axis [74,75]. In this manner, we could easily define the voxels corresponding to the vessels. Those voxels with only one neighborhood were considered as endpoints, while those others with two neighbors were treated as vessel points. Lastly, those voxels with more than two neighbors were labeled as branchpoints (usually three neighbors, but can be more). The parametrization of our vasculature structure permits breaking our vessel set into individual segments by eliminating the branchpoints that can be more easily analyzed.

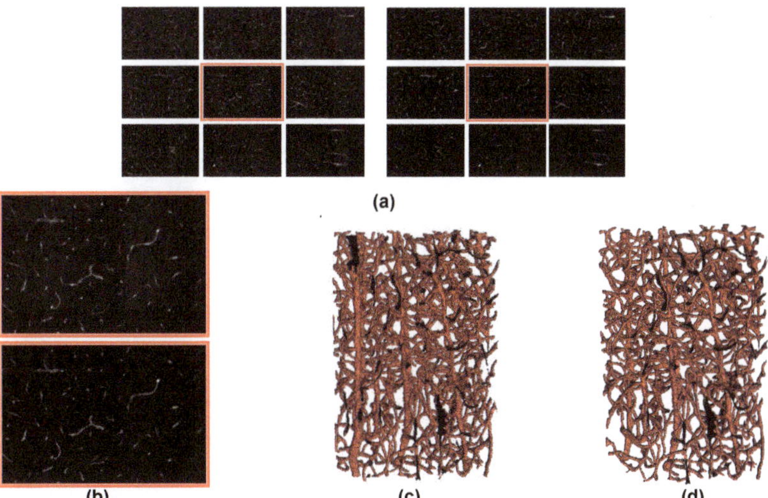

Figure 3. Image treatment: (**a**) Original images (**left**) and result of filtering (**right**). (**b**) Zoomed versions extracted from (**a**). Original slice (**up**), filtered slice (**down**). (**c**) Segmented volumen. (**d**) Result after artifacts erasing.

Regarding each segment, knowing the voxel size, it was possible to calculate some shape-related characteristics such as the longitude, the curvature or the tortuosity of the segment. Using the vessel set volume and the medial axis voxels data, we also calculated the radius for every skeleton voxel and then computed the mean radius for each segment. Once all the structures were parameterized, we straightforwardly constructed 1D models of the five vessel regions in which each voxel and segment are characterized for further analyses. A final representative model is depicted in the Figure 3d).

2.5. One-Dimensional (1D) Modeling

2.5.1. Governing Equations

The computational models of the previously created geometries were programmed in MaTLaB, and they were based on the hemodynamic network developed by A. R. Pries and T. W. Secomb [76]. In the literature, it is worldwide known that in microvascular networks, the velocities achieved by plasma and red blood cells (RBCs) are very low (60–1 mm/s) [77], translating into a very low Reynolds number (Re < 0.001 for all the analyzed regions) and leading to a laminar capillary flow. Hence, inertial forces have less influence than viscous forces. This fact, in addition with a low Womersley number ($Wo < 0.01$) indicating that the flow can be considered as no pulsatile, enabled a simplification of the Navier–Stokes equation into the Stokes equations (Equation (4)). As a result, we simplified the flow in the capillary bed as a ratio of the pressure drop in every capillary and its hydraulic resistivity:

$$\mu \nabla^2 \mathbf{v} + \nabla p = 0 \tag{4}$$

The model handled in this study, after its processing, became a 3D network built from nodes and cylindrical segments in which all the constitutive equations were solved. This model was composed by a collection of interconnected nodes and segments of the BBB microvascular geometry. In this vascular network, the nodes represented locations where vessels bifurcated or ended, and the segments represented the vessels. Each node was defined by coordinates and each segment was defined by nodes and diameters. Every segment was then divided into several intermediate segments, making possible the use of tortuous vessel length instead of simplifying the vessels as straight lines between two end points (Figure 4). Hence, the total length of the vessel was obtained as the sum of the lengths of consecutive intermediate segments (Equation (5), Figure 4).

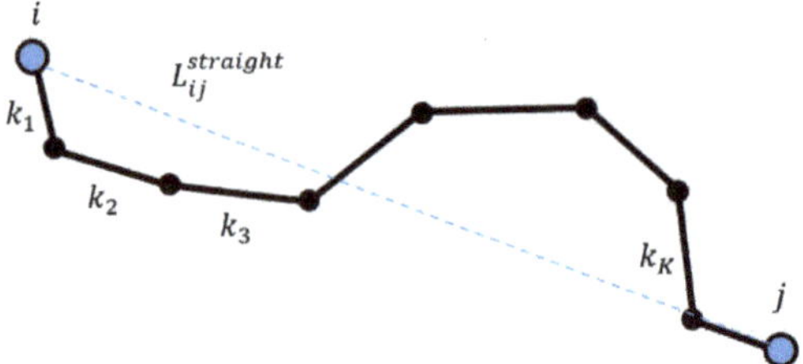

Figure 4. Schematic representation of the segments subdivisions.

$$L_{ij}^{total} = \sum_{i}^{k} L_{ij,k} \tag{5}$$

where L_{ij} was the total segment length divided by k intermediate segments. Initially, the mass flow entering a node was the same as the outflow of this node, fulfilling the continuity equation in every geometry node (Equation (6)).

$$\nabla \cdot \mathbf{v} = 0 \Rightarrow \dot{m}_{in} = \sum \dot{m}_{out} \tag{6}$$

On the other hand, the blood flow (Q) in every capillary segment was calculated as shown in Equation (7):

$$Q_{ij} = \frac{\Delta p_{ij}}{R_{ij}} \tag{7}$$

where Δp represented the pressure drop between the defining nodes of the segment ij and R represented the flow resistance of segment ij, given a cylindrical shape, which was calculated using the Hagen–Poiseuille Law (Equation (8)):

$$R_{ij} = \frac{128\mu_{ij}L_{ij}}{\pi D_{ij}^4} \tag{8}$$

The flow resistance of a segment R_{ij} depends on the diameter of the segment D_{ij} and on its length L_{ij}. In this study, the tortuous length of each vessel was taken into account instead considering only straight segments adding the tortuosity evaluated during the images' treatment. The flow resistance depends on the blood viscosity of the segment μ_{ij}, which varies considerably between segments due to the Fahraeus–Lindqvist effect. As known, the latter is caused by the biphasic nature of the blood and the small dimensions of the capillaries [78]. The effective viscosity μ_{eff} was calculated using the in vivo empirical description made by A. R. Pries [78]. This set of empirical equations takes into account the effects of the biphasic nature of the blood in the capillary bed and calculates its effective viscosity given a vessel diameter D, velocity and hematocrit H_D as follows:

$$\mu_{ij} = \mu_{ij}^{rel} \cdot \mu_{plasma} \tag{9}$$

$$\mu_{ij}^{rel} = \left[1 + (\mu_{0.45} - 1) \cdot \frac{(1 - H_D)^C - 1}{(1 - 0.45)^C - 1} \left(\frac{D}{D - 1.1} \right)^2 \right] \cdot \left(\frac{D}{D - 1.1} \right)^2 \tag{10}$$

$$\mu_{0.45} = 6 \cdot e^{-0.085D} + 3.2 - 2.44 \cdot e^{-0.06D^{0.045}} \tag{11}$$

$$C = 0.8 + e^{-0.075D} \cdot \left(-1 + \frac{1}{1 + 10^{-11} \cdot D^{12}} \right) + \frac{1}{1 + 10^{-11} \cdot D^{12}} \tag{12}$$

The hematocrit distribution in the bifurcations of the geometry was calculated using the phase separation law established by A. R. Pries and T. W. Secomb [76]. This law contains a set of empirical equations that define the hematocrit distribution in a bifurcation knowing the hematocrit in the mother branch and the flow and diameters of the daughter branches (Equations (13)–(15)):

$$FQ_E = 0, \quad if \quad FQ_B \leq X_0 \tag{13}$$

$$logit_{FQ_E} = A + B\,logit\left[\frac{FQ_B - X_0}{1 - 2X_0}\right], \quad if \quad X_0 \leq FQ_B \leq 1 - X_0 \tag{14}$$

$$FQ_E = 1, \quad if \quad 1 - X_0 \leq FQ_B \tag{15}$$

where FQ_B was the fractional blood flow in the daughter branch (ratio of the blood flow of the daughter branch and the mother branch) and FQ_E was the fractional erythrocyte flow in the daughter branch (ratio of the erythrocyte flow of the daughter branch and the mother branch). The relationship between erythrocyte flow, blood flow and the segment hematocrit was obtained using the following equation:

$$Q_E = Q_B \cdot H_D \tag{16}$$

The parameters A, B and X_0 were obtained as follows:

$$A = -13.29 \cdot \left(\frac{\frac{D_A^2}{D_B^2} - 1}{\frac{D_A^2}{D_B^2} + 1}\right) \cdot \frac{1 - H_D}{D_F} \tag{17}$$

$$B = 1 + \frac{6.98(1 - H_D)}{D_F} \tag{18}$$

$$X_0 = \frac{0.964(1 - H_D)}{D_F} \tag{19}$$

where D_A and D_B were the diameters of the daughter branches and D_F and H_D were the diameter and hematocrit of the mother branch.

2.5.2. Boundary Conditions

It is widely known that one of the most challenging parts in simulating microvascular networks is to establish the conditions in all the in/outflows that appear in the limits of the computational domain. These conditions are necessary for the solution of the 1D equations that describe the blood flow inside the microvasculature. In particular, flow, pressure and hematocrit conditions must be set and, depending on their values, the calculations predict accurate (or less accurate) physiologically meaningful results. In this work, as experimental measurements were not possible in murine brains, some approximations were taken, and literature data were adopted. Different authors used various solutions to this problem [5,24,76,77]. In this work, the solution presented by Lorthois et al. [5] was chosen, as it was simple and fast to implement and achieved valid predictions, comparing the obtained results. The used set of boundary conditions are taken from [77,79] and are described below:

(a) Pressures were imposed at the inlet and outlets. With that, there was no need to know the flow direction in all in- and outflows respectively, as the flow direction in the segments adjusted to fulfill the pressure boundary conditions.

(b) Boundary nodes (1 segment nodes) inside the geometries limits were assigned with a zero flow condition and with zero hematocrit. These nodes show the presence of *broken* vessels inside the geometry that could be produced during the segmentation. It is important to notice that these vessels have no physiological meaning but need to be treated.

(c) Three different sets of pressure boundary conditions were assigned depending on the segment to which the boundary node was attached to: venule, arteriole or capillary:

    1. At the arterial inflow, a pressure of 50 mmHg was given. The arterial pressure outflow was set to 40 or 45 mmHg depending on its nearness to the inflow. With that, the risk of a short circuit was eliminated.

2. At the venular outflows, a pressure of 10 mmHg was given.
3. In the capillary in/outflows, two cases were studied, following Lorthois and coworkers [5]:

Case 1: Zero flow condition: Flow is set to zero in all the capillary outflows. In this case, the flow goes from the arterial inlet passing through the whole geometry until it reaches a venular outlet. As reported [5], this condition would underestimate the flow in the geometry as it isolates it from its virtual neighbors.

Case 2: Constant pressure condition: A constant capillary boundary pressure was calculated so that the net capillary flow (the sum of the flow in all the inlets and outlets) was zero; thus, everything that enters through the arterioles exits through the venules. In other words, this pressure was adjusted such that the total flow entering the arteriolar network was the same as the total flow entering the venular network. In this way, the net flux to all the boundary capillary segments was zero. As a consequence, the net flux leaving the studied brain region through capillaries to supply neighboring areas was exactly compensated by the net flux arriving from neighboring areas through capillaries. As shown in the literature, this condition forces the flux lines to be perpendicular to the ends of the computational domain, maximizing the exchanges of fluid with the neighboring region. For this reason, this condition overestimates the flow in the geometry as it maximizes the flow exchange between the region itself and its virtual neighbors. [64].

(d) The boundary hematocrits values were set 0.45 at the arterial inlets, 0.4 at the arterial outlets, 0.2 at the venular outlets and a random value between 0.2 and 0.6 for the rest of the capillary boundary nodes, mimicking the chaotic and haeterogenic nature of this variable in the capillary beds [5,64].

The influence of imposing zero flow or a constant pressure at the capillary outlets was found to be limited. Similar results were found also by [5]. Finally, we chose the second option (constant pressure condition), as the first one (zero flow) tended to isolate the volume of the considered capillary regions.

3. Results

This paper focuses on the image-based circulatory network of the BBB and shows the versatility of the presented methodology for analyzing up to five cortical regions of the murine brain vasculature. The main purpose of the framework is to introduce a consistent methodology for elucidating the murine microvascular hemodynamics and other functions related to the BBB. The presented synthetic anatomical networks are easy to be treated using 1D hemodynamics. In this section, we illustrate some computational results in terms of flow, pressure, hematocrit and endothelial shear stress. The results took a few CPU minutes to be obtained and required around 20 GB of memory in serial execution on a HP Z440 Intel Xeon computer.

The network geometries are represented in Figure 5 and are colored by the values of the vessels diameter. These regions considered one inlet each but a different number of outlets, different vessel densities, mean diameters, curvatures and tortuosities and the number of bifurcations among other morphological differences. As visible from the figure, the morphology of the five regions is widely different. Of course, these topologies strongly influence the flow patterns and the associated nutrient transport in the surrounding tissue. For these reasons, it is relevant to show the different statistics associated to the flow simulations of the five cerebral regions. Frequency distributions of vessel diameters, length, surface area and volume that characterize the five networks reconstructed using the presented algorithm are shown in Figure 6. The geometries show a very good correlation in terms of segment diameters (in (μm)) and lengths (in (μm)), surface area (in (μm^2)) and total vascular volume (in (μm^3)) distributions, as shown in Figure 6 where the cumulative distribution function (CDF) of these variables is depicted. The presented curves match well the shape and order of magnitude of those presented by Linninger and coworkers [9] which were obtained using a mathematical synthesis of the cortical circulation for a whole mouse brain. The obtained relations between the frequency of appearance and diameters, lengths, surfaces and volumes are in agreement also with those found by other authors, showing that the used geometries are suitable for further use in the numerical simulations of the microvascular blood flow. This comparison ensures that our image-based modeling presents anatomically consistent microvasculature.

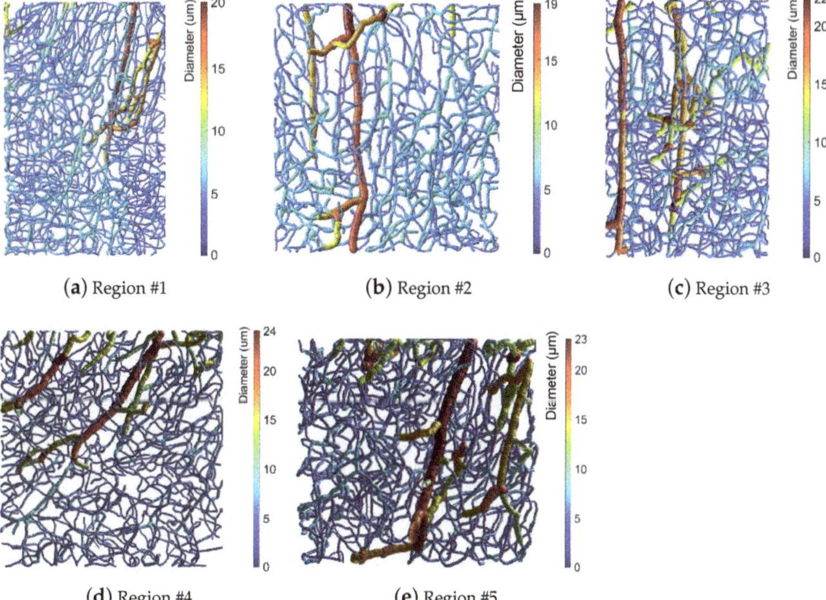

(**a**) Region #1 (**b**) Region #2 (**c**) Region #3

(**d**) Region #4 (**e**) Region #5

Figure 5. Morphology of the 5 cerebrovascular regions of the murine cortex considered in this study. The heat map represents by color the distribution of the value of the diameters within the microcirculatory synthetic network.

Additionally, the framework is capable of controlling the number of arteries and bifurcations and all the associated geometrical features that are quantified in the image data and included in the synthetic model. Previously published capillary networks use only straight segments with cylindrical shape for describing the microvasculature. However, real networks present curvature and tortuosity. Both variables were measured here directly from the images. In particular, the tortuosity was computed using the metric SOAM described by Bullit et al. [80]. We imposed the tortuosity measured directly from the images for mimicking imaged networks. Its CDF is depicted in the Figure 6e). Moreover, we provided a venous connection between arteriolar and capillary regions thanks to the double staining. Previous studies habitually neglected curvature and tortuosity, presenting straight vessel instead, and only a few consider venous drainage [9].

The five different geometries that have been analyzed in this work and depicted in Figure 5 are defined by the parameters summarized in Table 1.

Table 1. Morphological properties of the 5 considered cortical regions.

Region	Nodes	Boundary Nodes	Segments	Dimensions (μm)
#1	1250	210	1561	$698 \times 459 \times 310$
#2	751	136	932	$466 \times 432 \times 207$
#3	968	134	1265	$559 \times 365 \times 318$
#4	948	198	1164	$517 \times 532 \times 220$
#5	999	72	1292	$508 \times 383 \times 345$

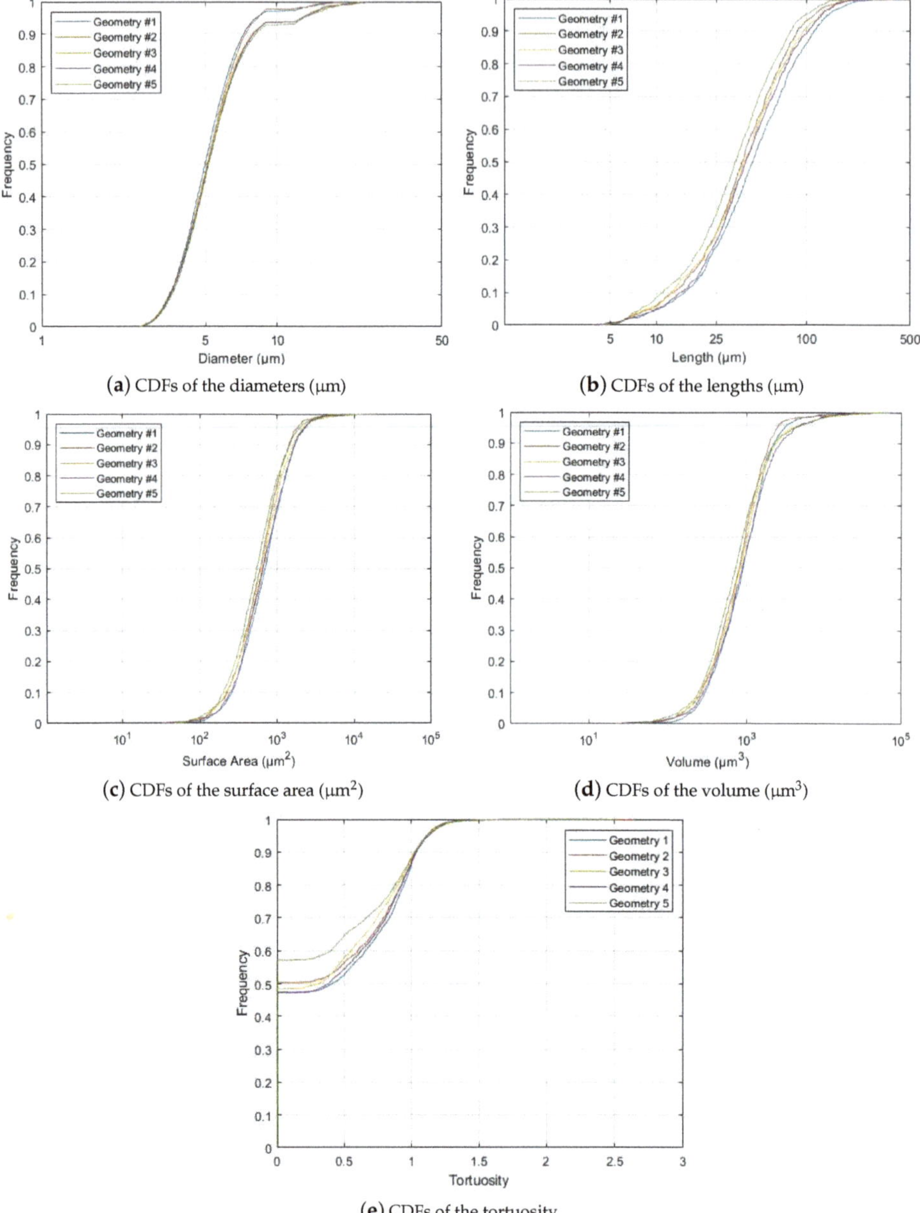

(**a**) CDFs of the diameters (μm)

(**b**) CDFs of the lengths (μm)

(**c**) CDFs of the surface area (μm^2)

(**d**) CDFs of the volume (μm^3)

(**e**) CDFs of the tortuosity

Figure 6. Statistical analysis of the considered regions: cumulative distribution functions of diameters (**a**), lengths (**b**), surface areas (**c**), volumes (**d**) and tortuosity (**e**).

The blood flow distributions of the five regions is depicted in Figure 7 in logarithmic scale for enhancing differences within the vessel segments. As the regions are of different size and present important morphological differences, the maximum and minimum of the scale is different for each geometry. The results of the simulations showed that a peak blood flow of 437.33 nL/min was found in Geometry #4, while the minimum blood flow was 99.65 nL/min and belonged to Geometry #3. Summarizing, we found a mean blood flow of 268.49 ± 168.84 nL/min. This value differed from the values by Hurtung and coworkers [24]. However, even though they have found a maximum blood flow of around 780 nLmin, they considered wider regions and scales than the ones used in this work. Of course, the comparison can be only performed qualitatively because it is about different samples with variable morphologies. The important variability of the blood flows found in the present work can be explained by the geometrical differences presented by the 5 regions. In some of them, the feeding arteriolar branch present 'shortcuts' to the outlets, having a preferential flow path of little resistance and increasing the blood flow. This happens, for example, in Geometries #4 and #5, indicated in Figure 7. Furthermore, there are slight differences in the feeding arteriolar trunk diameters, varying from 17.24 µm in Geometry #1 to 22.62 µm in Geometry #5, for instance. This causes less flow resistance for the same pressure loss between inflow and outflow, leading again to an increase of the blood flow.

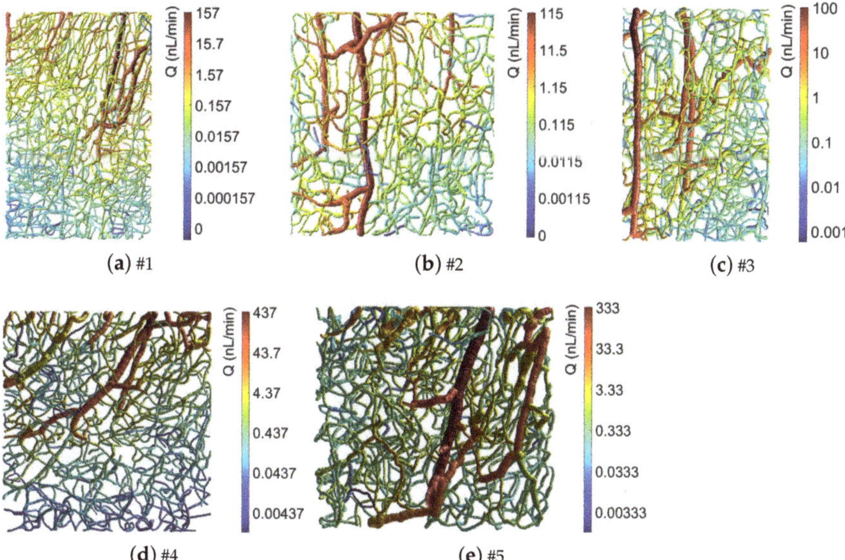

Figure 7. Computed blood flow (in [nL/min]) within the 5 cerebrovascular regions of the murine cortex. The heat map (in logarithmic scale) represents by colors the average blood flow distribution within the microcirculatory synthetic network.

Figure 8 shows the hematocrit distribution within the five regions. Initially, a maximum hematocrit of 80% has been set for any segment. The obtained distributions, as visible in the figure tend to be chaotic in all the regions. This happens because of the used geometries, as this distribution mostly depends on the asymmetry of the bifurcations inside the models due to the nature of the blood. As can be found in the literature, the hematocrit distribution differs from realistic to synthetic, symmetric, binary tree geometries. In these geometries, the hematocrit distribution is in fact mostly homogeneous [24]. The variations seen in Figure 8 are the result of the morphology of the five considered regions. The position of the venous drainage in the geometries can affect hugely the hematocrit distribution, as this is the location where the flow exits and where convergent bifurcations appear. Additionally, there are also some locations in the geometries where divergent bifurcations appear, leading to a decrease of the hematocrit in the segments until it reduces even to 0%, as seen for instance in Geometry #1. On the other hand, in all geometries, some vessels with slightly high hematocrit values can be seen. These increases of hematocrit depends both on the segment diameter and on the feeding segment hematocrit, being usually convergent bifurcations.

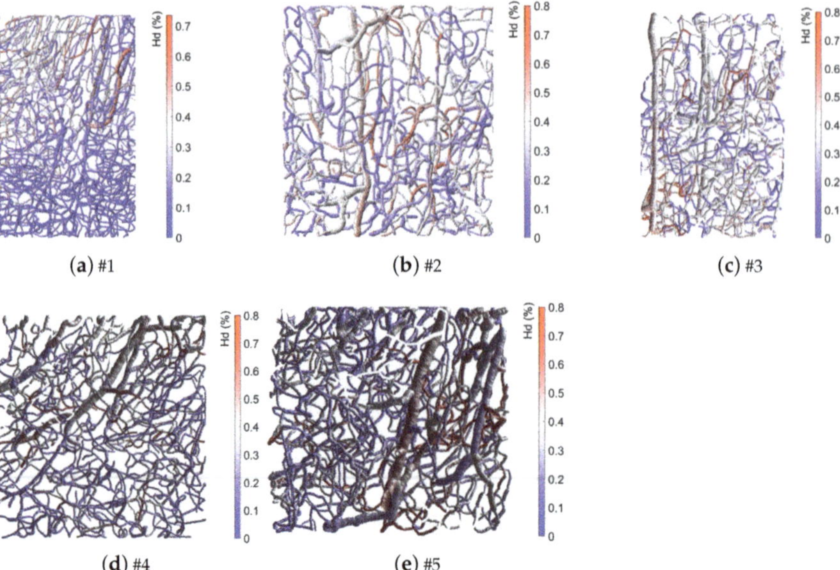

(a) #1   (b) #2   (c) #3

(d) #4   (e) #5

Figure 8. Computed hematocrit (in [%]) within the 5 cerebrovascular regions of the murine cortex. The heat map represents by colors the hematocrit percentage per segment within the microcirculatory synthetic network.

4. Discussion

The brain is the most complex organ of humans, but despite the extensive work dedicated in recent decades, still little is known about its functionalities, including the anatomy and the hemodynamics of its vasculature in comparison with all the other organs [81]. Several studies have attempted to describe the microvasculature structure and anatomical variations in the cerebral surface region often comparing humans and rats that present many similarities but also differences [82]. We have proposed a comprehensive framework based on tissue clarification, advanced microscopy and image treatment aimed at the analysis of the murine microvasculature that is feasible to be applied to humans. Through a mathematical algorithm, specific regions or even the entire murine brain geometry can theoretically be created for the analysis of its hemodynamics. The reconstruction of the entire anatomy from image data is difficult to be obtained as patient-specific data have a limited spatial resolution [9]. For this reason, the combination of anatomical images, from the tissue clarification to the obtention of 3D geometries, and mathematical modeling using advanced algorithms that allow the analysis of a consistent circulatory network is an efficient strategy, and it is the standard methodology in the literature. The advantage of the synthetic network is that it can be used for different purposes, for example for the simulation of blood flow and nutrients transport phenomena that can mimic the 3D vasculature. These simulations can cover regions of the in vivo data sets where imaging data are not consistent, as we have discussed in the Section 2.5.1 or regions not reconstructed [9]. Alternatively, the use of mathematical networks could also be capable of complementing real anatomical data serving as boundary conditions for 3D realistic anatomical microcirculatory models. The 1D modeling can be attached to 3D models replacing the limits of the computational vascular domains and can be used for applying the boundary conditions as elucidated by Linninger and coworkers [9]. Fractal networks have been often used in this sense for large and small arteries as well as for the cerebral vasculature [83–85].

In the past decade, synthetic vascular models have offered more and more an alternative to purely image-based approaches [65,81,86,87]. Unfortunately, binary trees can only approximate the real microvasculature because they only bifurcate in one direction and cannot take into account loops. For this reason, more recently, other authors start creating more complex vascular structures that could include anastomoses improving previous findings [9]. Our work demonstrated that the presented methodology offers such morphological structures as the obtained synthetic models faithfully represent the imaged cortical regions.

At the same time, researchers have progressively proposed improved mathematical algorithms providing increased models complexity yet providing accurate brain data-based networks. An example is the synthetic model introduced by Linninger et al., which simulates the cortical blood supply in a section of the human cortex. They provided a computational method for building realistic microcirculatory beds using Voronoi tessellation [66]. Due to the high computational costs, they later further extended this model using a single algorithm including arterial and venous trees with capillary connection [9]. Another example is the algorithm developed by Su et al. for creating a set of networks based on experimental statistics to bypass the complexities to reconstruct a cerebral microvascular network from real brain tissue data [81].

The principles of the modeling proposed in the present work are similar to those introduced by other studies in the literature [63]. The cerebral vasculature is represented by a network of bifurcating cylinders that provide a resistance to flow according to the Stokes equations. The proposed mathematical model was further used for studying the hemodynamic in the brain for showing the application of the developed methodology. Some computational results regarding the blood flow, the hematocrit and the endothelial shear stress distribution have been presented (see Figures 7–9) and demonstrate the feasibility, the utility and versatility of the presented framework. With the proposed framework, it is also possible to have a consistent quantification of the vascular morphology, providing data of the number of bifurcations, tortuosity, surface, vessel length and diameter, volume and volume density that can be used for characterizing the vascular structure and its functionality (see Figure 6). It is widely known that the neuronal tissue varies with the depth of the cerebral cortex so that the presented results may be used to help elucidate the relationship of the flow, pressure and shear stress characteristics with the depth in 1D realistic vascular networks as studied by other authors but still not yet fully understood [77]. The computational results support the hypothesis already diffused in the literature that the flow field and the hematocrit distribution are highly heterogenous in the microvasculature, suggesting that the oxygen and nutrients brain regulations depend on the cortical layer [25,77].

The presented simulations are based on a real anatomical data so that reconstructed geometries are controllable. However, the results leads to 1D flow and average values of velocity, WSS and other variables that approximate the real cerebrovascular hemodynamics. Of course, synthetic models are based on simplified geometries and simplified hemodynamic constraints as a boundary condition so that the resulting hemodynamic features are simplified as well [9]. For this reason, the results obtained in this study were compared with published results for demonstrating the consistency and robustness of the presented tool. Unfortunately, currently, an in vitro or an experimental validation is not feasible. Nevertheless, as stated in the literature [9], simplified hemodynamic models used in combination with synthetic vascular networks do not preclude rigorous blood flow simulations. In this sense, the advantage of the presented model is that one can control all geometrical parameters and preview the results in real time. Additionally, as explained before, the presented framework is feasible to be more and more complicated adding or improving model details and additional specific conditions. In conclusion, although it includes some simplifications, the presented mathematical model which incorporates anatomic-based morphometric properties can potentially be used for addressing open questions regarding healthy and diseased cortical blood flow in the cerebral microvasculature.

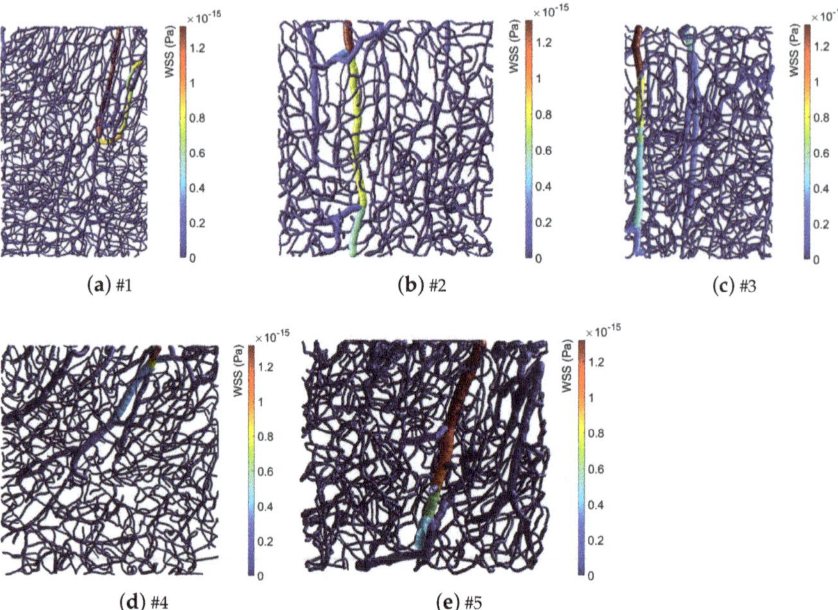

Figure 9. Computed endothelial shear stress of the 5 cerebrovascular regions of the murine cortex. The heat map represents by colors the average value of the shear stress per segment within the microcirculatory synthetic network.

5. Conclusions

We presented a comprehensive numerical tool for the generation and analysis of image-based artificial vascular networks. This novel methodology is based on tissue clearing, two-photons microscopy, image acquisition and treatment and 1D computational modeling. We have analyzed five cortical regions showing that the framework is capable of correctly synthesizing the cortex microvasculature from a morphological and hemodynamical point of view. Furthermore, the tissue clearing-based methodology is flexible and it can be applied to human brains that have bigger sizes. In contrast with previous studies, the methodology includes a physiological connection to the venous drainage and some morphological features such as curvature and tortuosity. The obtained results are in line with the literature so that the presented mathematical model allows studying the healthy cerebral microvasculature for computing the hemodynamics of the BBB. Lastly, the presented methodology is feasible to be applied as well to pathological cerebral microvasculature helping understanding the role of the hemodynamics in neurodegenerative diseases.

Author Contributions: Conceptualization, N.R., C.O.-d.-S., A.V. and M.M.; methodology, C.E., A.U., E.R.-d.-M., A.O., I.C.-D., N.R., A.V. and M.M.; software, S.M.; formal analysis, S.M., A.S., A.U., E.R.-d.-M. and A.O.; investigation, S.M., A.S., C.E., A.U., E.R.-d.-M., A.O., I.C.-D., N.R., C.O.-d.-S., A.V. and M.M.; resources, N.R., C.O.-d.-S., A.V. and M.M.; data curation, A.S., C.E., A.U., E.R.-d.-M., A.O., I.C.-D. and A.V.; writing—original draft preparation, S.M., A.S., I.C.-D. and M.M.; writing—review and editing, I.C.-D., N.R., C.O.-d.-S., A.V. and M.M.; visualization, S.M., A.S. and A.V.; supervision, N.R., C.O.-d.-S., A.V. and M.M.; project administration, N.R., C.O.-d.-S. and M.M.; funding acquisition, N.R., C.O.-d.-S. and M.M. All authors have read and agreed to the published version of the manuscript.

Funding: The study is financially supported by the Economic Department of the Navarre Government through the research project 0011-1411-2020-000025 (3D3B-AVATAR).

Institutional Review Board Statement: The study was conducted according to the guidelines of the Declaration of Helsinki, and approved by the Ethics Committee of the University of Navarra (076-19, 9 October 2019.

Data Availability Statement: Some of the presented data could be available from the corresponding author on reasonable request.

Acknowledgments: The support of the Instituto de Salud Carlos III (ISCIII) through the CIBER-BBN initiative is gratefully acknowledged.

Conflicts of Interest: The authors declare no conflict of interest.

References

1. Boileau, E.; Nithiarasu, P.; Blanco, P.J.; Muller, L.O.; Eikeland Fossan, F.; Hellevik, L.R.; Donders, W.P.; Huberts, W.; Willemet, M.; Alastruey, J. A benchmark study of numerical schemes for one-dimensional arterial blood flow modelling. *Int. J. Numer. Methods Biomed. Eng.* **2015**, *31*, e02732. [CrossRef]
2. Park, C.S.; Hartung, G.; Alaraj, A.; Du, X.; Charbel, F.T.; Linninger, A.A. Quantification of blood flow patterns in the cerebral arterial circulation of individual (human) subjects. *Int. J. Numer. Methods Biomed. Eng.* **2020**, *36*, e3288. [CrossRef] [PubMed]
3. Cassot, F.; Lauwers, F.; Fouard, C.; Prohaska, S.; Lauwers-Cances, V. A novel three-dimensional computer-assisted method for a quantitative study of microvascular networks of the human cerebral cortex. *Microcirculation* **2006**, *13*, 1–18. [CrossRef] [PubMed]
4. Lauwers, F.; Cassot, F.; Lauwers-Cances, V.; Puwanarajah, P.; Duvernoy, H. Morphometry of the human cerebral cortex microcirculation: general characteristics and space-related profiles. *Neuroimage* **2008**, *39*, 936–948. [CrossRef] [PubMed]
5. Lorthois, S.; Cassot, F.; Lauwers, F. Simulation study of brain blood ow regulation by intra-cortical arterioles in an anatomically accurate large human vascular network: Part I: methodology and baseline flow. *NeuroImage* **2011**, *54*, 1031–1042. [CrossRef]
6. Lorthois, S.; Cassot, F.; Lauwers, F. Simulation study of brain blood flow regulation by intra-cortical arterioles in an anatomically accurate large human vascular net- work. Part II: Flow variations induced by global or localized modifcations of arteriolar diameters. *NeuroImage* **2011**, *54*, 2840–2853. [CrossRef]
7. Schmid, F.; Reichold, J.; Weber, B.; Jenny, P. The impact of capillary dilation on the distribution of red blood cells in artificial networks. *Am. J. Physiol.—Heart Circ. Physiol.* **2015**, *308*, H733–H742. [CrossRef]
8. Schmid, F.; Tsai, P.S.; Kleinfeld, D.; Jenny, P.; Weber, B. Depth-dependent flow and pressure characteristics in cortical microvascular networks. *PLoS Comput. Biol.* **2017**, *13*, e1005392. [CrossRef]
9. Linninger, A.; Hartung, G.; Badr, S.; Morley, R. Mathematical synthesis of the cortical circulation for the whole mouse brain-part I. theory and image integration. *Comput. Biol. Med.* **2019**, *110*, 265–275. [CrossRef]
10. Kidoguchi, K.; andT. Mizobe, M.T.; Koyama, J.; Kondoh, T.; Kohmura, E.; Sakurai, T.; Yokono, K.; Umetani, K. In vivo X-ray angiography in the mouse brain using synchrotron radiation. *Stroke* **2006**, *37*, 1856–1861. [CrossRef]
11. Fang, Q.; Sakadzić, S.; Ruvinskaya, L.; Devor, A.; Dale, A.M.; D, A.B. Oxygen Advection and Diffusion in a Three Dimensional Vascular Anatomical Network. *Opt. Express.* **2008**, *16*, 17530–17541. [CrossRef]
12. Tsai, P.S.; Kaufhold, J.P.; Blinder, P.; Friedman, B.; Drew, P.J.; Karten, H.J.; Lyden, P.D.; Kleinfeld, D. Correlations of neuronal and microvascular densities in murine cortex revealed by direct counting and colocalization of nuclei and vessels. *J. Neurosci.* **2009**, *29*, 14553–14570. [CrossRef]
13. Gould, I.G.; Linninger, A.A. Hematocrit Distribution and Tissue Oxygenation in Large Microcirculatory Networks. *Microcirculation* **2015**, *12*, 1–18. [CrossRef]
14. Gould, I.G.; Tsai, P.; Kleinfeld, D.; Linninger, A. The capillary bed offers the largest hemodynamic resistance to the cortical blood supply. *J. Cereb. Blood Flow Metab.* **2017**, *37*, 52–68. [CrossRef]
15. Gagnon, L.; Smith, A.F.; Boas, D.A.; Devor, A.; Secomb, T.W.; Sakadzić, S. Modeling of Cerebral Oxygen Transport Based on In vivo Microscopic Imaging of Microvascular Network Structure, Blood Flow, and Oxygenation. *Front. Comput. Neurosci.* **2016**, *10*, 82. [CrossRef]
16. Hartung, G.; Badr, S.; Moeini, M.; Lesage, F.; Kleinfeld, D.; Alaraj, A.; Linninger, A. Voxelized simulation of cerebral oxygen perfusion elucidates hypoxia in aged mouse cortex. *PLoS Comput. Biol.* **2021**, *17*, 1008584. [CrossRef]
17. Chen, Q.; Jiang, L.; Li, C.; Hu, D.; Bu, J.; Cai, D.; Du, J. Haemodynamics-driven developmental pruning of brain vasculature in zebrafish. *PLoS Biol.* **2012**, *10*, e1001374. [CrossRef]
18. Blumers, A.L.; Yin, M.; Nakajima, H.; Hasegawa, Y.; Li, Z.; Karniadakis, G.E. Multiscale parareal algorithm for long-time mesoscopic simulations of microvascular blood flow in zebrafish. *Comput. Mech.* **2021**, *68*, 1131–1152. [CrossRef]
19. Roustaei, M.; In Baek, K.; Wang, Z.; Cavallero, S.; Satta, S.; Lai, A.; O'Donnell, R.; Vedula, V.; Ding, Y.; Marsden, A.L.; et al. Computational simulations of the 4D micro-circulatory network in zebrafish tail amputation and regeneration. *J. R. Soc. Interface* **2022**, *19*, 29210898. [CrossRef]

20. Anbazhakan, S.; Rios Coronado, P.E.; Sy-Quia, A.N.L.; Seow, A.; Hands, A.M.; Zhao, M.; Dong, M.L.; Pfaller, M.; Raftrey, B.C.; Cook, C.K.; et al. Blood flow modeling reveals improved collateral artery performance during the regenerative period in mammalian hearts. *Nat. Cardiovasc. Res.* **2022**, *1*, 775–790. [CrossRef]
21. Ronellenfitsch, H.; Katifori, E. Global optimization, local adaptation, and the role of growth in distribution networks. *Phys. Rev. Lett.* **2016**, *117*, 138301. [CrossRef] [PubMed]
22. Rocks, J.; Ronellenfitsch, H.; Liu, A.J.; Katifori, E. Limits of multifunctionality in tunable networks. *Proc. Natl. Acad. Sci. USA* **2019**, *116*, 2506–2511. [CrossRef] [PubMed]
23. Sangiorgi, S.; De Benedictis, A.; Protasoni, M.; Manelli, A.; Reguzzoni, M.; Cividini, A.; Dell'Orbo, C.; Tomei, G.; Balbi, S. Early-stage microvascular alterations of a new model of controlled cortical traumatic brain injury: 3D morphological analysis using scanning electron microscopy and corrosion casting. *J. Neurosurg.* **2013**, *118*, 763–774. [CrossRef] [PubMed]
24. Hartung, G.; Vesel, C.; Morley, R.; Alaraj, A.; Sled, J.; Kleinfeld, D.; Linninger, A. Simulations of blood as a suspension predicts a depth dependent hematocrit in the circulation throughout the cerebral cortex. *PLoS Comput. Biol.* **2018**, *14*, e1006549. [CrossRef]
25. Schmid, F.; Barrett, M.J.P.; Obrist, D.; Weber, B.; Jenny, P. Red blood cells stabilize flow in brain microvascular networks. *PLoS Comput. Biol.* **2019**, *15*, e1007231. [CrossRef]
26. Plouraboue, F.; Cloetens, P.; Fonta, C.; Steyer, A.; Lauwers, F.; Marc-Vergnes, J.P. X-ray high-resolution vascular network imaging. *J. Microsc.* **2004**, *215*, 139–148. [CrossRef]
27. Heinzer, S.; Kuhn, G.; Krucker, T.; Ulmann-Schuler, E.M.A.; Stampanoni, M.; Gassmann, M.; Marti, H.H.; Muller, R.; Vogel, J. Novel three-dimensional analysis tool for vascular trees indicates complete micro-networks, not single capillaries, as the angiogenic endpoint in mice overexpressing human VEGF(165) in the brain. *NeuroImage* **2008**, *39*, 1549–1558. [CrossRef]
28. Reichold, J.; Stampanoni, M.; Keller, L.; Buck, A.; Jenny, P.; Weber, B. Vascular graph model to simulate the cerebral blood flow in realistic vascular networks. *J. Cereb. Blood Flow Metab.* **2009**, *29*, 1429–1443. [CrossRef]
29. Waelchli, T.; Bisschop, J.; Miettinen, A.; Ulmann-Schuler, A.; Hintermueller, C.; Meyer, E.P.; Krucker, T.; Waelchli, R.; Monnier, P.P.; Carmeliet, P.; et al. Hierarchical imaging and computational analysis of three-dimensional vascular network architecture in the entire postnatal and adult mouse brain. *Nat. Protoc.* **2021**, *16*, 4564–4610. [CrossRef]
30. Reichold, J. Cerebral Blood Flow Modeling in Realistic Cortical Microvascular Networks. Ph.D. Thesis, Faculty of Science, ETH Zürich, Zürich, Switzerland, 2011.
31. Demene, C.; Tiran, E.; A.Sieu, L.; Bergel, A.; Gennisson, J.L.; Pernot, M.; De eux, T.; Cohen, I.; Tanter, M. 4D microvascular imaging based on ultrafast Doppler tomography. *NeuroImage* **2016**, *127*, 472–483. [CrossRef]
32. Hlushchuk, R.; Haberthuer, D.; Soukup, P.; Barré, S.F.; Khoma, O.Z.; Schittny, J.; Jahromi, N.H.; Bouchet, A.; Engelhardt, B.; Djonov, V. Innovative high-resolution microCT imaging of animal brain vasculature. *Brain Struct. Funct.* **2020**, *225*, 2885–2895. [CrossRef]
33. Ghavanati, S.; Yu, L.X.; Lerch, J.P.; Sled, J.G. A perfusion procedure for imaging of the muse cerebral vasculature by X-ray micro-CT. *J. Neurosci. Methods* **2014**, *221*, 70–77. [CrossRef]
34. Chung, K.; Wallace, J.; Kim, S.Y.; Kalyanasundaram, S.; Andalman, A.S.; Davidson, T.J.; Mirzabekov, J.J.; Zalocusky, K.A.; Mattis, J.; Denisin, A.K.; et al. Structural and molecular interrogation of intact biological systems. *Nature* **2013**, *497*, 332–337. [CrossRef]
35. Susaki, E.A.; Tainaka, K.; Perrin, D.; Kishino, F.; Tawara, T.; Watanabe, T.M.; Yokoyama, C.; Onoe, H.; Eguchi, M.; Yamaguchi, S.; et al. Whole-brain imaging with single-cell resolution using chemical cocktails and computational analysis. *Cell* **2014**, *157*, 726–739. [CrossRef]
36. Dodt, H.U.; Leischner, U.; Schierloh, A.; Jahrling, N.; Mauch, C.P.; Deininger, K.; Deussing, J.M.; Eder, M.; Zieglgansberger, W.; Becker, K. Ultramicroscopy: three-dimensional visualization of neuronal networks in the whole mouse brain. *Nat. Methods* **2007**, *4*, 331–336. [CrossRef] [PubMed]
37. Ertuerk, A.; Becker, K.; Jaehrling, N.; Mauch, C.P.; Hojer, C.D.; Egen, G.J.; Hellal, F.; Bradke, F.; Sheng, M.; Dodt, H.U. Three-dimensional imaging of solvent-cleared organs using 3DISCO. *Nat. Protoc.* **2012**, *7*, 1983–1995. [CrossRef]
38. Hama, H.; Kurokawa, H.; Kawano, H.; Ando, R.; Shimogori, T.; Noda, H.; Fukami, K.; Sakaue-Sawano, A.; Miyawaki, A. Scale: a chemical approach for fluorescence imaging and reconstruction of transparent mouse brain. *Nat. Neurosci.* **2011**, *14*, 1481–1488. [CrossRef]
39. Ke, M.T.; Fujimoto, S.; Imai, T. SeeDB: a simple and morphology-preserving optical clearing agent for neuronal circuit reconstruction. *Nat. Neurosci.* **2013**, *16*, 1154–1161. [CrossRef]
40. Renier, N.; Wu, Z.; Simon, D.J.; Yang, J.; Ariel, P.; Tessier-Lavigne, M. iDISCO: a simple, rapid method to immunolabel large tissue samples for volume imaging. *Cell* **2014**, *159*, 896–910 . [CrossRef]
41. Li, Y.; Xu, J.; Wan, P.; Yu, T.; Zhu, D. Optimization of GFP fluorescence preservation by a modified uDISCO clearing protocol. *Front. Neuroanat.* **2021**, *12*, 67. [CrossRef] [PubMed]
42. Chung, K.; Deisseroth, K. CLARITY for mapping the nervous system. *Nat. Methods* **2013**, *10*, 508–513. [CrossRef] [PubMed]
43. Tomer, R.; Ye, L.; Hsueh, B.; Deisseroth, K. Advanced CLARITY for rapid and high-resolution imaging of intact tissues. *Nat. Protoc.* **2014**, *9*, 1682–1697. [CrossRef]
44. D.Spence, R.; Kurtha, F.; Itoh, N.; Mongerson, C.R.L.; Wailes, S.H.; Peng, M.S.; MacKenzie-Graham, A.J. Bringing CLARITY to gray matter atrophy. *NeuroImage* **2014**, *101*, 625–632. [CrossRef]
45. Zheng, H.; Rinaman, L. Simplified CLARITY for visualizing immunofluorescence labeling in the developing rat brain. *Brain Structure and Function* **2016**, *2375–2383*, 221. [CrossRef]

46. Poplawsky, A.J.; Fukuda, M.; Bok-man Kang.; Hwan-Kim, J.; Suh, M.; S. G. Kim. Dominance of layer-specific microvessel dilation in contrast-enhanced high-resolution fMRI: Comparison between hemodynamic spread and vascular architecture with CLARITY. *NeuroImage* **2019**, *197*, 657–667. [CrossRef]

47. Martínez-Lorenzana, G.; Gamal-Eltrabily, M.; Tello-García, I.A.; Martínez-Torres, A.; Becerra-González, M.; González-Hernández, A.; Condés-Lara, M. CLARITY with neuronal tracing and immunofluorescence to study the somatosensory system in rats. *J. Neurosci. Methods* **2020**, *350*, 109048. [CrossRef]

48. Ren, Z.; Wu, Y.; Wang, Z.; Hu, Y.; Lu, J.; Liu, J.; Chen, Y.; Yao, M. CUBIC-plus: An optimized method for rapid tissue clearing and decolorization. *Biochem. Biophys. Res. Commun.* **2021**, *568*, 116–123. [CrossRef]

49. Tainaka, K.; Kubota, S.I.; Suyama, T.Q.; Susaki, E.A.; Perrin, D.; Ukai-Tadenuma, M.; Ukai, H.; Ueda, H.R. Whole-body imaging with single-cell resolution by tissue decolorization. *Cell* **2014**, *159*, 911–924. [CrossRef]

50. Susaki, E.A.; Tainaka, K.; Perrin, D.; Yukinaga, H.; Kuno, A.; Ueda, H.R. Advanced CUBIC protocols for whole-brain and whole-body clearing and imaging. *Nat. Protoc.* **2015**, *10*, 1709–1727. [CrossRef]

51. Pinheiro, T.; Mayor, I.; Edwards, S.; Joven, A.; Kantzer, C.G.; Kirkham, M.; Simon, A. CUBIC-f: An optimized clearing method for cell tracing and evaluation of neurite density in the salamander brain. *J. Neurosci. Methods* **2021**, *348*, 109002. [CrossRef]

52. Hasegawa, S.; Susaki, E.A.; Tanaka, T.; Komaba, H.; Wada, T.; Fukagawa, M.; R.Ueda, H.; Nangaku, M. Comprehensive three-dimensional analysis (CUBIC-kidney) visualizes abnormal renal sympathetic nerves after ischemia/reperfusion injury. *Kidney Int.* **2019**, *96*, 129–138. [CrossRef]

53. Murakami, T.C.; Mano, T.; Saikawa, S.; Horiguchi, S.A.; Shigeta, D.; Baba, K.; Sekiya, H.; Shimizu, Y.; Tanaka, K.F.; Kiyonari, H.; et al. A three-dimensional single-cell-resolution whole-brain atlas using CUBIC-X expansion microscopy and tissue clearing. *Nat. Neurosci.* **2018**, *21*, 625–637. [CrossRef] [PubMed]

54. Matsumoto, K.; Mitani, T.T.; Horiguchi, S.A.; Kaneshiro, J.; Murakami, T.C.; Mano, T.; Fujishima, H.; Konno, A.; Watanabe, T.M.; Hirai, H.; et al. Advanced CUBIC tissue clearing for whole-organ cell profiling. *Nature Protocols* **2021**, *14*, 3506–3537. [CrossRef] [PubMed]

55. Karc, R.; Neumann, F.; Neumann, M.; Schreiner, W. Staged growth of optimized arterial model trees. *Ann. Biomed. Eng.* **2000**, *28*, 495–511. [CrossRef] [PubMed]

56. Karch, R.; Neumann, F.; Podesser, B.K.; Neumann, M.; Szawlowski, P.; Schreiner, W. Fractal properties of perfusion heterogeneity in optimized arterial trees: a model study. *J. Gen. Physiol.* **2003**, *122*, 307–322. [CrossRef]

57. Schreiner, W.; Karch, R.; Neumann, M.; Neumann, F.; Roedler, S.M.; Heinze, G. Heterogeneous perfusion is a consequence of uniform shear stress in optimized arterial tree models. *J. Theor. Biol.* **2003**, *3*, 285–301. [CrossRef]

58. Sakadzic, S.; Roussakis, E.; Yaseen, M.A.; Mandeville, E.T.; Srinivasan, V.J.; Arai, K.; Ruvinskaya, S.; Devor, A.; Lo, E.H.; Vinogradov, S.A.; et al. Two-photon high-resolution measurement of partial pressure of oxygen in cerebral vasculature and tissue. *Nat. Methods* **2010**, *7*, 755–759. [CrossRef]

59. Blinder, P.; Shih, A.Y.; Rafie, C.; Kleinfeld, D. Topological basis for the robust distribution of blood to rodent neocortex. *Proc. Natl. Acad. Sci. USA* **2010**, *107*, 12670–12675. [CrossRef]

60. Keller, A.L.; Schuz, A.; Logothetis, N.K.; Weber, B. Vascularization of cytochrome oxidase-rich blobs in the primary visual cortex of squirrel and macaque monkeys. *J. Neurosci.* **2011**, *31*, 1246–1253. [CrossRef]

61. Kasischke, K.A.; Lambert, E.M.; Panepento, B.; Sun, A.; Gelbard, H.A.; Burgess, R.W.; Foster, T.H.; .; Nedergaard, M. Two-photon NADH imaging exposes boundaries of oxygen diffusion in cortical vascular supply regions. *J. Cereb. Blood Flow Metab.* **2011**, *31*, 68–81. [CrossRef]

62. Sherwin, S.J.; Franke, V.; Peiró, J.; Parker, K. One-dimensional modelling of a vascular network in space-time variables. *J. Eng. Math.* **2003**, *47*, 217–250. [CrossRef]

63. Boas, D.A.; Jones, S.R.; Devor, A.; Huppert, T.J.; Dale, A.M. A vascular anatomical network model of the spatio-temporal response to brain activation. *NeuroImage* **2008**, *40*, 1116–1129. [CrossRef] [PubMed]

64. Lorthois, S.; Cassot, F. Fractal analysis of vascular networks: insights from morphogenesis. *J. Theor. Biol.* **2010**, *262*, 614–633. [CrossRef] [PubMed]

65. Peyrounette, M.; Davit, Y.; Quintard, M.; Lorthois, S. Multiscale modelling of blood ow in cerebral microcirculation: details at capillary scale control accuracy at the level of the cortex. *PLoS ONE* **2018**, *13*, e0189474. [CrossRef] [PubMed]

66. Linninger, A.A.; Gould, I.G.; Marinnan, T.; Hsu, C.Y.; Chojecki, M.; Alaraj, A. Cerebral Microcirculation and Oxygen Tension in the Human Secondary Cortex. *Ann. Biomed. Eng.* **2013**, *41*, 2264–2284. [CrossRef] [PubMed]

67. Hsu, C.Y.; Schneller, B.; Alaraj, A.; Flannery, M.; Zhou, X.J.; Linninger, A. Automatic recognition of subject-specifc cerebrovascular trees. *Magn. Reson. Med.* **2016**, *77*, 398–410. [CrossRef]

68. Hsu, C.Y.; Ghaffari, M.; Alaraj, A.; Flannery, M.; Zhou, X.J.; Linninger, A. Gap-free segmentation of vascular networks with automatic image processing pipeline. *Comput. Biol. Med.* **2017**, *82*, 29–39. [CrossRef]

69. Buades, A.; Coll, B.; Morel, J.M. A Non-Local Algorithm for Image Denoising. *IEEE Comput. Soc. Conf. Comput. Vis. Pattern Recognit.* **2005**, *2*, 60–65.

70. Hasinoff, S.W. Photon, Poisson Noise. In *Computer Vision*; Ikeuchi K., Ed.; Springer: Boston, MA, USA, 2014.

71. Zana, F.; Klein, J.C. Segmentation of vessel-like patterns using mathematical morphology and curvature evaluation. *IEEE Trans. Image Process.* **2001**, *10*, 1010–1019.

72. Bradley, D.; Roth, G. Adapting Thresholding Using the Integral Image. *J. Graph. Tools* **2007**, *12*, 13–21. [CrossRef]

73. Soille, P. *Morphological Image Analysis: Principles and Applications*; Springer: Berlin/Heidelberg, Germany, 1999.
74. Lee, T.C.; Kashyap, R.L.; Chu, C.N. Building skeleton models via 3-D medial surface/axis thinning algorithms. *Comput. Vis. Graph. Image Process.* **1994**, *56*, 462–478. [CrossRef]
75. Kerschnitzki, M.; Kollmannsberger, P.; Burghammer, M.; Duda, G.N.; Weinkamer, R.; Wagermaier, W.; Fratzl, P. Architecture of the osteocyte network correlates with bone material quality. *J. Bone Miner. Res.* **2013**, *28*, 1837–1845. [CrossRef]
76. Pries, A.R.; Secomb, T.W. *Blood Flow in Microvascular Networks. In Microcirculation*; Elsevier: Amsterdam, The Netherlands, 2008; pp. 3–36.
77. Schmid, F. Cerebral Blood Flow Modeling with Discrete Red Blood Cell Tracking Analyzing Microvascular Networks and Their Perfusion. Ph.D. Thesis, Faculty of Science. ETH Zurich, Zürich, Switzerland, 2017.
78. Pries, A.R.; Secomb, T.W.; Gebner, T.; Sperandio, M.B.; Gross, J.F.; Gaehtgens, P. Resistance to Blood Flow in Microvessels In Vivo. *Circ. Res.* **1994**, *75*, 904–915. [CrossRef]
79. Shapiro, H.M.; Stromberg, D.D.; Lee, D.R.; Wiederhielm, C.A. Dynamic pressures in the pill arterial microcirculation. *Am. J. Physiol.-Leg. Content* **1971**, *221*, 279–283. [CrossRef]
80. Bullit, E.; Gerig, G.; Pizer, S.M.; Lin, W.; Aylward, S.R. Measuring tortuosity of the intracerebral vasculature from MRA images. *IEEE Trans. Med. Imaging* **2003**, *22*, 1163–1171. [CrossRef]
81. Su, S.W.; Catherall, M.; Payne, S. The influence of network structure on the transport of blood in the human cerebral microvasculature. *Microcirculation* **2012**, *19*, 175–187. [CrossRef]
82. Lee, R.M. Morphology of cerebral arteries. *Pharmacol. Ther.* **1995**, *66*, 149–173. [CrossRef]
83. Olufsen, M.S. A structured tree outflow condition for blood flow in larger systemic arteries. *Am. J. Physiol.* **1999**, *276*, H257–H268. [CrossRef]
84. Olufsen, M.S.; Peskins, C.S.; Kim, W.Y.; Pedersen, E.M.; Nadim, A.; Larsen, J. Numerical simulation and experimental validation of blood flow in arteries with structured tree outflow conditions. *Ann. Biomed. Eng.* **2000**, *28*, 1281–1299. [CrossRef]
85. Malvè, M.; Chandra, S.; García, A.; Mena, A.; Martínez, M.A.; Finol, E.A.; Doblaré, M. Impedance-based outflow boundary conditions for human carotid haemodynamics. *Comput. Methods Biomech. Biomed. Eng.* **2014**, *17*, 1248–1260. [CrossRef]
86. El-Bouri, W.K.; Payne, S.J. Multi-scale homogenization of blood ow in 3-dimensional human cerebral microvascular networks. *J. Theor. Biol.* **2015**, *380*, 40–47. [CrossRef] [PubMed]
87. Bui, A.V.; Manasseh, R.; K. Liffman, I.D.Šutalo. Development of optimized vascular fractal tree models using level set distance function. *Med. Eng. Phys.* **2010**, *32*, 790–794. [CrossRef] [PubMed]

 *mathematics*

Article

Quantitative Criteria for the Degree of Pathological Remodeling of the Aortic Duct

Eugene Talygin [1,*], Alexander Gorodkov [1,2], Teona Tibua [1] and Leo Bockeria [1]

[1] Bakulev National Medical Research Center for Cardiovascular Surgery, Russian Academy of Medical Sciences, 121552 Moscow, Russia

[2] Lomonosov Institute of Fine Chemical Technologies, Russian Technological University, Vernadsky Prospect, 86, 119571 Moscow, Russia

* Correspondence: skalolom@gmail.com

Citation: Talygin, E.; Gorodkov, A.; Tibua, T.; Bockeria, L. Quantitative Criteria for the Degree of Pathological Remodeling of the Aortic Duct. *Mathematics* **2022**, *10*, 4773. https://doi.org/10.3390/math10244773

Academic Editor: Mauro Malvè

Received: 30 October 2022
Accepted: 10 December 2022
Published: 15 December 2022

Publisher's Note: MDPI stays neutral with regard to jurisdictional claims in published maps and institutional affiliations.

Abstract: Analysis of the properties of the aorta was carried out by numerous researchers using several parameters. However, the general laws of change in the dynamic geometry of the aortic flow channel in connection with the hydrodynamics of the swirling blood flow have not been studied properly. Therefore, at present, attempts to correct various diseases are carried out based on the location of the aneurysm, and not in accordance with the general patterns of changes in the dynamic geometry of the entire aortic channel. For a proper understanding of the aortic flow channel remodeling mechanisms, it is necessary to determine the quantitative parameters that formalize the geometry of this channel. The geometric shape of the aorta primarily depends on the hydrodynamics of the flow inside the aortic flow channel, which is the only source of force impact on its walls. The main result of the present study was that we obtained the new quantitative parameters that characterize the normal aorta and the degree of its shape deviations caused by pathological changes of the aortic duct. These parameters were calculated based on the software processing of the three-dimensional aortic reconstruction in normal conditions and in the case of differently localized aortic aneurysm.

Keywords: potential swirling flow; navier-stokes equations; unsteady swirling flow; tornado-like jets

MSC: 76Z05

1. Introduction

Any perturbations in the blood stream, which is a biologically active fluid flowing in a channel with biologically active walls, inevitably lead to the activation of the body's defense systems and/or damage to the walls of the flow channel. Therefore, the main hypothesis is that the blood flow in the central parts of the circulatory system (heart and great vessels) is carried out without the formation of separation and stagnant zones, that is, it is a potential flow. In such a flow, by definition, any types of interaction are minimized both in the flow core and on the walls of the flow channel.

The cellular composition of the walls [1,2], their biomechanical properties [1], the distribution of velocities and shear stresses [3,4], and metabolism [5,6] were studied. Many model studies of the aorta have been carried out [7]. However, by present days there are no proper quantitative criterions that allow one to formalize the degree of aortic duct pathological remodeling. In the present study these parameters have been obtained by considering analytical solutions for the velocity vector of swirling blood flow in the heart and great vessels.

2. Materials and Methods

In previously published works [8–10], it has been shown that the dynamic geometry of the heart and great vessels corresponds with a high degree of accuracy to the direction of streamlines of swirling flows described by exact solutions of the Navier-Stokes and continuity equations for a class of self-organizing tornado-like flows of a viscous fluid. These solutions were obtained in 1986 by G.I. Kiknadze and Yu.K. Krasnov [11] and are generally expressed by relation (1).

$$\begin{cases} u_r = C_0(t)r + \dfrac{C_1}{r} \\ u_z = -2C_0(t)z + C_2(t) \\ u_\varphi(r,t) = \dfrac{1}{r} * \left(A_1 + A_2 \Gamma \left(1 + \dfrac{C_1}{2v}, \alpha(0) * r^2 \right) \right), \end{cases} \tag{1}$$

Here $C_0(t)$ and $C_2(t)$ are time-dependent functions, A_1, A_2, and C_1 are constants, and $\alpha(t)$ is a function that has the following form:

$$\alpha(t) = \frac{e^{-2 \int_0^t C_0(\tau_1)d\tau_1}}{B_1 - A \int_0^t e^{-2 \int_0^\tau C_0(\tau_2)d\tau_2} d\tau_1}$$

Here B_1—is the arbitrary constant.

In the steady state, the swirling flow under consideration can be described by the relations for the Burgers vortex [12]:

$$\begin{cases} u_r = -C_0 * r \\ u_z = 2C_0 * z \\ u_\varphi = \dfrac{\Gamma_0}{2\pi r} * \left(1 - e^{-\frac{C_0^2 * r}{2v}} \right) \end{cases} \tag{2.1}$$

However, it is known that the blood flow is roughly unsteady and occurs in a pulsating regime. Previously, it has been shown [13,14] that the geometry of the flow channel during the entire cardiac cycle corresponds to the direction of the streamlines described by these solutions. Therefore, we used these solutions as quasi-stationary, if only $C_0(t)$ and $\Gamma_0(t)$ depend on time. In this case, relation (2.1) may be transformed as follows:

$$\begin{cases} u_r = -C_0(t) * r \\ u_z = 2C_0(t) * z \\ u_\varphi = \dfrac{\Gamma_0(t)}{2\pi r} * \left(1 - e^{-\frac{C_0^2(t) * r}{2v}} \right) \end{cases} \tag{2.2}$$

Here, only the azimuthal velocity component depends on the viscosity; therefore, the modulus of this component decreases with the evolution of the flow.

The geometry of the flow channel must correspond to the direction of the streamlines of the swirling flow inside. Expression (2.2) includes two functions of time—$C_0(t)$ and $\Gamma_0(t)$. To use these solutions as quasi-stationary in a non-stationary pulsating flow, the cardiac cycle was considered as a sequence of discrete states, in each of which the flow is described by relations (2.1). In this case, the instantaneous values of $C_0(t)$ and $\Gamma_0(t)$ are described by the measured geometric characteristics of the flow channel.

The instantaneous position of the streamlines of the considered flow in the longitudinal-radial projection is described by the following relation:

$$zr^2 = \beta(t) \tag{3}$$

$\beta(t)$—a time-dependent function, the instantaneous value of which is determined by the choice of values C_0 and Γ_0 at a given time.

Streamlines in a fixed position of the flow channel of the heart and aorta in the axial-radial projection are described by the following expression:

$$\varphi = \frac{\Gamma_0}{2\pi C_0} * \left(\frac{1}{r} + \sqrt{\frac{\pi C_0}{2v}} * \int_{0}^{\sqrt{\frac{C_0}{2v}}r} e^{-t^2} dt - \frac{1}{r} * e^{-\frac{C_0 r^2}{2v}} \right) + \varphi_0 \tag{4}$$

As has been shown already, the aorta retains the shape of a converging canal throughout the entire cardiac cycle [13]. According to our assumptions, the pulse oscillations of the aortic wall are much less than the geometric transformations that occur in the process of remodeling. Therefore, the aortic flow channel was normally considered as a tube with almost constant geometric characteristics. As a result, fluctuations in the values of the characteristics C_0 and Γ_0 were considered negligible.

Since the streamlines are an evolution of the hyperbolic helix (4), continuous flow along such streamlines is possible only in a channel that has the form of a lower order hyperbolic right-handed helix (with fewer turns per aortic length). To approximate the shape of the aortic flow channel, a flat hyperbolic spiral of the following form was used:

$$r = \left(\frac{a}{\varphi} \right)^{power} + bias$$

Here (r, φ) are the radial and longitudinal coordinates, and $(a, power, bias)$ are the parameters by which the approximation is carried out.

As a result, the obtained quantitative characteristics of the shape of the aorta are composed of two components, which are the values taken by the constants C_0 and Γ_0 and the parameters describing the helical properties of the aortic flow channel.

Let us consider 2 sections of the aortic flow channel outlying at a distance z_1 and z_2 respectively from the plane of the aortic valve. The corresponding aortic radiuses on these sections are r_1 and r_2. For $z_1 < z_2$ it follows that $r_1 > r_2$ due to the convergence of the flow channel. Since the specific spatial location of the point of origin of the cylindrical coordinate system in which the parameters of the aortic flow channel were calculated is unclear, a fictitious point of origin was introduced into consideration, which is located at a distance z0 from the plane of the aortic valve inside the cavity of the left ventricle. This point can change its position during the entire cardiac cycle, however, within the framework of this work, it was assumed to be immobile. It can be considered as the point of origin of the swirling blood flow. Then relation (3) for the sections under consideration can be written as follows:

$$(z_0 + z_1) * r_1^2 = (z_0 + z_2) * r_2^2 \Leftrightarrow z_0 = \frac{z_2 r_2^2 - z_1 r_1^2}{r_1^2 - r_2^2} \tag{5}$$

From (2.2) it can be derived that the contribution of the azimuthal component in the total swirling flow velocity vector decreases with the distance from the origin point. Thus, if a certain section of the aortic flow channel S_1 is closer to the point of origin than the section S_2, the corresponding value of the modulus of the azimuth component can be expressed as follows: $u_{\phi,1} > u_{\phi,2}$. Accordingly, the minimum value of $|u_\phi|$ occurs near the end of the aorta. For simplicity it can be stated as $|u_\phi| = 0$. Then the total velocity vector through the section S_2 at the end of the aorta will be written as follows:

$$u_2 = \sqrt{u_z^2 + u_r^2} = C_0 \sqrt{4(z_0 + z_2)^2 + r_2^2} \Leftrightarrow C_0 = \frac{u_2}{\sqrt{4(z_0 + z_2)^2 + r_2^2}} \tag{6}$$

Considering the section S_1, the derived expression for the linear velocity of the swirling flow through this section can be written as follows:

$$u_1 = \sqrt{u_z^2 + u_r^2 + u_\varphi^2} \Leftrightarrow u_\varphi^2 = u_1^2 - u_z^2 - u_r^2 \Leftrightarrow \left(\frac{\Gamma_0}{2\pi r} * \left(1 - e^{-\frac{C_0^2 * r_1}{2v}} \right) \right)^2$$
$$= u_1^2 - 4C_0^2 (z_0 + z_1)^2 - C_0^2 r_1^2$$

Expanding the brackets and simplifying the expression above, we get:

$$\Gamma_0 = \frac{\sqrt{4\pi r_1^2 u_1^2 - 4\pi r_1^2 C_0^2 \left(4(z_0 + z_1)^2 + r_1^2 \right)}}{\left(1 - e^{-\frac{C_0^2 * r_1}{2v}} \right)} \tag{7}$$

For the correct reconstruction of a parametric spatial curve approximating the shape of the aortic flow channel, it is necessary to reconstruct the central line of the aorta first.

The central line is a spatial curve drawn between two planes cutting through an extended cavity in such a way that the distance from this line to the boundaries of the cavity is the maximum.

To construct the central line, 2 planes were used that cut the aortic flow channel perpendicular to the direction of the swirling blood flow. For all studied aortas, one of these planes is the plane that includes the aortic valve. The algorithm for choosing another plane depends on the type of aorta. For a relatively healthy aorta, the second plane should coincide with the section of the aorta in a bifurcation zone; for an aorta with a pathological disorder of the vascular bed, the plane was chosen approximately at the level of 2/3 of the aorta length, counting from the aortic valve. This is due to a large distortion of the geometry of the aortic flow channel in the abdominal region, which is associated with serious errors in the reconstruction of the required central line. Then, points were fixed on the selected planes in the central region of the cutting planes (one for each plane). These points (labeled as p_1 and p_2) were used to construct the required center line.

To determine the central line, it is necessary to determine the trajectory $C = C(s)$ connecting the selected points p_1 and p_2. This trajectory should ensure the minimization of the value of the following functional:

$$I_{centerline}(C) = \int_{C^{-1}(p_1)}^{C^{-1}(p_2)} F(C(s)) ds \tag{8}$$

In the written expression, $F(x)$ is a certain scalar field, the value of which at points lying closer to the center of the cavity is less than at points far from the center. The simplest example is a function whose value at a point is inversely proportional to the distance from this point to the boundary of the cavity. Such a function can be represented by the following relation:

$$DT(x) = \min_{y \in d\Omega} (|x, y|) \tag{9}$$

In this relation, $|(x, y)|$ is the Euclidean distance from the point x to the point y, $d\Omega$ is the boundary of the cavity Ω, corresponding to the radius of the channel at this point. Choosing the scalar field $F(x) = DT^{-1}(x)$ defined by expression (8), the central lines will need to be located on the medial axis of the given cavity Ω. The medial axis of the cavity is defined as the set of ball centers inscribed in a given cavity, where the size of the inscribed ball will be the maximum only if it in turn is not inscribed in another such ball. The construction of such a medial axis and the central line associated with it is carried out using the Voronoi diagram [15]. In the three-dimensional case, the Voronoi diagram is a surface composed of convex polygons whose vertices are the centers of the maximum inscribed balls.

The construction of the Voronoi diagram was performed by Delaunay triangulation [16], which was accompanied by the removal of polygons that partially fell out of the given cavity. This method allowed us to reformulate the problem described by expression (6) in the form of an eikonal equation, a non-linear partial differential equation:

$$|\nabla T(x) = F(x)| \tag{10}$$

The boundary condition for the written equation is $T(p_1) = 0$. Such an equation can be solved using the fast sweep method. As soon as the solution of equation (8) was obtained for the entire Voronoi diagram, the backpropagation method was used to construct a trajectory from point p_1 to point p_2 in the direction of the maximum decrease in the value of the scalar field (i.e., in the direction of the medial axis of the cavity).

Central lines have been constructed using The Vascular Modelling Toolkit (vmtk).

All required data engineering and math computations has been done using standard python packages—pandas, numpy, scipy, scikit-learn and vmtk.

3. Results

3.1. Computation of C_0 and Γ_0

Eighteen aortic conditions were studied in 14 patients, of which 4 had normal aortas, 6 had lesions located in the proximal regions, and 4 had lesions localized in the distal regions. Three-dimensional reconstructions of the aortas were obtained using MSCT.

Each reconstruction is a surface in STL format. The surfaces presented in this format are a set of conjugated triangles with indication of the normal to them. The following algorithm was used to calculate the C_0 and Γ_0 values for these reconstructions:

1. The initial matrix with the coordinates of the triangles that make up the surface of the aortic duct was presented as a matrix A with dimensions [num_rows, num_features]. Here num_rows is the total number of triangles that make up the surface, and num_features is the number of points representing each such triangle (num_features = 9 for 3D space).

2. The computed matrix A was projected onto a plane by the PCA method (there was a decrease in the dimension from 9 to 2). Next, the length of the 2d-projection of the aortic flow channel was measured, and on the distance from the beginning of the aorta by 10% and 90% of the entire length of the aorta, the aortic radius was measured. Thus, two pairs of values (z_1, r_1) and (z_2, r_2) were obtained, which are important for C_0 and Γ_0 computation according to expressions (5) and (6).

Normally, for patients without obvious aortic pathology, the streamlines can be plotted as follows (Figure 1a,b).

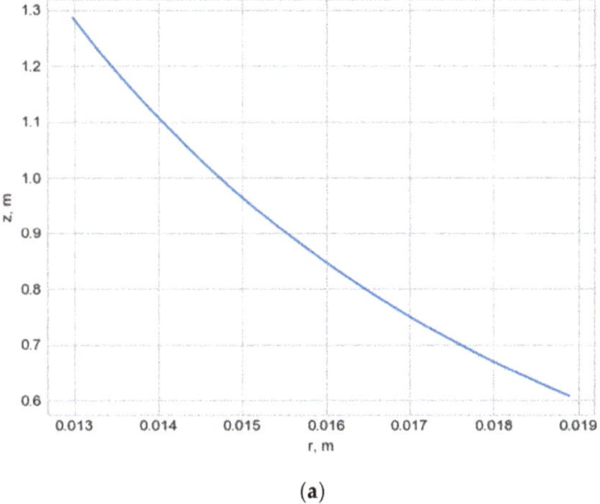

(a)

Figure 1. *Cont.*

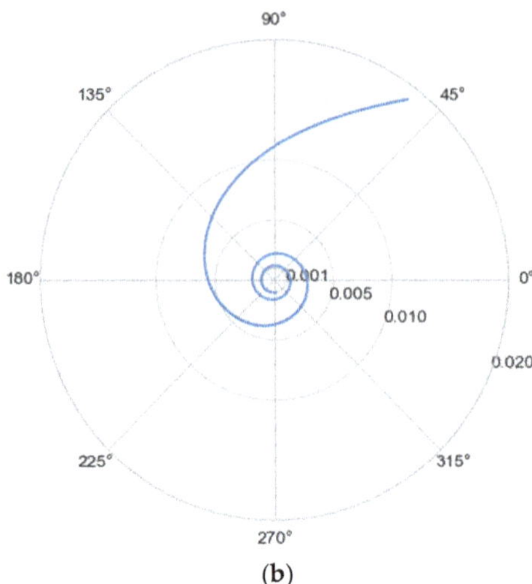

(**b**)

Figure 1. (**a**) Streamlines of the swirling flow in the normal aorta in longitudinal-radial projection. (**b**) Streamlines of the swirling flow in the normal aorta in the axial-radial projection.

Figure 1b shows a plot of C_0 and Γ_0 values depending on the location of aortic pathology. Table 1 shows the corresponding values of the parameters C_0 and Γ_0.

Table 1. Values of constants C_0 and Γ_0 for the studied aortas, 'loc'—localization of aortic lesion (down—distal aneurysm, up—proximal aneurysm, norm—no pathology).

Name	C_0 (s^{-1})	Γ_0 (m^2/s)	loc
lar_s	0.301	0.046	up
hom_a	0.357	0.058	up
are_d	0.537	0.073	down
ino_d	0.407	0.070	down
mir_s	0.460	0.037	norm
lar_d	0.307	0.048	up
she_a	0.243	0.071	down
mal_a	0.135	0.076	down
zag_a	0.368	0.041	norm
mir_d	0.427	0.037	norm
are_s	0.378	0.068	down
pav_a	0.218	0.071	down
bor_a	0.545	0.089	down
gor_d	0.314	0.041	norm
poz_a	0.255	0.044	norm
bar_a	0.564	0.039	up
ino_s	0.396	0.072	down
bru_a	0.484	0.052	up

As can be seen in Figure 2, a pair of C_0 and Γ_0 values can serve as a quantitative criterion for identifying aortic pathology. All three considered cases (pathology in the descending aorta, pathology in the ascending aorta and the norm) are linearly separable.

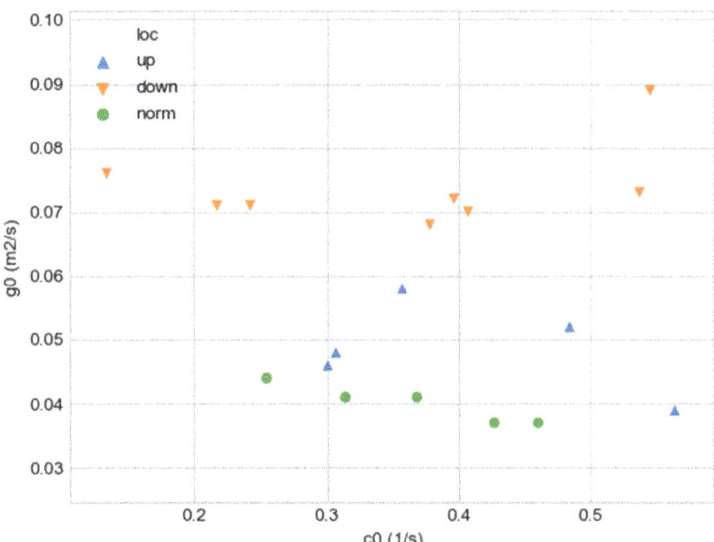

Figure 2. Comparison of C_0 and Γ_0 values for aortas from Table 1. Blue triangles indicate aortas with pathology in the proximal sections, orange triangles—aortas with pathology in the distal sections, and green circles—aortas without severe pathology.

In the case of a pathological disturbance of the vascular bed in the descending section, the value of the circulation Γ_0 of the swirling flow increases significantly. At the same time, a C_0 raising can be observed followed by the increase in the transverse gradients of the blood flow velocity. Fluctuations of C_0 and Γ_0 values may reflect the action of compensatory and regulatory mechanisms of the cardiovascular system. However, the actions of these mechanisms are inevitably associated with excessive energy consumption to maintain the flow structure and can also lead to an increased force impact on the aortic wall.

In the case of pathology in the ascending region, one can observe a slight increase in the Γ_0 value and a relatively small (compared with the pathology of the descending region) increase in C_0 value.

However, the parameters C_0 and Γ_0 do not unequivocally allow the establishment of the fact of pathological remodeling of the aortic duct. In Figure 1a,b, the dots representing aortas with distal damage lie very close to the dots corresponding to the normal aorta.

3.2. Approximation of the Aorta Flow Channel by a Parametric Spiral

Figure 3a,b show the result of plotting the central line for an aorta with no obvious pathological disorders and an aorta with pathology, respectively.

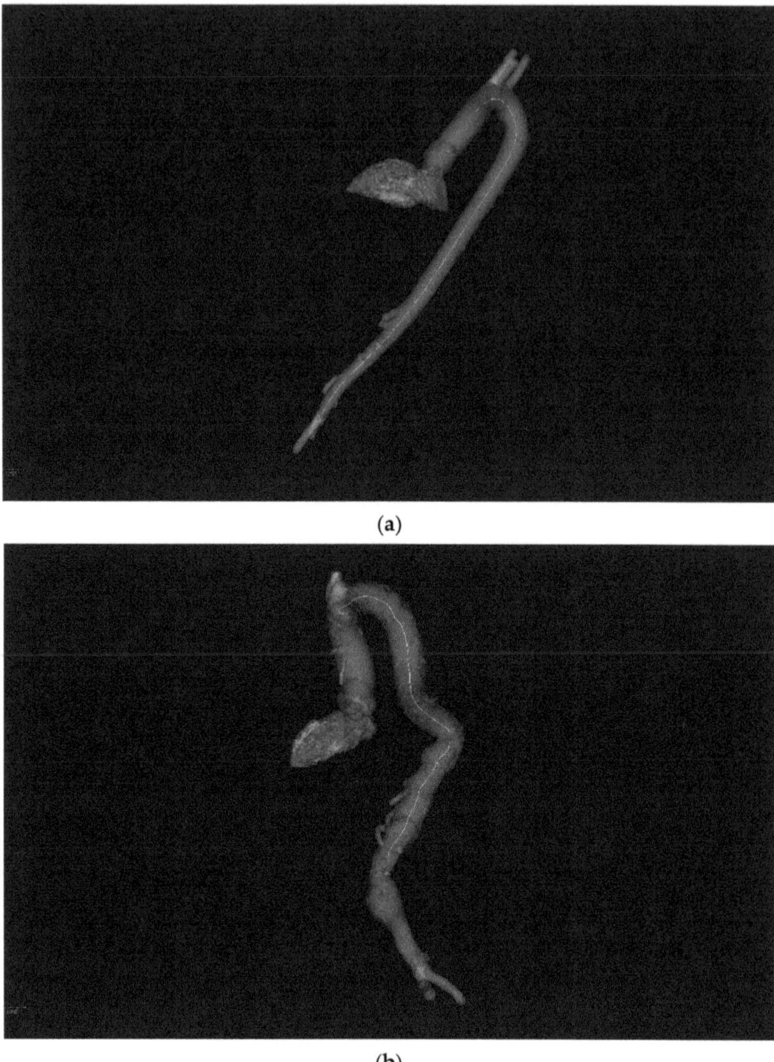

(a)

(b)

Figure 3. (**a**) Three-dimensional STL—reconstruction of the flow channel of the aorta without severe pathology. A central line is built inside the flow channel. (**b**) Three-dimensional STL—reconstruction of the flow channel of the aorta with severe pathology of the distal sections. A central line is built inside the flow channel.

Each center line is a matrix with dimensions [num_points, 3], where num_points is the number of points in the center line, and 3 is the number of spatial dimensions in the Cartesian coordinate system. The resulting lines must be approximated by some spiral curves to obtain the characteristic parameters. The search for spirals was carried out in the class of hyperbolic spirals described by the following relation in the polar coordinate system:

$$r = \left(\frac{a}{\varphi}\right)^{power} + bias \tag{11}$$

In relation (11), the unknown parameters are $(a, \ power, \ bias)$.

The hyperbolic class of spirals was chosen since the streamlines of the swirling blood flow in the aorta in the radial-axial projection are described by a hyperbolic spiral, and the similarity principle indicates that the flow channel, in which the flow evolves without the formation of separation and stagnant zones, must have similar geometry.

For each central line, an approximating spiral was constructed in accordance with the following algorithm:

1. The original [num_points, 3] centerline was projected onto a plane using the PCA method to obtain a [num_points, 2] matrix, where 2 is the (x, y) coordinates. The resulting projection is labeled centerline_proj.
2. The coordinates of the resulting line centerline_proj were converted to the polar coordinate system (r, φ) using the following expressions:

$$r = \sqrt{x^2 + y^2}$$
$$\varphi = atan2(y, \ x)$$

Here $atan2(y, \ x)$—2-argument arctangent used to translate Cartesian coordinates into polar coordinates. This arctangent can be stated as follows:

$$atan2(y, \ x) = \begin{cases} 2arctan\left(\dfrac{y}{\sqrt{x^2+y^2}+x}\right), \ if \ x > 0 \ and \ y \neq 0, \\ \pi, \ if \ x < 0 \ and \ y = 0, \\ undefined, \ if \ x = 0 \ and \ y = 0 \end{cases}$$

Using the least squares method, the parameters $(a, \ power, \ bias)$ from expression (11) were selected in such a way that the polar representation of the centerline_proj line is most accurately described by the parametric hyperbolic spiral (11).

The obtained parameters of the approximating spiral $(a, \ power, \ bias)$, the coefficient of determination R^2, and the standard deviation (mae) of the real line centerline_proj from the approximating spiral were entered in the resulting table.

Based on the calculated five parameters $(a, \ power, \ bias, \ R^2, \ mae)$. 2 synthetic parameters were calculated by the PCA method (*feature_1, feature_2*). These parameters store all the necessary information about the quantitative differences in the parameters $(a, \ power, \ bias, \ R^2, \ mae)$ for the normal aorta and in the presence of a pathological change in the vascular bed and allow visual interpretation of these differences.

The application of the formulated algorithm for one central line looks like this:

The plotting of the initial three-dimensional central line centerline (white line) is performed in Figure 4.

Figure 4. STL-reconstruction of the aortic flow channel for a patient without severe pathology and the central line inside the channel.

The graph of the line centerline_proj was plotted, with the projection of the central line onto the plane in Cartesian and polar coordinate systems (Figure 5).

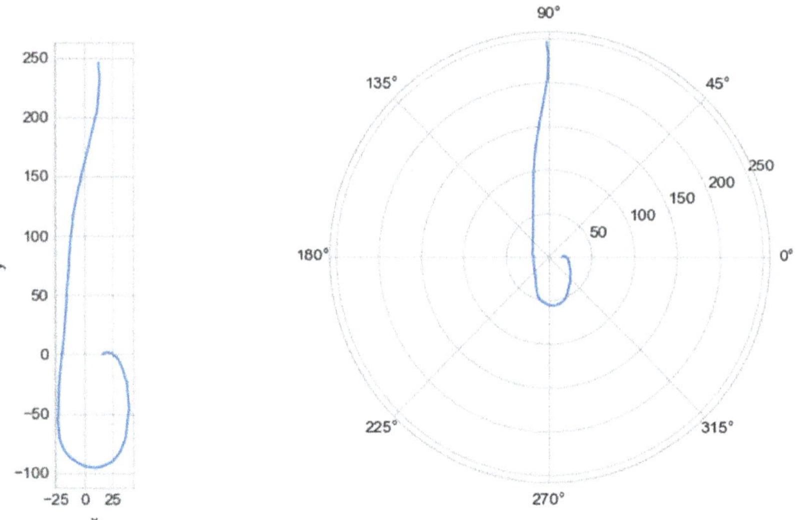

Figure 5. On the left—the projection of the central line in the Cartesian coordinate system, on the right—in the polar coordinate system.

Using the least squares method, the parameters of the approximating spiral were calculated. In Figure 6, the original curve and its approximation are plotted in the polar coordinate system.

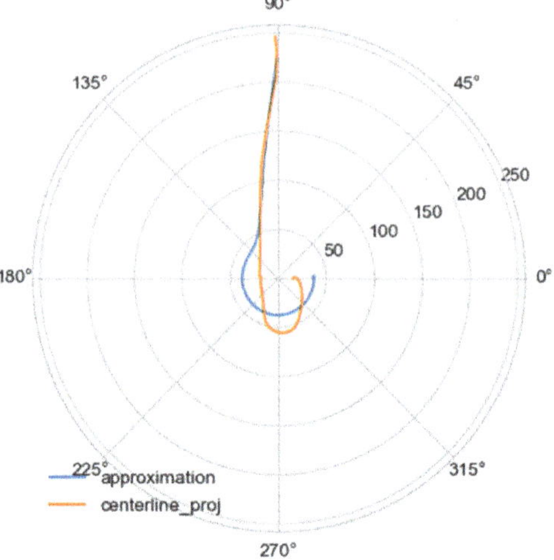

Figure 6. Comparison of a flat projection of the central line of the aorta and its approximation by a spiral in polar coordinates. The orange line is the projection of the central line, the blue line is the fitting curve.

Approximating spirals for all central lines were constructed using an identical algorithm. The results obtained are shown in Table 2.

Table 2. Comparison of the geometric characteristics of the approximation of the central line of the aorta.

Name	a	Power	bias	mae	R^2	loc	feature_1	feature_2
are_s	2.896	7.893	-16.042	14.877	0.885	down	8.541	1.131
ino_s	4.008	5.256	-3.502	22.912	0.798	down	-3.933	9.622
ino_d	3.894	5.215	1.000	23.025	0.779	down	-8.834	9.775
she_a	3.125	6.702	0.993	15.979	0.894	down	-8.471	2.855
are_d	2.579	9.392	1.000	14.694	0.881	down	-8.522	0.384
bor_a	2.703	8.728	-131.587	14.359	0.925	down	124.059	-0.879
mal_a	3.009	7.190	1.000	13.648	0.892	down	-8.493	0.580
pav_a	0.426	17.357	1.000	18.678	0.810	down	-8.627	-0.195
mir_s	2.509	11.196	1.000	8.989	0.973	norm	-8.566	-5.461
bar_a	2.276	12.468	3.263	7.553	0.968	norm	-10.852	-7.325
zag_a	2.626	9.523	1.000	8.622	0.973	norm	-8.543	-4.959
mir_d	2.470	11.583	7.610	9.403	0.971	norm	-15.179	-5.227
gor_d	2.097	17.613	1.000	10.575	0.951	norm	-8.654	-7.230
poz_a	2.735	9.002	0.931	14.640	0.908	up	-8.447	0.539
hom_a	2.790	8.040	1.270	18.492	0.806	up	-8.760	4.365
lar_s	2.441	11.655	1.034	17.337	0.908	up	-8.580	1.574
lar_d	2.475	10.675	1.000	15.498	0.918	up	-8.538	0.452

(a, power, bias)—parameters of the approximating hyperbolic spiral for the projection of the central line from expression (11), mae—the value of the standard deviation of the approximating curve from the projection of the central line, R^2—coefficient of determination for a specific approximation, feature_1, feature_2—derived parameters, obtained from $\left(\text{a, power, bias, } R^2\text{, mae}\right)$ by PCA method.

As can be seen from the table, the value of the coefficient of determination R^2 for aortas without noticeable remodeling is higher, and the value of the standard deviation of the approximation *mae* is lower than for aortas with pathological disorders of the vascular bed. For normal aortas, the coefficient of determination exceeds 0.95, which indicates high approximation accuracy. For aortas with severe pathology of the vascular bed, the coefficient of determination exceeds 0.77, which indicates the significance of the chosen approximation method.

In Figure 6 was depicted approximation for normal aorta denoted as mir_s in Table 2. Other approximations are depicted on Figures 7–22 The notations and lines color are the same, as on Figure 6.

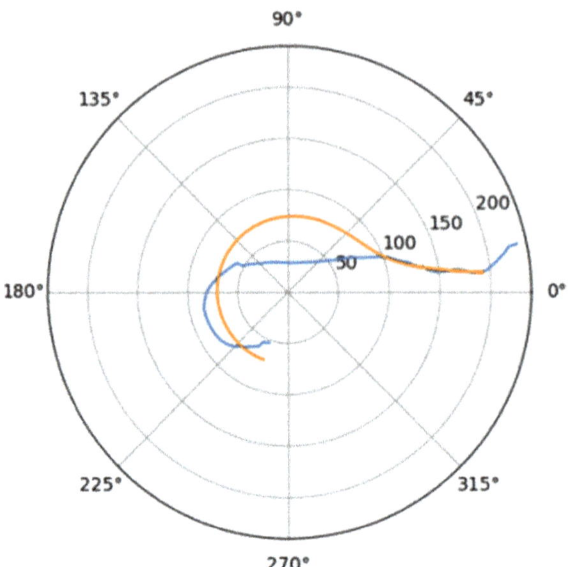

Figure 7. Approximation for are_s.

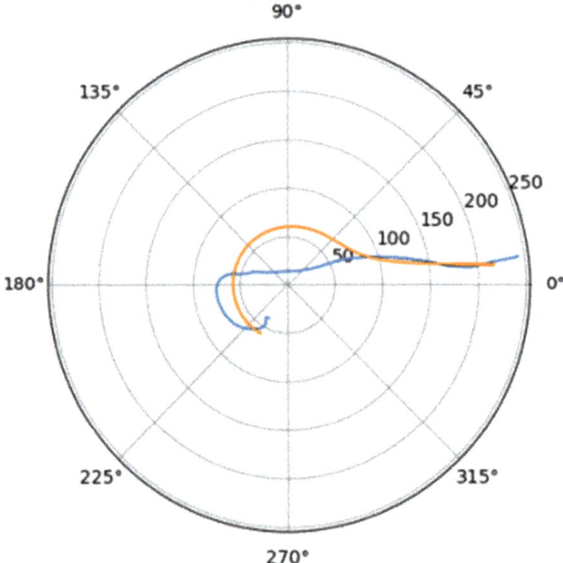

Figure 8. Approximations for ino_s.

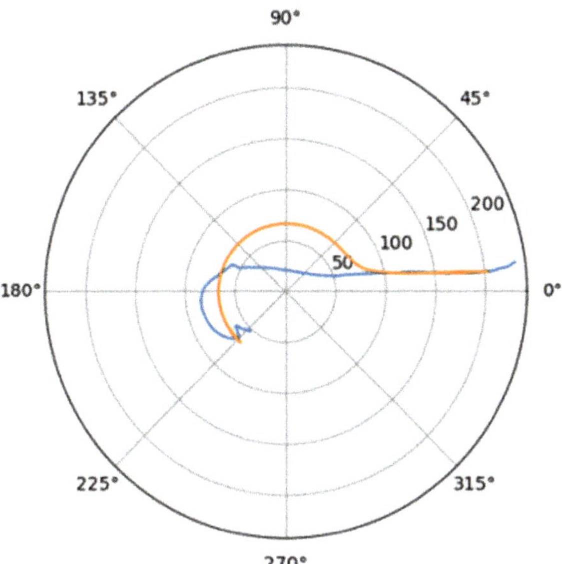

Figure 9. Approximation for ino_d.

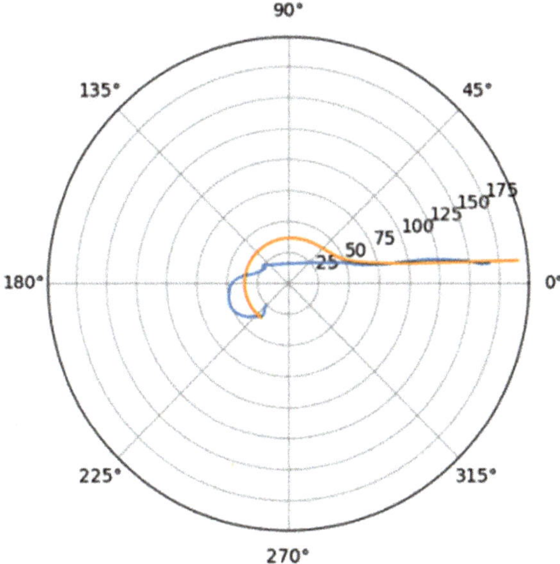

Figure 10. Approximations for she_a.

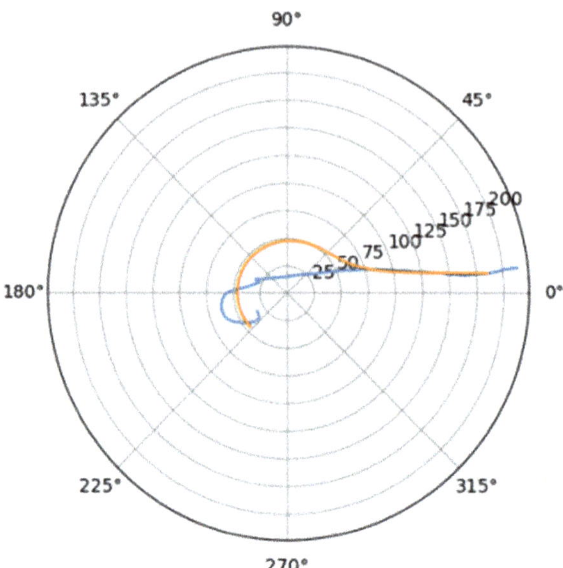

Figure 11. Approximation for are_d.

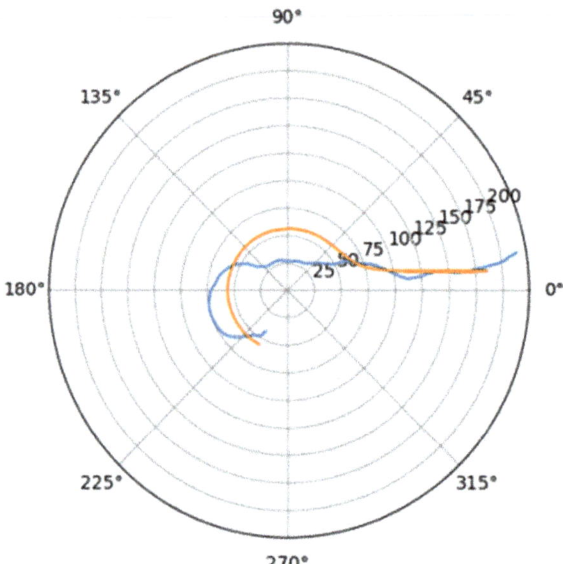

Figure 12. Approximations for bor_a.

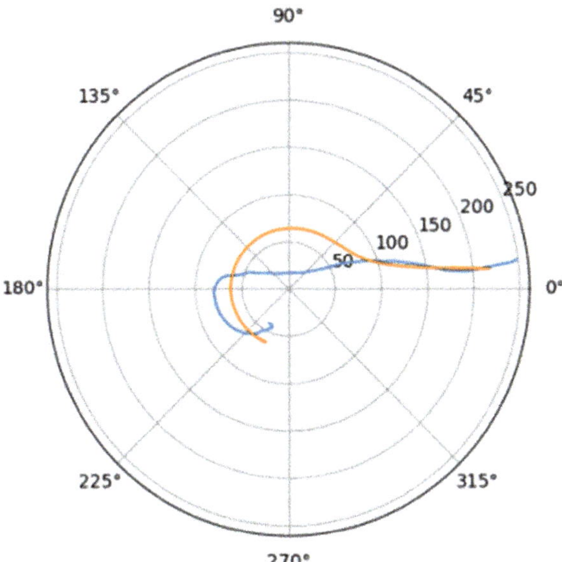

Figure 13. Approximation for mal_a.

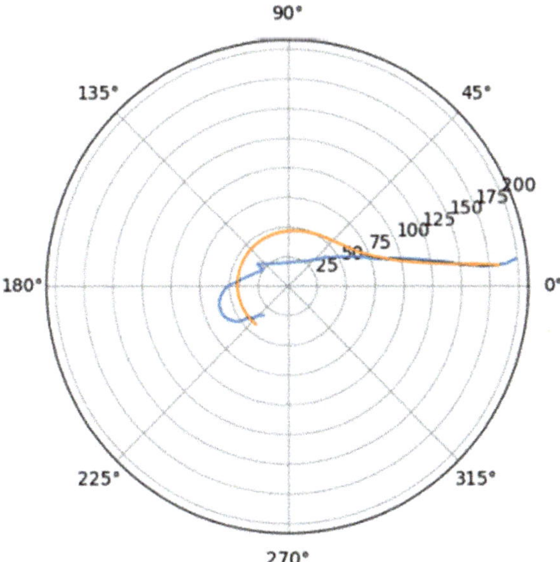

Figure 14. Approximations for pav_a.

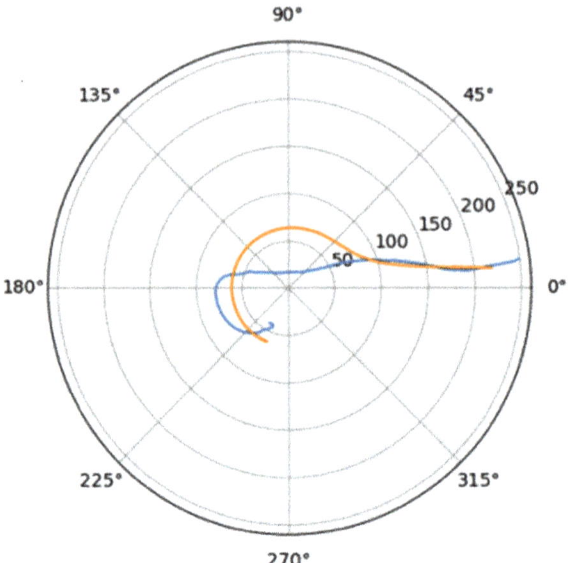

Figure 15. Approximation for bar_a.

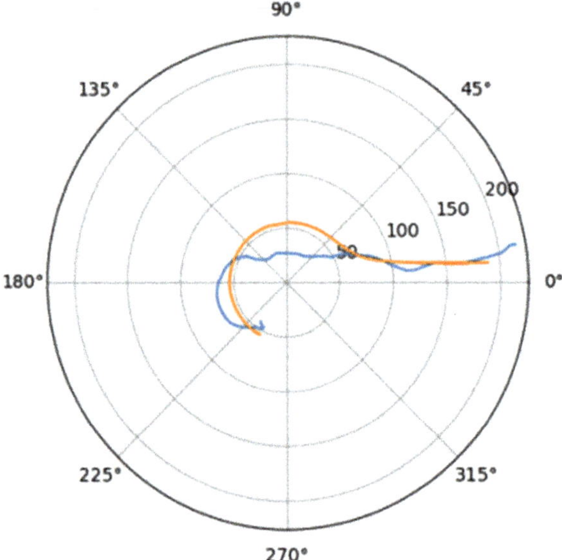

Figure 16. Approximations for zag_a.

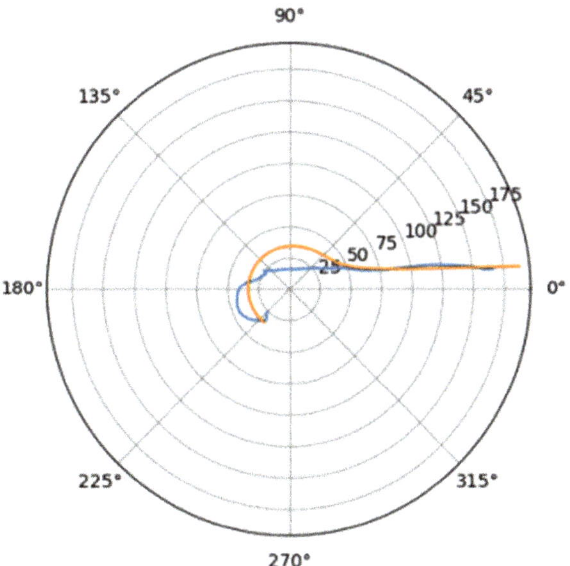

Figure 17. Approximation for mir_d.

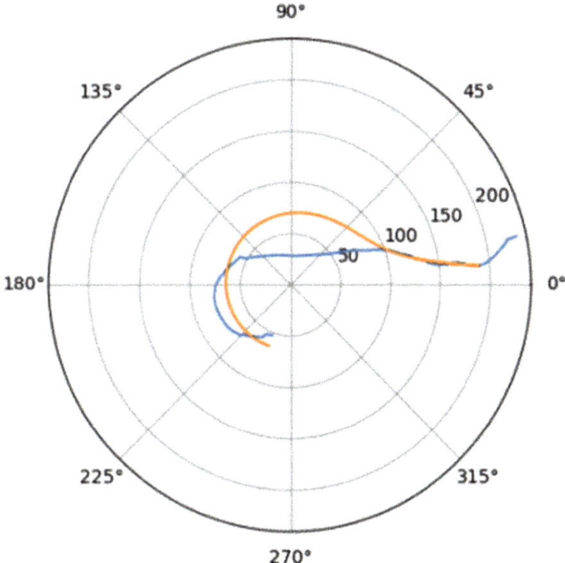

Figure 18. Approximations for gor_d.

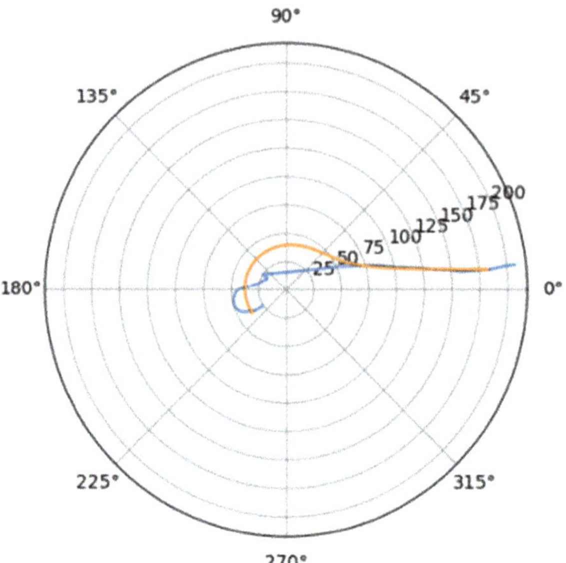

Figure 19. Approximation for poz_a.

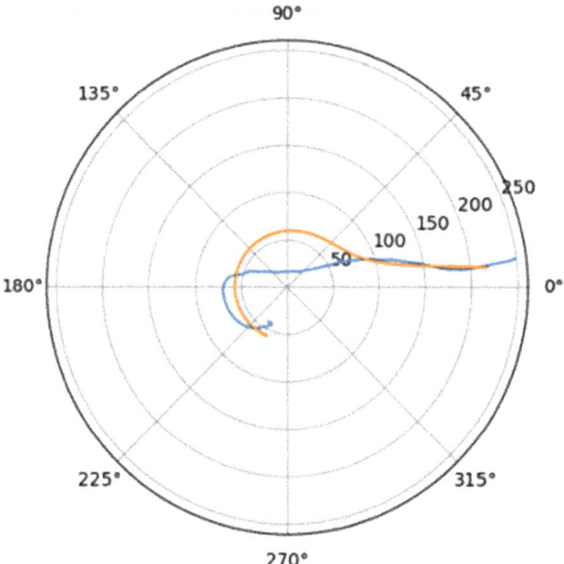

Figure 20. Approximations for hom_a.

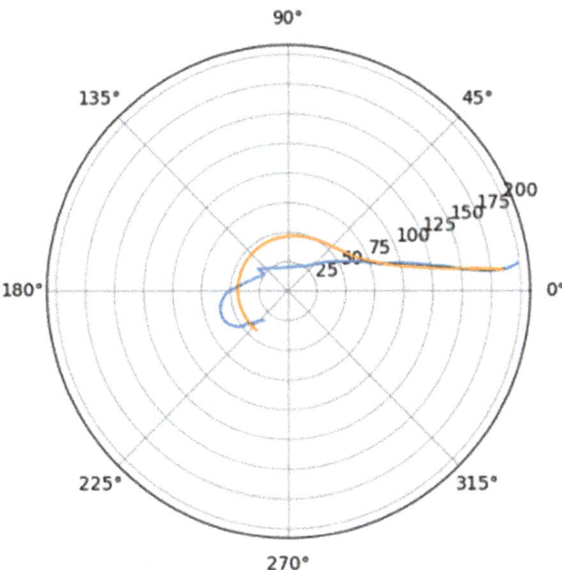

Figure 21. Approximation for lar_s.

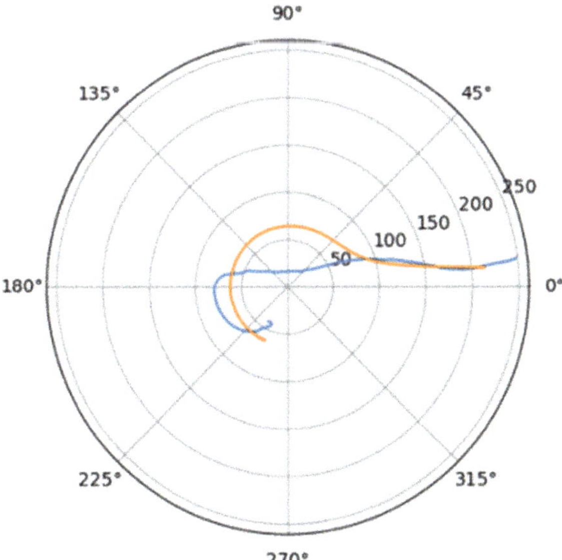

Figure 22. Approximations for lar_d.

The values of the derived quantitative features feature_1 and feature_2 make it possible to unambiguously separate aortas without a noticeable pathological disorder and aortas with a violation of the geometry of the vascular bed (Figure 23).

The values of quantitative features (*feature_1, feature_2*), which were calculated by approximating the central line of the aortic flow channel with a hyperbolic spiral, make it possible to clearly separate the aorta without pronounced pathological remodeling and the aorta with pathological disturbance of the vascular bed. However, the obtained parameters do not allow one to reliably divide aortas according to the type of pathological remodeling (lesion in the proximal or distal sections).

As one can see, there is one obvious outlier on a plot from Figure 23. It is caused by a severely damaged aortic duct in the distal regions. As a result, the proposed algorithm can't properly handle such altered geometry and issues biased values for features.

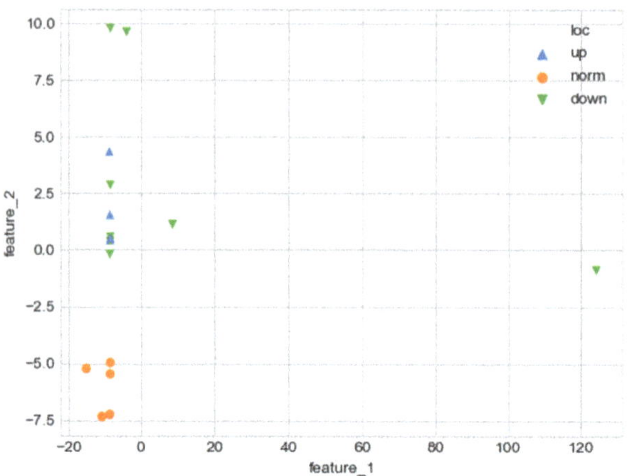

Figure 23. Comparison of quantitative characteristics of geometry for aortas. Blue triangles indicate aortas with pathology in the proximal sections, green triangles—aortas with pathology in the distal sections, and orange circles—aortas without severe pathology.

With a pathological violation of the geometry of the flow channel of the aorta, an area is formed in the local section of the channel, the radius of which significantly exceeds the radius of the same section in the norm. As follows from relations (2.1), for the volume of the swirling flow that fills this additional "pathological" region, at constant values of C_0 and Γ_0 the radial and azimuthal velocity components will be higher than in the normal case. An increase in the total linear velocity vector in the presence of severe pathological remodeling of the aortic duct was experimentally confirmed. However, if we assume that the parameters C_0 and Γ_0 for a pathological case normally coincide, the increase in the total vector of blood flow velocity in pathology will exceed the experimentally observed changes. Therefore, the value of Γ_0 in case of a pathological violation of the geometry of the flow channel will exceed the normal values to compensate for the increase in the radial size of the area with the aneurysm. The value of C_0 with pathology will increase slightly. This will lead to an insignificant increase in the radial and longitudinal departure velocity, which will cause an increase in the energy spent to maintain the evolution of the swirling flow. However, at the same time, an increase in the value of C_0 leads to a decrease in the viscous radius of the swirling flow (the region in which the influence of viscosity is strong). This will inevitably lead to a decrease in the energy spent on maintaining the twist. As a result, in case of a pathological violation of the geometry of the flow channel, we will get a small increase in the energy spent on maintaining the swirling blood flow; however, this value lies within the limits that approximately correspond to indirect observations.

4. Discussion

The geometric configuration of the aorta undergoes significant changes in various pathological conditions, such as hypertension, atherosclerosis, some infectious diseases, etc. At the same time, it is difficult to determine the stage at which changes in the geometry of the aortic flow channel are still compensatory in nature, and at which they are a manifestation of decompensation. There is no formal approach to the analysis of the geometric configuration of the aorta because the fields of force impact on the aortic wall, both in normal conditions and in the development of pathology, have not been sufficiently studied. The only source of force can be the flow of blood in the lumen of the aorta. However, there is still a discussion about the structure of this flow. Previously published papers have argued that the swirling flow pattern is of fundamental importance for adequate unseparated blood flow along the aorta. The shape of the lumen (longitudinal-radial size and distribution of elasticity along the aorta) corresponds with high accuracy to the directions of streamlines of the swirling flow described earlier using quasi-stationary equations (2.2). In the present study, an analysis of the spatial geometric configuration of the aorta in normal conditions and in lesions of predominantly proximal or distal sections was carried out. This analysis showed that the factors of the flow structure (C_0 and Γ_0) and the parameters of approximation of the projection of the aorta on the frontal plane of the human body by a hyperbolic spiral together make it possible to separate these states according to a formal feature.

As a result, new quantitative parameters characterizing the degree of pathological remodeling of the aortic duct and a method for their calculation were proposed. This method is based on a complex analysis of the geometry of the aortic flow channel. This analysis makes it possible to determine the hydrodynamic parameters of the swirling blood flow inside the aortic flow channel, the geometric correspondence of the shape of the flow channel to the direction of the swirling flow streamlines, and to formalize the geometry of the flow channel itself.

The field of shear stresses arising on the channel walls in a swirling flow regime differs significantly from the stresses arising in a fluid flow in laminar or turbulent regimes. Therefore, to analyze the impact of the flow on the walls of blood vessels, it is necessary to consider the twisted structure of the flow and to analyze the shape of the aorta parameters characterizing the degree of this twist—C_0 and Γ_0. The shape of the aorta is also determined by the nature of the swirling blood flow, so the channel geometry was approximated to a hyperbolic spiral. Only in a channel whose shape corresponds to a hyperbolic spiral is it possible to preserve the potentiality of the flow.

In this work, we used the geometric characteristics of the canal associated with the swirling blood flow in the canal lumen (C_0 and Γ_0), which made it possible to divide the aortas into 2 groups: the first group included normal aortas and aortas with pathology localized in the distal sections, and the second group included aorta with localization of pathology in the proximal sections. The construction of the central line and the approximation of this line to a hyperbolic spiral made it possible to further divide the aorta into 2 groups—the first included normal aorta, the second—aorta with lesions in the proximal and distal sections. At the same time, the error of the results does not exceed the threshold value adopted in experimental medicine (5%). The registered shifts in the values of the given parameters may reflect the type and severity of the pathology and may be considered predictors of critical conditions leading to the formation of aneurysms, dissections, and ruptures of the aortic wall.

5. Conclusions

New quantitative parameters were obtained, reflecting the degree of pathological remodeling of the aortic duct, as well a method for their calculation. These parameters can be useful for analyzing the mechanisms of aortic remodeling, determining the limits of compensatory changes in its geometry, and assessing the risks of decompensation leading to the formation of an aneurysm, dissection, or rupture of the aortic wall.

Author Contributions: Conceptualization, E.T. and A.G.; Methodology, E.T. and A.G.; Investigation, E.T, T.T. and A.G.; Math Derivation, E.T.; Supervision and Project Administration, A.G and L.B.; Writing, E.T. and A.G.; Translation, A.G. All authors have read and agreed to the published version of the manuscript.

Funding: This research was funded by Russian Scientific Foundation (grant N22-15-00148).

Data Availability Statement: Not applicable.

Conflicts of Interest: The authors declare no conflict of interest.

References

1. Wang, S.; Tokgoz, A.; Huang, Y.; Zhang, Y.; Feng, J.; Sastry, P.; Sun, C.; Figg, N.; Lu, Q.; Sutcliffe, M.P.F.; et al. Bayesian Inference-Based Estimation of Normal Aortic, Aneurysmal and Atherosclerotic Tissue Mechanical Properties: From Material Testing, Modeling and Histology. *IEEE Trans. Biomed. Eng.* **2019**, *66*, 2269–2278. [CrossRef]
2. Grotenhuis, H.B.; de Roos, A. Structure and function of the aorta in inherited and congenital heart disease and the role of MRI. *Heart* **2011**, *97*, 66–74. [CrossRef]
3. Gallo, D.; de Santis, G.; Negri, F.; Tresoldi, D.; Ponzini, R.; Massai, D.; Deriu, M.A.; Segers, P.; Verhegghe, B.; Rizzo, G.; et al. On the Use of In Vivo Measured Flow Rates as Boundary Conditions for Image-Based Hemodynamic Models of the Human Aorta: Implications for Indicators of Abnormal Flow. *Ann. Biomed. Eng.* **2012**, *40*, 729–741. [CrossRef] [PubMed]
4. Salmasi, M.Y.; Pirola, S.; Sasidharan, S.; Fisichella, S.M.; Redaelli, A.; Jarral, O.A.; O'Regan, D.P.; Oo, A.Y.; Moore, J.E., Jr.; Xu, X.Y.; et al. High Wall Shear Stress can Predict Wall Degradation in Ascending Aortic Aneurysms: An Integrated Biomechanics Study. *Front. Bioeng. Biotechnol.* **2021**, *9*, 750656. [CrossRef]
5. Howard, D.P.J.; Banerjee, A.; Fairhead, J.F.; Handa, A.; Silver, L.E.; Rothwell, P.M. Age-specific incidence, risk factors and outcome of acute abdominal aortic aneurysms in a defined population. *Br. J. Surgery* **2015**, *102*, 907–915. [CrossRef]
6. Celi, S.; Vignali, E.; Capellini, K.; Gasparotti, E. On the Role and Effects of Uncertainties in Cardiovascular in silico Analyses. *Front. Med. Technol.* **2021**, *3*, 748908. [CrossRef] [PubMed]
7. Li, X.; Zhao, G.; Zhang, J.; Duan, Z.; Xin, S. Prevalence and trends of the abdominal aortic aneurysms epidemic in general population–a meta-analysis. *PloS ONE* **2013**, *8*, e81260. [CrossRef] [PubMed]
8. Talygin, E.A.; Zazybo, N.A.; Zhorzholiany, S.T.; Krestinich, I.M.; Mironov, A.A.; Kiknadze, G.I.; Bokerya, L.A.; Gprpdkov, A.Y.; Makarenko, V.N.; Alexandrova, S.A. Quantitative Evaluation of Intracardiac Blood Flow by Left Ventricle Dynamic Anatomy Based on Exact Solutions of Non-Stationary Navier-Stocks Equations for Selforganized tornado-Like Flows of Viscous Incompresssible Fluid. *Uspekhi Fiziol. Nauk.* **2016**, *47*, 48–68. [PubMed]
9. Bockeria, L.A.; Gorodkov, A.Y.; Nikolaev, D.A.; Kiknadze, G.I.; Gachechiladze, I.A. Swirling flow velocity field analysis using the MR-velocimetry. Bulleten NCSSKh im. Bakuleva. *Serdechno-Sosud. Zabol.* **2003**, *4*, 70–74. (In Russian)
10. Talygin, E.A.; Zhorzholiani, S.T.; Agafonov, A.V.; Kiknadze, G.I.; Gorodkov, A.Y.; Bokeriya, L.A. Quantitative Evaluation of Disorders of the Swirled Blood Flow Structure in the Aorta with Pathological Alteration of Its Channel Geometry Using Numerical Simulation of the Aorta. *Hum. Physiol.* **2019**, *45*, 527–535. [CrossRef]
11. Kiknadze, G.I.; Krasnov, Y.K. Evolution of a spout-like flow of viscous fluid. *Sov. Phys. Dokl.* **1986**, *31*, 799–801.
12. Burgers, J.M. A mathematical model illustrating the theory of turbulence. In *Advances in Applied Mechanics*; Elsevier: Amsterdam, The Netherlands, 1948; Volume 1, pp. 171–199.
13. Zhorzholiani, S.T.; Mironov, A.A.; Talygin, E.A.; Tsyganokov, Y.M.; Agafonov, A.M.; Kiknadze, G.I. Analysis of Dynamic Geometric Configuration of the Aortic Channel from the Perspective of Tornado-Like Flow Organization of Blood Flow. *Bull. Exp. Biol. Med.* **2018**, *164*, 514–518. [CrossRef] [PubMed]
14. Zhorzholiani, S.h.T.; Talygin, E.A.; Krasheninnikov, S.V.; Tsigankov, Y.M.; Agafonov, A.V.; Gorodkov, A.Y.; Kiknadze, G.I.; Chvalun, S.N.; Bokeria, L.A. Elasticity change along the aorta is a mechanism for supporting the physiological self-organization of tornado-like blood flow. *Hum. Physiol.* **2018**, *44*, 532–540. [CrossRef]
15. Boyd, S.; Vandenberghe, L. *Convex Optimization*; Cambridge University Press: Cambridge, UK, 2004.
16. Okabe, A.; Boots, B.; Sugihara, K. *Spatial Tessellations: Concepts and Applications of Voronoi Diagrams*; Wiley: New York, NY, USA, 1992.

Article

Chaotic Model of Brownian Motion in Relation to Drug Delivery Systems Using Ferromagnetic Particles

Saša Nježić [1], Jasna Radulović [2], Fatima Živić [2,*], Ana Mirić [3], Živana Jovanović Pešić [2], Mina Vasković Jovanović [2] and Nenad Grujović [2]

1   Faculty of Medicine, University of Banja Luka, Save Mrkalja 14, 78000 Banja Luka, Bosnia and Herzegovina
2   Faculty of Engineering, University of Kragujevac, Sestre Janjic 6, 34000 Kragujevac, Serbia
3   Institute for Information Technologies—National Institute of the Republic of Serbia, University of Kragujevac, Jovana Cvijica bb, 34000 Kragujevac, Serbia
*   Correspondence: zivic@kg.ac.rs

Abstract: Deterministic and stochastic models of Brownian motion in ferrofluids are of interest to researchers, especially those related to drug delivery systems. The Brownian motion of nanoparticles in a ferrofluid environment was theoretically analyzed in this research. The state of the art in clinical drug delivery systems using ferromagnetic particles is briefly presented. The motion of the nanoparticles in an external field and as a random variable is elaborated by presenting a theoretical model. We analyzed the theoretical model and performed computer simulation by using Maple software. We used simple low-dimensional deterministic systems that can exhibit diffusive behavior. The ferrofluid in the gravitational field without the presence of an external magnetic field in the xy plane was observed. Control parameter p was mapped as related to the fluid viscosity. Computer simulation showed that nanoparticles can exhibit deterministic patterns in a chaotic model for certain values of the control parameter p. Linear motion of the particles was observed for certain values of the parameter p, and for other values of p, the particles move randomly without any rule. Based on our numerical simulation, it can be concluded that the motion of nanoparticles could be controlled by inherent material properties and properties of the surrounding media, meaning that the delivery of drugs could possibly be executed by a ferrofluid without an exogenous power propulsion strategy. However, further studies are still needed.

Keywords: Brownian motion; chaotic model; ferrofluid; targeted drug delivery; exogenous power propulsion strategy

MSC: 62P10; 62P30; 62P35

Citation: Nježić, S.; Radulović, J.; Živić, F.; Mirić, A.; Jovanović Pešić, Ž.; Vasković Jovanović, M.; Grujović, N. Chaotic Model of Brownian Motion in Relation to Drug Delivery Systems Using Ferromagnetic Particles. *Mathematics* **2022**, *10*, 4791. https://doi.org/10.3390/math10244791

Academic Editor: Efstratios Tzirtzilakis

Received: 2 November 2022
Accepted: 7 December 2022
Published: 16 December 2022

Publisher's Note: MDPI stays neutral with regard to jurisdictional claims in published maps and institutional affiliations.

1. Introduction

Models of Brownian motion, both stochastic and deterministic, have been of interest to researchers for a long time [1]. Lucretius considered in 60 BC that the movement of dust particles in the air is caused by the movement of small invisible particles. The chaotic movements of coal particles in alcohol were described by Ingenhousz in 1785. In 1827, the botanist Brown observed the movement of pollen in water under a microscope. A mathematical description of Brown's movement was given in 1880 by Thiele and in 1900 by Bachelier. In 1905, Einstein developed a stochastic theory of Brownian motion [2], which Perrin experimentally proved in 1909. Langevin in 1908 used a stochastic differential equation to describe changes in macroscopic variables. In 1965, Mori described transport, collective motion and Brownian motion within statistical-mechanical theory [3]. Saffman and Delbruck investigated Brownian motion in 1975 in biological membranes. Caldeira and Legget investigated the quantum Brownian motion of 1983 [4].

Fujisaka, Grossmann, Thomae and Geisel from 1982 to 1985 wanted to form a dynamic theory of Brownian motion [5]. In 1998, Gaspard and his associates experimentally proved microscopic chaos [6]. In 2005, Cecconi attempted to determine the microscopic nature of diffusion by data analysis [7]. Since 2011, Brownian motion in superfluids has been considered. There are various studies of Brownian motion and stochastic and chaotic models are observed, and the nature of this movement is a difficult question and the answer is not determined by the character of the model [8,9].

The first definition of Brownian motion was related to stochastic process [10], in relation to a wide range of different real stochastic processes, and represented by the Wiener process, which describes continuous-time stochastic process with real values [11]. Brownian motion can be observed as stochastic or deterministic in chaos theory, based on the deterministic equations that describe stochastic phenomena [11], but the governing parameters that might provide a full replication of the experiment are difficult to determine or define. Recent computer simulation experiments have shown the possibility to model the chaotic system as a stochastic one, by controlling simulation parameters and initial conditions [11]. Such an approach has enabled research on how to govern the chaotic system (and determine governing parameters) through the study of particle trajectories that have random motion, that is, by using deterministic equations to reproduce random behavior [11].

The generation of deterministic Brownian motion is possible through additional degrees of freedom in the Langevin equation of the phenomenological system of particle mixing and agitation in fluids [12]. Another study [13] replicated Brownian motion by using a fully deterministic set of differential equations and applied it to a real problem of electronic circuit implementation. Their deterministic model showed that some variables within the model can enable modeling of the circuit dynamics as a stochastic Brownian behavior [13]. There are numerous real systems that exhibit Brownian behavior, and modeling such systems by deterministic systems (without random components) is an important area in recent research, including drug delivery systems [13]. An analysis of 126 different combinations of governing parameters is given in [13], and around 10% of those cases involved deterministic Brownian-like motion. They obtained stochastic or deterministic Brownian motion based on the initial setup conditions (assigned initial parameters values for circuit implementation) [13].

Ferrofluids are a suspension of small particles of 10 nm, each of which contains one permanent ferromagnetic domain [14]; thus, each particle is a permanent magnet, which, in the absence of an external magnetic field, rotates randomly under the action of Brown's forces, which are strong due to the small particle size [15]. In ferrofluids, dipoles exist without fields and rotate randomly by Brownian motion [16]. Ferrofluids are interesting magnetic fluids that can be controlled by an external magnetic field [17]. There are several applications for ferrofluids in industrial as well as technological fields such as magnetic memory, inkjet printers, magnetic seals, etc. [18]. They are known for their biomedical applications such as magnetic resonance contrast agents [19], hyperthermia [20], targeted drug delivery to tumor and cancer cells, antibacterial activity, etc. [21,22].

Numerical computational simulations have emerged in the past decade as powerful tools for the analysis and prediction of the material physical behavior at macro/micro and nano scales, with extensive research on applied models and software solutions. Thermal conductivity of fluids is the most influential factor for the fluid behavior, and different approaches to estimate or predict it for novel nanofluids have been studied using mathematical models [23–25] or experimentally based models [26]. For example, models related to the rheological properties of hybrid non-Newtonian nanofluids are important, since it is proven that with the increase in volume fractions of nanoparticles, the effect of temperature increase is more influential, leading to the non-Newtonian behavior of the nanofluid and also having a strong effect on viscosity [26].

A mathematical model was developed, using SigmaPlot software, to study the thermal conductivity of nanofluids, through the study of different volume fractions of ternary hybrid nanofluids and mono and binary hybrid nanofluids [23]. This model correlated volume fractions of different nanofluids and resulting thermal conductivity in order to provide a tool for the estimation of thermal conductivity of a ternary hybrid nanofluid [23].

Thermal properties of DNA structure in water fluid were estimated by using equilibrium and non-equilibrium molecular dynamics approaches and LAMMPS (Large-scale Atomic/Molecular Massively Parallel Simulator) software [24]. Another work on the prediction of the thermal conductivity of hybrid Newtonian nanofluid proposed an algorithm to solve the problem in the Artificial Neural Network (ANN) [25] that considered the volume fraction of nanoparticles and temperature.

This paper studies the possibility to use and influence Brownian motion to produce patterned trajectories of particles in a diffusive motion of the ferrofluid, aiming to assist in more efficient drug delivery nanofluid systems. A chaotic model of Brownian motion was theoretically analyzed and simulated by using Maple software. The chaotic model was mapped with an introduced control parameter, p, which depends on the viscosity coefficient and particle mass and size, in analogy with the Langevin equation. The ferrofluid in the gravitational field without the presence of an external magnetic field in a two-dimensional model was observed.

2. Materials and Methods

2.1. Brownian Motion

Numerous experiments show that there is a constant internal movement in every substance. This internal movement is in fact the movement of the molecules that make up the observed substance. This movement of molecules is unregulated, never stops and depends only on temperature. The phenomenon discovered by Brown directly indicates the stochastic nature of the movement of molecules, where the same initial condition will not replicate the resulting motion (the same trajectory in time). Using a microscope, it was observed that very small particles floating in a liquid are in a state of continuous stochastic motion, and the smaller the particles, the faster they move [27]. This motion, called Brownian motion, never stops, does not depend on any external cause and is a manifestation of the particles' motion due to colliding with surrounding molecules of fluid and internal energy of matter: the potential energy of all the particles and thermal energy of moving particles (kinetic energy), which is correlated to the temperature and number of particles (mass). When they collide with a solid body, liquid molecules, which are constantly moving, are subjected to a certain amount of movement. If the body is in a liquid and has larger dimensions, the number of molecules that come across it from all sides is also very large, and their shocks are compensated at any time and the body remains practically motionless [28].

If the body is small, such compensation may be incomplete: it can accidentally hit one side of the body with a much larger number of molecules than the other, causing the body to move [28]. It is a movement performed by Brown's particles under the action of chaotic blows of molecules. Brown's particles have several billion times the mass of individual molecules and their velocities are very low compared to the speeds of molecules, but their movement can be observed with a microscope. In this way, too, a substance not only has a granular structure—that is, it consists of individual separate parts—but it also consists of particles that are constantly moving.

2.2. Ferrofluid

A magnetic colloidal particle, also known as a ferrofluid, is a colloidal suspension of single-domain magnetic particles, typically about 10 nm in size, dispersed in a liquid carrier [15,16]. The liquid carrier can be polar or non-polar [14]. Since the 1960s, when ferrofluid was initially synthesized, its technical and medical applications have not stopped growing [18]. Ferrofluid differs from ordinary magnetorheological fluids used for shock absorbers, brakes and clutches, formed by micron-sized particles dispersed in oil [18]. In magnetorheological fluid, the application of a magnetic field increases the viscosity, so that for a sufficiently strong field, it can behave like a solid [16,18]. On the other hand, ferrofluid retains its fluidity even if it is exposed to a strong magnetic field. Ferrofluids are optically isotropic, but in the presence of an external magnetic field, they show induced birefringence [29]. Wetting of certain substrates can also cause bifurcation in thin ferrofluid layers. To avoid agglomeration, the magnetic particle should be coated with a shell of a suitable material [30,31]. In relation to the coating, ferrofluid is divided into two main groups: surfactant, if the coating is a surfactant molecule, and ionic, if it is an electric shell [30,31].

Colloidal suspensions of magnetic particles in liquids that have the ability to magnetize in an external magnetic field are called ferrofluids [32]. These are magnetic materials in liquid form. The liquid can be water or an organic solution in which ferromagnetic or ferrimagnetic particles are dispersed. The particles are most often hematite, Fe_2O_3, or magnetite, Fe_3O_4, and they need to be stabilized due to high surface energy by adding a polymer or ionic component (surfactant). Usually, such stable particles are about 10 nm in diameter, and their surface energy is reduced by long-chain surfactants which, thanks to the long chains, prevent agglomeration, or the same charge on the surface of magnetic particles leads to a mutual repulsion, preventing agglomeration [30]. Ferrofluid particles do not precipitate even for a long time, they do not agglomerate and they do not separate from liquids even by applying an extremely strong magnetic field. The combination of the liquid phase and the magnetic behavior makes it possible to manipulate the fluid by changing its position using an external magnetic field [32].

To avoid agglomeration, ferrofluid particles are coated [30,31]. Depending on the coating, ferrofluids are divided into two groups: surfactant-coated ferrofluids and electrostatically stabilized ferrofluids [31]. Surfactant-coated ferrofluids contain magnetic particles coated with amphiphilic molecules such as oleate to prevent aggregation. Spherical repulsion between particles acts as a physical barrier that keeps the particles in solution and stabilizes the colloid. If the particles are dispersed in a nonpolar phase, such as oil, the polar head of the surfactant is attached to the surface of the particles, and the hydrophobic chain is in contact with the liquid (Figure 1a). If the particles are dispersed in the polar phase such as water, a two-layer coating of the particles is required to form a hydrophilic layer around them (Figure 1b).

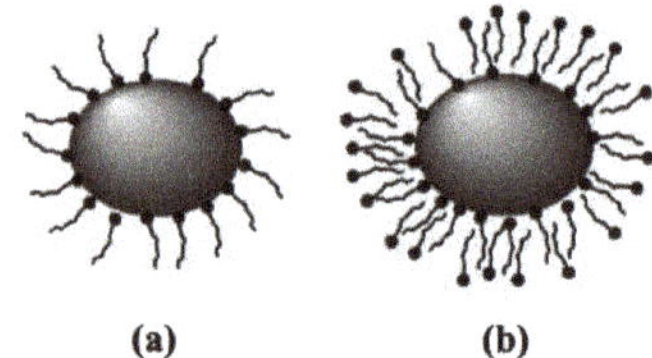

Figure 1. Single layer (**a**) Two-layer. (**b**) Coated magnetic particles.

One of the fastest developed areas of research is one in which nanotechnology, biology and medicine intertwine. According to many experts, the application of nanotechnology in medicine, better known as nanomedicine, will lead to a revolution in the field of targeted drug delivery systems [33,34], disease diagnosis, bioengineering and the improvement of contrast agents in magnetic resonance imaging. The term "teragnostics" is usually used in this context, which, as the name itself implies, is a synthesis of diagnostics and therapy. The localization of ferrofluids by the applied magnetic field gives an interesting application of ferrofluids in medicine. A lot of research has been dedicated to the use of ferrofluids as a system for targeted delivery of drugs used in chemotherapy [35]. A drug is injected into tumor carcinomas and retained there for some time by a magnetic field. The amount of medicine needed is much less than the amount of medicine that would be needed to distribute the medicine throughout the body. After turning off the magnetic field, the drug will disperse in the body, but since it is a much smaller amount, there are practically no side effects.

The ability of ferrofluid to absorb the energy of electromagnetic waves at a frequency different from the frequency at which water absorbs energy allows heating of the localized part of the tissue where the ferrofluid is injected (for example, tumors) without heating the surrounding tissues. This phenomenon is called hyperthermia. Hyperthermia is one of the methods used in cancer therapy that is based on increasing the temperature of tumor tissue above 41 °C. As a result, the function of tumor cells is disturbed and they die. Magnetic hyperthermia is based on the effect of releasing heat when magnetic nanoparticles are found in a changing magnetic field. Magnetic nanoparticles can be successfully localized in tumor tissue, which allows heating only in the desired place. In therapy, these particles are most often used in colloidal form. It is possible to bind chemotherapy drugs or radionuclides to these particles and thus achieve a combined effect.

2.3. State of the Art in Clinical Drug Delivery Systems Using Ferromagnetic Particles

Drug delivery systems based on nanotechnology have improved the delivery of drugs due to their changes in pharmacokinetics, enabling a longer half-life of the drug in the bloodstream and reducing toxicity [36]. Magnetic nanoparticles play an important role in the diagnosis and treatment of diseases such as cancer, heart and neurological diseases [37]. These particles are often used in the targeted delivery of medically active substances because they deliver the drug to the desired place via tissue magnetic absorption or strong ligand–receptor interaction [38].

The drug–carrier complex can be administered intravenously or by arterial injection [39]. It can also be administered orally, but the main problem with such administration is the delivery of peptides and proteins due to their breakdown in gastric acid, low absorption and first-pass metabolism through the liver [40].

If a drug is delivered by a magnetic field, the gradient of the external magnetic field associated with the magnetic field within human tissues enables the transfer and accumulation of magnetic nanoparticles in the body [41]. However, there are a number of intracellular and extracellular barriers that can be limiting factors. One of the possible solutions is covering the surface of nanoparticles with biocompatible materials (different organic and inorganic compounds) [37]. Coating the surface of nanoparticles increases the half-life of the drug by delaying clearance [38]. Macrophages take up uncoated nanoparticles at a rate that depends on their functional surface, size and hydrophilicity, followed by clearance in the liver and spleen. Plasma proteins bind to the surface of nanoparticles, accelerating phagocytosis. Coatings enable the slowing down of detection by macrophages and thus reduce clearance. For this purpose, the most often used is polyethylene glycol (PEG), the attachment of which provides a "stealth" protective effect. PEG is suitable for this purpose because it shows low toxicity and immunogenicity and is excreted by the kidneys [42]. In addition, surface coating enables covalent binding of biomolecules such as antibodies and proteins and their transport to the target tissue. It is necessary that these coatings be sensitive to the change in pH value, which would enable the controlled release of the drug [43]. Drug release can be stimulated by chemical radiation, mechanical forces and magnetic hyperthermia [38].

Figure 2 shows the structure of a magnetic nanoparticle carrying the active molecule on the surface.

The most important characteristics of nanoparticles used for drug delivery are intrinsic magnetic properties, the shape and size of nanoparticles, non-toxicity, stability in water, surface charge and coating. Many magnetic materials with ideal magnetic properties, such as cobalt or chromium, are very toxic and cannot be used in medicine, while materials based on iron oxide (magnetite or maghemite) are safe. In addition to low toxicity, these nanoparticles show high stability against degradation [38]. The size of the particles is such that they allow entry into biological structures, and it varies from 3 to 30 nm [41].

This method can be applied to solid tumor mass, neoplasms with metastases and tumors that are in the early phase of cell growth. Treatment would involve the application of particles that specifically recognize clusters of cancer cells, carrying a medically active substance that will act [44].

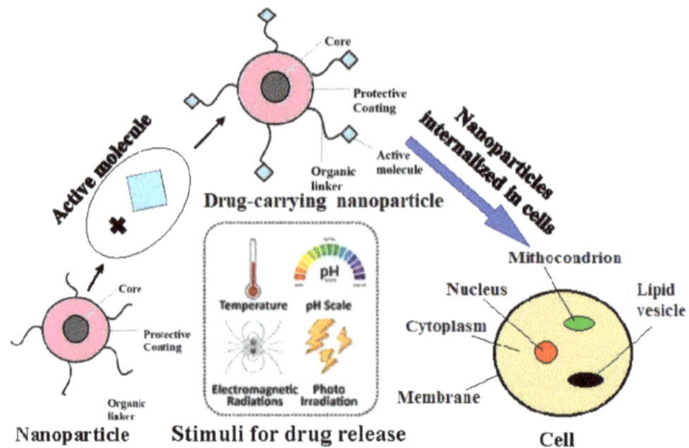

Figure 2. Structure of a magnetic nanoparticle with active molecule on the surface (drug-carrying nanoparticle) and the influential factors on its behavior within a drug delivery system, including stimuli for drug release and cell structure that interact with such nanoparticles.

2.4. Motion of Particles in an External Field and as a Random Variable

Nanoparticles move after collision with smaller water molecules. Smaller water molecules come across the nanoparticle from all directions and collide with it, thus imparting their momentum to it. When at some point after such a collision a particle receives a momentum in that direction, it starts to move in that direction, until another water molecule collides with it and gives it its momentum by collision. If in the field of classical mechanics, this mode of motion is, in principle, deterministic and there is a corresponding Hamiltonian [45]. In reality, due to the large number of water molecules that interact with each other and with the nanoparticle, and due to the unknown initial conditions, such a system cannot be described using the Hamiltonian in practice, i.e., it is impossible to say with certainty how much and in which direction the particle will move in at some point. On the contrary, a shift in any direction is equally probable (this can be seen intuitively from the symmetry of the system). That is why the tool of statistical physics is used to study such a system. Due to the large number of water molecules that make up the system, instead of observing the microscopic effect on the nanoparticle of each of them, one can observe their "total" macroscopic effect after some time. After repeating such an experiment several times with the same initial conditions (as much as it can be controlled, e.g., placing a nanoparticle with the same physical values in the same place), a different value is obtained for the total macroscopic displacement of the nanoparticle after a certain time. This gives an empirical distribution of the probability of the total movement of a nanoparticle, and therefore, instead of looking at exactly how much a nanoparticle will move at a given moment, one can look at the probability that the particle will move by a certain value at a given moment.

A fluid containing a large number of particles (Brownian particles) and moving in one dimension is observed. The density is given by $n(x, t)$, where n is the particle density, x is the position coordinate and t is the time. Brownian motion over time makes the density uniform. The flow of particles during diffusion j_d is defined by:

$$j_d(x,t) = -D\frac{\partial n(x,t)}{\partial x} \tag{1}$$

The flow of particles causes a change in density in time according to the continuity equation:

$$\frac{\partial n(x,t)}{\partial t} = -\frac{\partial j_d(x,t)}{\partial x} = D\frac{\partial^2 n(x,t)}{\partial x^2} \tag{2}$$

In the equation, D [m^2/s] is the diffusion constant (it depends on the type of particle material), and the given equation is the diffusion equation. The flow of particles is opposite to the direction of the density gradient and the flow direction is from the area of higher density to the area of lower density.

A uniform force field (gravitational field) acts on the Brownian particles. The stated field accelerates the particles until the velocity reaches a certain limiting velocity, ug, and where the sum of the forces acting on the particle, the force field, F, and the friction force is equal to zero, $F = m\gamma ug$, where m is the mass of the particle and γ [1/s] is the velocity gradient. Now, the particle moves with a velocity u_g, which is determined by the force F and the friction that the fluid acts on the particles. Now, the flux of particles is j_F in the gravitational field:

$$j_F(x,t) = n(x,t)u_g = \frac{nF}{m\gamma} \tag{3}$$

where m is the mass of the particle and $m\gamma$ is the mass flux [kg/s] and n is the density, while F is the strength of the force field. The total flow of particles is written:

$$j = j_d + j_F \tag{4}$$

Now, the diffusion equation is:

$$\frac{\partial n(x,t)}{\partial t} = -\frac{\partial j(x,t)}{\partial x} = D\frac{\partial^2 n(x,t)}{\partial x^2} - \frac{F}{m\gamma}\frac{\partial n(x,t)}{\partial x} \tag{5}$$

For the diffusion coefficient, Einstein finds that the Brownian particle is related to its mobility μ [s/kg] via the equation:

$$D = \mu kT \tag{6}$$

where k is the Boltzmann constant and T is the temperature. This equation is derived as follows. For an arbitrary distribution of the density of Brownian particles after a long time, the flow of particles will equalize and an equilibrium state is obtained (it does not change with time), $n(x, t) = n(x)$. This is the sedimentation equilibrium, which is independent of time t_0, and then the particle density is where the position coordinate is x_0:

$$n(x) = n(x_0)e^{\frac{F(x-x_0)}{kT}} \tag{7}$$

Now, the first and second derivatives of the particle flow are performed and the total flow is zero and is obtained as:

$$\frac{D}{kT} = \frac{1}{m\gamma} \tag{8}$$

Accordingly, Einstein's relation is obtained as:

$$D = \frac{kT}{m\gamma} = \mu kT, \tag{9}$$

where $\mu = 1/m\gamma$ is the mobility of the particles, which is equal to the ratio of the force F and the limiting velocity ug. The force field can also be of electric potential if the Brownian particle is charged.

If the density is high, then the interaction of Brownian particles is ignored and the density is defined as follows. The probability that a Brownian particle is located at the position coordinate x at time t is defined if it was at time t_0 at x_0 with $V(x, t|x_0, t_0)$. Now, the density is $n(x, t)$, the integral of the product $n(x_0,t_0)$ and the probability of passing over all possible values of x_0. Moreover, since there are many Brownian particles of density at some x, there is a certain probability that the Brownian particles arrived at a certain x in some time $t - t_0$ from all other particles of the liquid as:

$$n(x,t) = \int n(x_0,t_0)V(x,t/x_0,t_0)dx_0, \tag{10}$$

where $V(x,t/x_0,t_0)$ is conditional probability.

The given equation is included in the diffusion equation as:

$$\int n(x_0,t_0)\frac{\partial V(x,t/x_0,t_0)}{\partial t}dx_0 = D \int n(x_0,t_0)\frac{\partial^2 V(x,t/x_0,t_0)}{\partial x^2}dx_0, \tag{11}$$

where V is the probability of the position x of the particle at time t.

Both equations are shifted to one side and equalized to zero:

$$\int \left[n(x_0,t_0)\left(\frac{\partial V(x,t/x_0,t_0)}{\partial t} - D\frac{\partial^2 V(x,t/x_0,t_0)}{\partial x^2} \right) \right]dx_0 = 0 \tag{12}$$

The equation will be correct regardless of the value of dx_0, which is the derivative of the position coordinate. Accordingly, the probability $V(x,t/x_0,t_0)$ fulfills the diffusion equation as:

$$\frac{\partial V(x,t/x_0,t_0)}{\partial t} = D\frac{\partial^2 V(x,t/x_0,t_0)}{\partial x^2} \tag{13}$$

Since the equation is valid regardless of the selected initial position coordinate x_0 and initial time t_0, it is written further as:

$$\frac{\partial V(x,t)}{\partial t} = D\frac{\partial^2 V(x,t)}{\partial x^2} \tag{14}$$

It is assumed that the initial condition is that at the initial time t_0, all particles are at the same coordinate position x_0, and the following can be stated:

$$V(x_0,t_0) = \delta(x - x_0), \tag{15}$$

where the right side of the equation is the Dirac delta function.

So, the solution for conditional probability, V, is:

$$V(x_0,t_0/x,t) = \frac{1}{\sqrt{4\pi D(t - t_0)}}e^{-\frac{(x-x_0)^2}{4D(t-t_0)}} \tag{16}$$

The resulting solution is a Gaussian distribution with expected value x_0 and variance $4D(t - t_0)$, and the resulting equation is the Green's function [46].

Now, the time interval within which Brownian motion is observed can be divided into $t_0, \ldots, t_i, \ldots, t_N$. $\Delta X(t_i)$ is the displacement of the Brownian particle between time $t_i - 1$ and t_i. Next, $X(t_i)$ is the position of the particle at time t_i, and we put $X(t_0 = 0) = 0$. The Brownian particle is surrounded on all sides by an average equal number of particles (which is shown by the symmetry of the system), and since there is no total flow of water in which the particles are located (the average total velocity of water molecules is zero), the expected probability of the Brownian particle motion is zero because it is equally likely that it can be hit by a particle from any direction. Accordingly, it can be stated:

$$\langle \Delta X(t_i) \rangle = 0 \tag{17}$$

Now, the position of the particle is:

$$X(t_N) = \sum_{i=0}^{N} \Delta X(t_i) \tag{18}$$

The probability of the position is zero:

$$\langle X(t_N)\rangle = \langle \sum_{i=0}^{N} \Delta X(t_i)\rangle = \sum_{i=0}^{N}\langle \Delta X(t_i)\rangle = 0 \tag{19}$$

The autocovariance of the displacement of the Brownian motion is:

$$\langle \Delta X(t_i)\Delta X(t_j)\rangle = 0, i \neq j \tag{20}$$

So, the covariance between two variables is observed. Autocovariance is a measure of the covariance between the value of a stochastic variable at some time t and its value at some other time. Correlation is the covariance divided by the product of the variances of both variables (normalized to the interval from -1 to 1). The correlation between two variables measures how much one variable changes as the other variable changes, and it is a measure of their mutual linear dependence. Two variables with a correlation of -1 change exactly the opposite (when one increases, the other always decreases); when the correlation is 0, there is no linear dependence of one variable on the other; and when the correlation is 1, when one increases, the other always increases. Furthermore, it is valid for the variance if there is no autocorrelation between the shifts:

$$\langle X^2\rangle = \sum_{i=0}^{N}\langle \Delta X(t_i)^2\rangle \tag{21}$$

It is assumed that they are all $\langle \Delta X(t_i)^2\rangle$ equal and their value is ΔX^2, so:

$$\langle X^2\rangle = N\langle \Delta X^2\rangle = t\frac{\Delta X^2}{\Delta t} \tag{22}$$

The time Δt is the time between collisions between water molecules and Brownian particles, where a water molecule hits a Brownian particle and then it moves by $\pm\Delta X$. When the time Δt has passed, the Brownian particle collides with the water molecule again and moves by $\pm\Delta X$, and so on. The speed of all particles has been replaced with the average speed, which is the most probable in thermal equilibrium. The displacement of the Brownian particle $\Delta x = x(t) - x(0)$ is related to the diffusion coefficient, as:

$$D = \frac{1}{2\Delta t}, \tag{23}$$

and in the following way:

$$\langle X^2\rangle = 2Dt \tag{24}$$

Based on the obtained equation and Einstein's relation [2], it is concluded that the collision time between two molecules is inversely proportional to the friction constant and temperature. A higher friction constant (viscosity coefficient) actually means that collisions with a Brownian particle are more frequent, and a higher speed (temperature) of the particle leads to more frequent collisions.

2.5. Maple

Maple, a computer algebra software, is general purpose software for symbolism, numeric, graphics and simulation. The symbolic approach implies the exact treatment of numbers, symbols, expressions and formulas. The numerical approach involves approximating decimal numbers (the number of digits can be large).

Maple's library contains over three thousand functions: calculation of derivatives, solving algebraic and differential equations, operations with matrices, factorization of polynomials, data processing, creation of FORTRAN or C-code, Fourier transform. Maple's document combines text, commands for calculation, results and graphics, and can be translated into LATEX code. An integral part of Maple is a high-level programming language that allows the users to work with their own procedures. They also have packages of special functions for linear algebra, statistics, geometry and combinatorics.

Maple is used to perform a computer experiment (simulation) when performing an imaginary experiment. Computer simulation is complementary to theory and experiment. The model of a system is complex and no analytical solution can be found, so the numerical method and simulation are used. The computer experiment is based on equations. In the dynamic model, there is a connection between applied mathematics, computer science and applied science. One way to explain the motion of ferrofluids in a gravitational field without the presence of an external magnetic field is:

1. Within the theory, a mathematical model is constructed.
2. Applied mathematics itself provides basic algorithms, computer science provides a scientific program and computer science provides system software.
3. A computer prediction is obtained which is experimentally verified.

The motion of a ferrofluid in the gravitational field is observed, with the modeling concept based on the previously elaborated theoretical model. Its chaotic and deterministic behavior in the system is studied.

3. Results

3.1. Dynamic Model

We analyzed the theoretical model as previously presented and used Maple software to perform the computational experiment. The ferrofluid in the gravitational field without the presence of an external magnetic field in the xy plane was observed and the initial condition on the x and y axes for the ferrofluid is given. It performs 400 collisions with fluid molecules. For the values of the control parameter p that depend on the viscosity coefficient and particle mass and size (in analogy with β coefficient in Equation (26), as shown in the following equations), we obtained a path on the basis of which the movement can be characterized. The character of the movement itself depends on the value of the control parameter. The numerical model can conduct a large number of different computer simulations in a short time. We started with simple low-dimensional deterministic systems that can exhibit diffusive behavior. Chaotic behavior is possible to be associated with diffusion in simple low-dimensional models, supporting the idea that chaos was at the very origin of diffusion [47].

Deterministic diffusion is a phenomenon also present in chaotic maps on the line. Many researchers are dealing with this phenomenon [5]. In 1908, Langevin used a stochastic differential equation to describe slow changes in macroscopic variables. A Stokes viscous force and a fluctuating random force with a Gaussian distribution act on the particle. Einstein views Brownian motion as diffusion. The Langevin equations are defined as stochastic equations [48]. Fujisaka and Grossmann worked on the dynamical theory of Brownian motion [5]. A one-dimensional discrete-time dynamical system example can be given by Equation (25):

$$x(t+1) = [x(t)] + F(x(t) - [x(t)]),\qquad(25)$$

where $x(t)$ is the position of the particle x as a function of time t that is performed by the diffusion in the real axis. The brackets [ ] denote an integer number of arguments. $F(u)$ is a map defined at interval [0, 1]. Based on the Langevin equation, we can observe the Brownian motion of a particle of mass m in a two-dimensional model as follows:

$$\frac{d^2\varsigma}{dt^2} = -\beta\frac{d\varsigma}{dt} + f_\varsigma,\qquad(26)$$

where:

$$\beta = \frac{6\pi\eta r}{m}, f_\varsigma = \frac{F_\varsigma}{m}, \varsigma = x, y,\qquad(27)$$

where η is fluid viscosity, r is particle radius and m is particle mass. In Equation (26), the Stokes viscous force and a fluctuating random force with a Gaussian distribution act on the particle [12].

3.2. Chaotic Model

Considering the equations given in a previous section, the chaotic model can be mapped as given in Equation (28). Equation (28) is derived from Equation (25) and describes the Brownian motion of particles in a two-dimensional chaotic model, where variable t is substituted by $\zeta = x, y$.

$$\xi(t+1) = [\xi(t)] + F(\xi(t) - [\xi(t)]); \ \xi = x, y, \tag{28}$$

where $[\zeta]$ is the integer part of ζ while

$$F(u) = \begin{cases} 2(1+q)u, 0 \le u \le \frac{1}{2} \\ 2(1+q)(u-1)+1, \frac{1}{2} \le u \le 1 \end{cases}, \tag{29}$$

where *F(u)* is a map defined on the interval [0, 1] that fulfills the following properties:

(i). The map, *u(t + 1) = F (u(t))* (mod 1) is chaotic.
(ii). *F(u)* must be larger than 1 and smaller than 0 for some values of u, so there exists a non-vanishing probability to escape from each unit cell (a unit cell of real axis is every interval $C\ell \equiv [\ell, \ell + 1]$, with $\ell \in Z$); ℓ is a number from the group of integer numbers Z.
(iii). *Fr(u) = 1 − Fl(1 − u)*, where *Fl* and *Fr* define the map in $u \in [0, \frac{1}{2}]$ and $u \in [\frac{1}{2}, 1]$, respectively. This anti-symmetry condition with respect to *u = 1/2* is introduced to avoid a net drift.

If the theoretical model presented in our study is mapped by the chaotic model, it can be stated as in Equation (30). With the introduction of the previous *F(u)* sinusoidal function, Equation (28) can be accordingly stated as Equation (30).

$$\zeta(t+1)=\zeta(t)+p\sin(2\pi\zeta(t)), \tag{30}$$

where $\zeta = x, y$ is a time-dependent coordinate t and p is a control parameter that depends on the viscosity coefficient of the fluid.

When a series of computer experiments are performed where the parameter p changes, it is observed that the Brownian particle can be in a stochastic or chaotic motion. Legitimacy can be derived from the following. The particle motion has deterministic patterns for the following values of the parameter p:

$$p = N + \frac{1}{2}, \ N = 0, \frac{1}{2}, 1, \frac{3}{2}, 2, \frac{5}{2} \dots \tag{31}$$

Computer experiments showed that the ferrofluid can exhibit different modes of deterministic dynamics within the two-dimensional model, depending on the initial value of the parameter p, as shown in Figures 3–8. Figure 3 shows the linear trajectory of the ferrofluid. Figures 4 and 5 are classic examples of the chaotic motion of a particle. Figures 6–8 show the transition from a deterministic to chaotic state of the system. This is demonstrated by the computer experiment where for the same initial condition (input value of the parameter p), the same patterns were obtained each time the computer simulation was repeated.

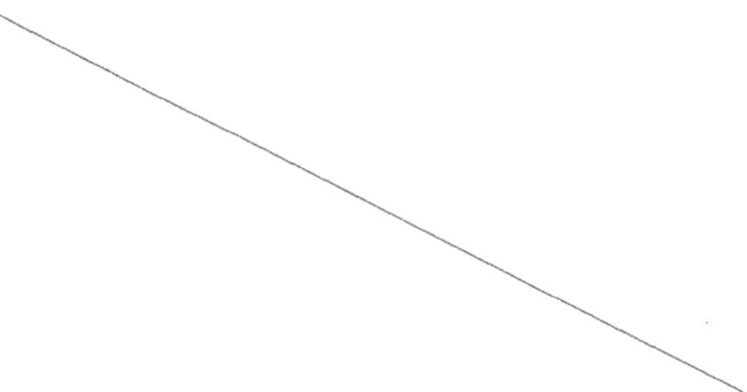

Figure 3. The linear trajectory of the particle in a two-dimensional chaotic model for the value of parameter p = 1.0.

Figure 4. The trajectory of the particle in a two-dimensional chaotic model for the value of parameter, $p = 0.9$.

Figure 5. Trajectory of the particle in a two-dimensional chaotic model for the value of parameter, $p = 0.8$.

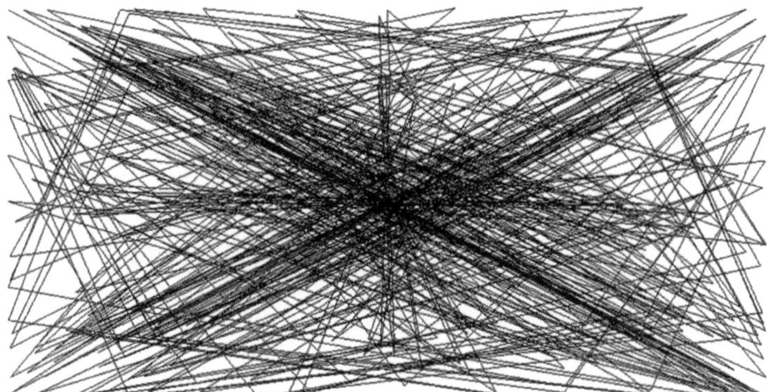

Figure 6. Trajectory of the particle in a two-dimensional chaotic model for the value of parameter, $p = 0.7$.

Figure 7. Trajectory of the particle in a two-dimensional chaotic model for the value of parameter, $p = 0.6$.

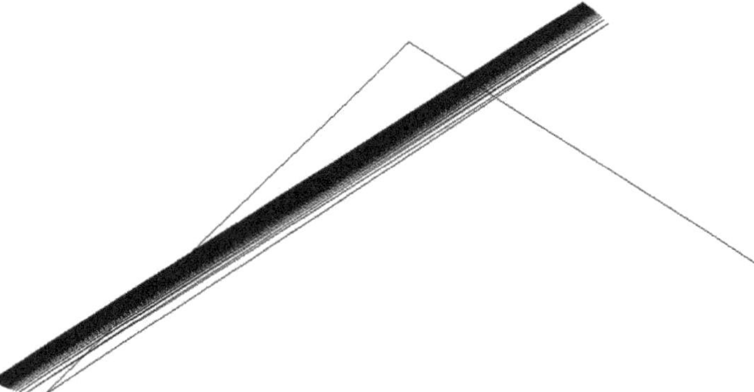

Figure 8. The trajectory of the particle in a two-dimensional chaotic model for the value of parameter, $p = 0.5$.

The condition of the system is described by the vector $\xi(t)$ of d dimension—Equation (30). The trajectory is discretized in time where the discretization step is τ and the vector of dimension d is introduced in Equation (32) with the associated string (of m length) given in Equation (33).

$$\Xi^m(t) = (\xi(t), \xi(t + \tau), \dots, \xi(t + m\tau - \tau)) \tag{32}$$

$$W^m(\varepsilon, t) = (i(\varepsilon, t), i(\varepsilon, t + \tau), \ldots, i(\varepsilon, t + m\tau - \tau)) \tag{33}$$

where $i(\varepsilon, t + j\tau)$ denotes the cell of $\zeta(t + j\tau)$, with a length of ε.

The value of 0.3060 for Kolmogorov–Sinai entropy (equal to the sum of positive Lyapunov exponents) was obtained, calculated according to the following equations:

$$h_{KS} = \lim_{\varepsilon \to 0} h(\varepsilon, \tau) \tag{34}$$

$$h(\varepsilon, \tau) = \frac{1}{\tau} \lim_{m \to \infty} \frac{1}{m} H_m(\varepsilon, \tau) \tag{35}$$

$$H_m(\varepsilon, \tau) = - \sum_{W^m(\varepsilon)} P(W^m(\varepsilon)) \ln P(W^m(\varepsilon)) \tag{36}$$

Variable h_{KS} has a value between zero and infinity, thus proving that the system is chaotic.

4. Discussion

The behavior of a particle (ferrofluid) moving in a fluid under the influence of a gravitational field without the presence of an external magnetic field is observed. Likewise, the delivery of drugs to the body could be possible without the presence of any electric or magnetic field, but only under the influence of the gravitational field. Accordingly, Brownian motion is studied, which, under the influence of the gravitational field, can be stochastic, deterministic or chaotic. Different models of the aforementioned movement have been observed, showing stochastic and chaotic movement [10–13]. Based on the model in our study, it can be observed that the particle moves randomly for certain values of the control parameter p and exhibits linearity in motion for other values of the parameter. The control parameter affects the movement of the particle. Linear motion of the particles was observed for certain values of the parameter p (as shown in Equation (31) and Figure 3). For other values of the parameter p, the particles move randomly without any rule (Figures 4 and 5). It can be noticed from Figures 4–8 that even the chaotic motion can exhibit patterns of a deterministic movement for certain material properties of the particles (and the surrounding fluid, as well as their interrelated properties) that will result in the desired nanofluid behavior.

The control parameter, p, is related to the friction constant and viscosity coefficient. Friction constant in a fluid motion has a direct relation with Reynolds number that further determines whether the laminar or turbulent flow of fluid will occur. Parameter p can be further correlated to the Peclet number in a microfluidic setup, thus indicating advectively dominated distribution or diffuse fluid flow. Changes in the parameter p are associated with changes in the viscosity coefficient and particle mass and size. The rheological behavior of nanofluids is complex because the increase in volume fractions of nanoparticles in a fluid may result in non-Newtonian nanofluid, with more pronounced temperature effects on viscosity changes [26].

If we compare Figures 3–8, it can be seen that trajectory shapes were significantly changed for slight changes in parameter p: a value of 1 resulted in a fully linear trajectory, while a value of 0.9 produced a fully random path. A further decrease in p to the lowest value, 0.5, again introduced patterns of linearity within chaotic motion. Since parameter p is related to the viscosity and particle radius and mass, it could be assumed that such transitional behavior with changes in p can be attributed to a complex phenomenon underlining dependencies between viscosity and volume fractions of particles in nanofluids, consistent with [26]. The rheological behavior of nanofluids is dramatically different for Newtonian and non-Newtonian nanofluids and dynamic transitions among these two modes, as influenced by the changes in viscosity are still not fully clarified. Figure 3 (with the highest p value) and Figure 8 (with the lowest p value) exhibit certain similarities in trajectory pattern, since both of these have fully linear parts of trajectories, in accordance with values in Equation (31). For values of p in between these numbers (Equation (31)), fully random trajectories were generated by simulation. However, for both of these regimes (random and linear trajectories), a decrease in p value produced a closer path (denser total trajectory), in accordance with the fact that viscosity decreases with the decrease in parameter p.

Accordingly, we could tailor the trajectory path of the particle in the liquid, regardless of the exogenous power propulsion strategy (e.g., external magnetic field), by tailoring the values of the parameter p which is related to viscosity and volume fractions of nanoparticles in a fluid. This means that it could be possible to realize targeted drug delivery by designing the system of nanoparticles in a fluid media at certain temperatures, consistent with recent research articles [49]. Research showed that there is dependence between the motility of particles and the density of neighbors, which has been a foundation for designing self-organizing nanofluids for drug delivery by tailoring the active Brownian motion of the particle [50]. The density of trajectories in our simulation significantly changed with changes in parameter p (Figures 3–8), in accordance with research [50] that showed a different size, density and shape of nano-cluster aggregates due to changes in Brownian motion.

Fine tailoring of the Brownian motion can produce different desired effects, including tailoring of the time and amount of the drug release [51]. On the other hand, drug delivery systems based on micro/nanomotors have been designed to overcome the influence of the Brownian motion through the control of nanoparticles' motion by some exogenous force (like external magnetic field) [49]. If the immobilization of nanoparticles increases, heating efficiency decreases [51]. How is this related to the confinement of the space within which the particles' trajectories can appear (as in the case of path shown at Figure 8) has not been the study yet, even though there are some studies related to the nanoparticle motion in a cylindrical tube and associated effects of the boundaries, curvature, size and density of the particle, including the influence of the Brownian dynamics [52] and transport phenomena in confined flows of nanoparticles [53]. Tuning of polymer amphiphilicity can increase the efficiency of drug delivery systems [54]. Amphiphilicity has direct influence on the particle collision modes, thus indicating that chaotic models of Brownian motion might exhibit patterns in particle trajectories for certain conditions.

Néel relaxation of magnetic nanoparticles has been studied, but the study on the correlation of Brownian motion to another magnetic relaxation mechanism is recent, showing the influence of Brownian relaxation on nanocage size [55]. There are complex interactions in the coupling of Brownian and Néel relaxation processes [56], which produces a highly nonlinear field-dependent magnetization response, including the pronounced influence of the size of nanoparticles clusters [57].

There is a correlation between the magnetization curve of the ferromagnetic particles system and Langevin curve [58]. If we observe single-domain ferromagnetic particles, their magnetic behavior at elevated temperatures can be correlated to the atomistic Langevin paramagnetism [59]. On the other hand, changes in temperature result in viscosity changes; thus, it is reasonable to expect that we could apply our model to a colloidal suspension of single-domain magnetic particles—ferrofluid, as described in previous chapters. The magnitude of the uniform magnetization vector for a single-domain ferromagnetic particle is proven to be constant with the direction of fluctuation based on a random motion of particles due to the heat changes (thermal agitation) [60]. Accordingly, the deterministic stochastic processes might be representative of such a process, meaning that the Langevin equation is relevant [60]. In the case of our model, we assume that parameter p has a correlation to the viscosity coefficient, particle radius and its mass, which further influences the degree of deterministic behavior of the chaotic system, as shown in Figures 3–8. However, further study is needed in relation to additional parameters that describe the magnetic behavior of ferrofluids.

Based on the above results, it can be concluded that the delivery of drugs could be executed without the presence of an external magnetic or electric field. Patterns of deterministic trajectories can be designed by predefined values of the parameter p in a computer simulation, which can further lead to the design of the nanoparticle system for targeted drug delivery without an exogenous power propulsion strategy. However, complex relations between different influential factors need further study, including further development of the theoretical model for the motion of nanoparticles in an external field and fluid environment. The significance of such a dynamic model for the development of drug delivery systems is related to the possibility to control the motion of the drug-containing nanoparticles, through the design of the inherent material properties of the particles and surrounding media. The possibility to use and influence Brownian motion to produce patterned particles' trajectories in diffusive motion of the ferrofluid aiming to assist in more efficient drug delivery systems of ferromagnetic nanofluids would support significant advancements in medical treatments.

5. Conclusions

A chaotic model of Brownian motion was theoretically analyzed and simulated using Maple software. The chaotic model was mapped and control parameter p was introduced, which depends on the viscosity coefficient and particle mass and size, in analogy with the Langevin equation. The ferrofluid in the gravitational field without the presence of an external magnetic field in a two-dimensional mathematical model was observed. It performed 400 collisions with fluid molecules.

Computer simulation showed that nanoparticles can exhibit deterministic patterns in a chaotic model for certain material properties of the particles (and the surrounding fluid, as well as their interrelated properties) that could result in the controlled nanofluid behavior. Trajectory shapes were significantly changed for slight changes in the parameter p: a value of 1 resulted in a fully linear trajectory, while a value of 0.9 produced a fully random path. The lowest value of p (0.5) introduced patterns of linearity within chaotic motion, with noticed changes in the shape and density of trajectories. Since parameter p is related to the fluid viscosity and particle radius and mass, it could be assumed that such transitional behavior with changes in the parameter p can be attributed to a complex phenomenon underlining dependencies between viscosity and volume fractions of particles in nanofluids. For values of p in between the designated numbers (Equation (31)), fully random trajectories were generated by simulation. Accordingly, we could tailor the trajectory path of the particle in the liquid, regardless of the exogenous power propulsion strategy (e.g., external magnetic field), by tailoring the values of the parameter p, which is related to viscosity and volume fractions of nanoparticles in a fluid.

Fine tailoring of the Brownian motion can produce different desired effects, including tailoring of the time and amount of the drug release. Patterns of deterministic trajectories can be designed by predefined values of the parameter p in a computer simulation, which can further lead to the design of the nanoparticle system for targeted drug delivery without an exogenous power propulsion strategy (e.g., external magnetic field). However, complex relations between different influential factors need further study, including further development of the theoretical model that will consider magnetic properties of the nanoparticles in a ferrofluid.

Author Contributions: Conceptualization, S.N., J.R., F.Ž., M.V.J. and N.G.; methodology, S.N., J.R., F.Ž., A.M. and N.G.; software, S.N., M.V.J. and N.G.; validation, S.N., F.Ž. and Ž.J.P.; formal analysis, J.R., F.Ž., A.M., M.V.J. and N.G.; investigation, S.N., J.R. and F.Ž.; resources, S.N., F.Ž. and Ž.J.P.; data curation, S.N., F.Ž. and Ž.J.P.; writing—original draft preparation, S.N., F.Ž., A.M. and Ž.J.P.; writing—review and editing, S.N., J.R., F.Ž., A.M., Ž.J.P., M.V.J. and N.G.; visualization, S.N., F.Ž. and A.M.; supervision, J.R., F.Ž. and N.G.; funding acquisition, F.Ž. All authors have read and agreed to the published version of the manuscript.

Funding: This paper is funded through the EIT's HEI Initiative SMART-2M project, supported by EIT RawMaterials, funded by the European Union.

Data Availability Statement: Not applicable.

Conflicts of Interest: The authors declare no conflict of interest. The funders had no role in the design of the study; in the collection, analyses, or interpretation of data; in the writing of the manuscript; or in the decision to publish the results.

References

1. Mörters, P.; Peres, Y.; Schramm, O.; Werner, W. *Brownian Motion*; Cambridge series in statistical and probabilistic mathematics; Cambridge University Press: Cambridge, UK; New York, NY, USA, 2010; ISBN 9780521760188.
2. Topping, J. Investigations on the Theory of the Brownian Movement. *Phys. Bull.* **1956**, *7*, 281. [CrossRef]
3. Mori, H. Transport, Collective Motion, and Brownian Motion. *Prog. Theor. Phys.* **1965**, *33*, 423–455. [CrossRef]
4. Caldeira, A.O.; Leggett, A.J. Path Integral Approach to Quantum Brownian Motion. *Phys. A Stat. Mech. Its Appl.* **1983**, *121*, 587–616. [CrossRef]
5. Fujisaka, H.; Grossmann, S. Chaos-Induced Diffusion in Nonlinear Discrete Dynamics. *Z. Phys. B Condens. Matter* **1982**, *48*, 261–275. [CrossRef]
6. Gaspard, P.; Briggs, M.E.; Francis, M.K.; Sengers, J.V.; Gammon, R.W.; Dorfman, J.R.; Calabrese, R.V. Experimental Evidence for Microscopic Chaos. *Nature* **1998**, *394*, 865–868. [CrossRef]
7. Cecconi, F.; Cencini, M.; Falcioni, M.; Vulpiani, A. Brownian Motion and Diffusion: From Stochastic Processes to Chaos and Beyond. *Chaos* **2005**, *15*, 026102. [CrossRef]
8. Cencini, M.; Falcioni, M.; Olbrich, E.; Kantz, H.; Vulpiani, A. Chaos or Noise: Difficulties of a Distinction. *Phys. Rev. E* **2000**, *62*, 427–437. [CrossRef]
9. Peredo-Ortíz, R.; Hernández-Contreras, M.; Hernández-Gómez, R. Magnetic Viscoelastic Behavior in a Colloidal Ferrofluid. *J. Chem. Phys.* **2020**, *153*, 184903. [CrossRef]
10. Bass, R.F. *Stochastic Processes*, 1st ed.; Cambridge University Press: Cambridge, UK, 2011; ISBN 9781107008007.
11. Martín-Pasquín, F.J.; Pisarchik, A.N. Brownian Behavior in Coupled Chaotic Oscillators. *Mathematics* **2021**, *9*, 2503. [CrossRef]
12. Huerta-Cuellar, G.; Jiménez-López, E.; Campos-Cantón, E.; Pisarchik, A.N. An Approach to Generate Deterministic Brownian Motion. *Commun. Nonlinear Sci. Numer. Simul.* **2014**, *19*, 2740–2746. [CrossRef]
13. Echenausía-Monroy, J.L.; Campos, E.; Jaimes-Reátegui, R.; García-López, J.H.; Huerta-Cuellar, G. Deterministic Brownian-like Motion: Electronic Approach. *Electronics* **2022**, *11*, 2949. [CrossRef]
14. Dhiman, J.S.; Sood, S. Linear and Weakly Non-Linear Stability Analysis of Oscillatory Convection in Rotating Ferrofluid Layer. *Appl. Math. Comput.* **2022**, *430*, 127239. [CrossRef]
15. Rickert, W.; Winkelmann, M.; Müller, W.H. Modeling the Magnetic Relaxation Behavior of Micropolar Ferrofluids by Means of Homogenization. In *Theoretical Analyses, Computations, and Experiments of Multiscale Materials*; Giorgio, I., Placidi, L., Barchiesi, E., Abali, B.E., Altenbach, H., Eds.; Springer International Publishing: Cham, Switzerland, 2022; Volume 175, pp. 473–486. ISBN 9783031045479.
16. Xu, H.; Dai, Q.; Huang, W.; Wang, X. The Supporting Capacity of Ferrofluids Bearing: From the Liquid Ring to Droplet. *J. Magn. Magn. Mater.* **2022**, *552*, 169212. [CrossRef]
17. Ivanov, A.O.; Camp, P.J. Effects of Interactions, Structure Formation, and Polydispersity on the Dynamic Magnetic Susceptibility and Magnetic Relaxation of Ferrofluids. *J. Mol. Liq.* **2022**, *356*, 119034. [CrossRef]
18. Yang, W.; Zhang, Y.; Yang, X.; Sun, C.; Chen, Y. Systematic Analysis of Ferrofluid: A Visualization Review, Advances Engineering Applications, and Challenges. *J. Nanopart. Res.* **2022**, *24*, 102. [CrossRef]
19. Klein, Y.P.; Abelmann, L.; Gardeniers, H. Ferrofluids to Improve Field Homogeneity in Permanent Magnet Assemblies. *J. Magn. Magn. Mater.* **2022**, *555*, 169371. [CrossRef]
20. Déjardin, J.-L.; Kachkachi, H. Time Profile of Temperature Rise in Assemblies of Nanomagnets. *J. Magn. Magn. Mater.* **2022**, *556*, 169354. [CrossRef]
21. Alla, S.K.; Yeddu, V.; Prasad Rao, E.V.; Mandal, R.K.; Prasad, N.K. Synthesis and Characterization of Manganese Substituted Cerium Oxide Nanoparticles by Microwave Refluxing Method. *MSF* **2015**, *830–831*, 608–611.
22. Larson, R.G. *The Structure and Rheology of Complex Fluids*; Oxford University Press: Oxford, UK; New York, NY, USA, 1999; pp. 801–802. [CrossRef]
23. Boroomandpour, A.; Toghraie, D.; Hashemian, M. A Comprehensive Experimental Investigation of Thermal Conductivity of a Ternary Hybrid Nanofluid Containing MWCNTs- Titania-Zinc Oxide/Water-Ethylene Glycol (80:20) as Well as Binary and Mono Nanofluids. *Synth. Met.* **2020**, *268*, 116501. [CrossRef]
24. Jolfaei, N.A.; Jolfaei, N.A.; Hekmatifar, M.; Piranfar, A.; Toghraie, D.; Sabetvand, R.; Rostami, S. Investigation of Thermal Properties of DNA Structure with Precise Atomic Arrangement via Equilibrium and Non-Equilibrium Molecular Dynamics Approaches. *Comput. Methods Programs Biomed.* **2020**, *185*, 105169. [CrossRef]
25. He, W.; Ruhani, B.; Toghraie, D.; Izadpanahi, N.; Esfahani, N.N.; Karimipour, A.; Afrand, M. Using of Artificial Neural Networks (ANNs) to Predict the Thermal Conductivity of Zinc Oxide–Silver (50%–50%)/Water Hybrid Newtonian Nanofluid. *Int. Commun. Heat Mass Transf.* **2020**, *116*, 104645. [CrossRef]
26. Yan, S.-R.; Toghraie, D.; Abdulkareem, L.A.; Alizadeh, A.; Barnoon, P.; Afrand, M. The Rheological Behavior of MWCNTs–ZnO/Water–Ethylene Glycol Hybrid Non-Newtonian Nanofluid by Using of an Experimental Investigation. *J. Mater. Res. Technol.* **2020**, *9*, 8401–8406. [CrossRef]

27. Landers, J.; Salamon, S.; Webers, S.; Wende, H. Microscopic Understanding of Particle-Matrix Interaction in Magnetic Hybrid Materials by Element-Specific Spectroscopy. *Phys. Sci. Rev.* **2021**, *0*, 20190116. [CrossRef]
28. Itzykson, C.; Drouffe, J.-M. *Statistical Field Theory. Volume 1. From Brownian Motion to Renormalization and Lattice Gauge Theory*; Cambridge University Press: Cambridge, UK, 1989; ISBN 9780511622779.
29. Rablau, C.; Vaishnava, P.; Sudakar, C.; Tackett, R.; Lawes, G.; Naik, R. Magnetic-Field-Induced Optical Anisotropy in Ferrofluids: A Time-Dependent Light-Scattering Investigation. *Phys. Rev. E* **2008**, *78*, 051502. [CrossRef] [PubMed]
30. Rigoni, C.; Beaune, G.; Harnist, B.; Sohrabi, F.; Timonen, J.V.I. Ferrofluidic Aqueous Two-Phase System with Ultralow Interfacial Tension and Micro-Pattern Formation. *Commun. Mater* **2022**, *3*, 26. [CrossRef]
31. Scherer, C.; Figueiredo Neto, A.M. Ferrofluids: Properties and Applications. *Braz. J. Phys.* **2005**, *35*, 718–727. [CrossRef]
32. Berger, P.; Adelman, N.B.; Beckman, K.J.; Campbell, D.J.; Ellis, A.B.; Lisensky, G.C. Preparation and Properties of an Aqueous Ferrofluid. *J. Chem. Educ.* **1999**, *76*, 943. [CrossRef]
33. Wahajuddin, A. Superparamagnetic Iron Oxide Nanoparticles: Magnetic Nanoplatforms as Drug Carriers. *Int. J. Nanomed.* **2012**, *7*, 3445–3471. [CrossRef]
34. Chourpa, I.; Douziech-Eyrolles, L.; Ngaboni-Okassa, L.; Fouquenet, J.-F.; Cohen-Jonathan, S.; Soucé, M.; Marchais, H.; Dubois, P. Molecular Composition of Iron Oxide Nanoparticles, Precursors for Magnetic Drug Targeting, as Characterized by Confocal Raman Microspectroscopy. *Analyst* **2005**, *130*, 1395. [CrossRef]
35. Kandasamy, G.; Sudame, A.; Maity, D.; Soni, S.; Sushmita, K.; Veerapu, N.S.; Bose, S.; Tomy, C.V. Multifunctional Magnetic-Polymeric Nanoparticles Based Ferrofluids for Multi-Modal in Vitro Cancer Treatment Using Thermotherapy and Chemotherapy. *J. Mol. Liq.* **2019**, *293*, 111549. [CrossRef]
36. Katz, E. Synthesis, Properties and Applications of Magnetic Nanoparticles and Nanowires—A Brief Introduction. *Magnetochemistry* **2019**, *5*, 61. [CrossRef]
37. Stergar, J.; Ban, I.; Maver, U. The Potential Biomedical Application of NiCu Magnetic Nanoparticles. *Magnetochemistry* **2019**, *5*, 66. [CrossRef]
38. Kianfar, E. Magnetic Nanoparticles in Targeted Drug Delivery: A Review. *J. Supercond. Nov. Magn.* **2021**, *34*, 1709–1735. [CrossRef]
39. Price, P.M.; Mahmoud, W.E.; Al-Ghamdi, A.A.; Bronstein, L.M. Magnetic Drug Delivery: Where the Field Is Going. *Front. Chem.* **2018**, *6*, 619. [CrossRef] [PubMed]
40. Cheng, J.; Teply, B.A.; Jeong, S.Y.; Yim, C.H.; Ho, D.; Sherifi, I.; Jon, S.; Farokhzad, O.C.; Khademhosseini, A.; Langer, R.S. Magnetically Responsive Polymeric Microparticles for Oral Delivery of Protein Drugs. *Pharm. Res.* **2006**, *23*, 557–564. [CrossRef] [PubMed]
41. McBain, S.C.; Yiu, H.H.P.; Dobson, J. Dobson Magnetic Nanoparticles for Gene and Drug Delivery. *Int. J. Nanomed.* **2008**, *3*, 169–180. [CrossRef]
42. Ferrari, M. Nanovector Therapeutics. *Curr. Opin. Chem. Biol.* **2005**, *9*, 343–346. [CrossRef]
43. Yamaoka, T.; Tabata, Y.; Ikada, Y. Distribution and Tissue Uptake of Poly(Ethylene Glycol) with Different Molecular Weights after Intravenous Administration to Mice. *J. Pharm. Sci.* **1994**, *83*, 601–606. [CrossRef]
44. Arruebo, M.; Fernández-Pacheco, R.; Ibarra, M.R.; Santamaría, J. Magnetic Nanoparticles for Drug Delivery. *Nano Today* **2007**, *2*, 22–32. [CrossRef]
45. Grigolini, P.; Rocco, A.; West, B.J. Fractional Calculus as a Macroscopic Manifestation of Randomness. *Phys. Rev. E* **1999**, *59*, 2603–2613. [CrossRef]
46. Arfken, G.B.; Weber, H.-J. *Mathematical Methods for Physicists*, 6th ed.; Elsevier: Boston, MA, USA, 2005; ISBN 9780120598762.
47. Dettmann, C.P.; Cohen, E.G.D. Note on chaos and diffusion. *J. Stat. Phys.* **2001**, *103*, 589–599. [CrossRef]
48. Naqvi, K.R. The Origin of the Langevin Equation and the Calculation of the Mean Squared Displacement: Let's Set the Record Straight. *arXiv* **2005**. [CrossRef]
49. Zhang, W.; Zhang, Z.; Fu, S.; Ma, Q.; Liu, Y.; Zhang, N. Micro/Nanomotor: A Promising Drug Delivery System for Cancer Therapy. *ChemPhysMater* **2022**, *In Press, Corrected Proof.* S2772571522000444. [CrossRef]
50. Bäuerle, T.; Fischer, A.; Speck, T.; Bechinger, C. Self-Organization of Active Particles by Quorum Sensing Rules. *Nat. Commun.* **2018**, *9*, 3232. [CrossRef]
51. Perera, A.S.; Jackson, R.J.; Bristow, R.M.D.; White, C.A. Magnetic Cryogels as a Shape-Selective and Customizable Platform for Hyperthermia-Mediated Drug Delivery. *Sci. Rep.* **2022**, *12*, 9654. [CrossRef]
52. Vitoshkin, H.; Yu, H.-Y.; Eckmann, D.M.; Ayyaswamy, P.S.; Radhakrishnan, R. Nanoparticle Stochastic Motion in the Inertial Regime and Hydrodynamic Interactions Close to a Cylindrical Wall. *Phys. Rev. Fluids* **2016**, *1*, 054104. [CrossRef]
53. Radhakrishnan, R.; Farokhirad, S.; Eckmann, D.M.; Ayyaswamy, P.S. Nanoparticle Transport Phenomena in Confined Flows. In *Advances in Heat Transfer*; Elsevier: Amsterdam, The Netherlands, 2019; Volume 51, pp. 55–129. ISBN 9780128177006.
54. Horvat, S.; Yu, Y.; Böjte, S.; Teßmer, I.; Lowman, D.W.; Ma, Z.; Williams, D.L.; Beilhack, A.; Albrecht, K.; Groll, J. Engineering Nanogels for Drug Delivery to Pathogenic Fungi Aspergillus Fumigatus by Tuning Polymer Amphiphilicity. *Biomacromolecules* **2020**, *21*, 3112–3121. [CrossRef]
55. Kang, M.A.; Fang, J.; Paragodaarachchi, A.; Kodama, K.; Yakobashvili, D.; Ichiyanagi, Y.; Matsui, H. Magnetically Induced Brownian Motion of Iron Oxide Nanocages in Alternating Magnetic Fields and Their Application for Efficient SiRNA Delivery. *Nano Lett.* **2022**, *22*, 8852–8859. [CrossRef]

56. Ilg, P.; Kröger, M. Dynamics of Interacting Magnetic Nanoparticles: Effective Behavior from Competition between Brownian and Néel Relaxation. *Phys. Chem. Chem. Phys.* **2020**, *22*, 22244–22259. [CrossRef]
57. Trisnanto, S.B.; Takemura, Y. Effective Néel Relaxation Time Constant and Intrinsic Dipolar Magnetism in a Multicore Magnetic Nanoparticle System. *J. Appl. Phys.* **2021**, *130*, 064302. [CrossRef]
58. Elmore, W.C. The Magnetization of Ferromagnetic Colloids. *Phys. Rev.* **1938**, *54*, 1092–1095. [CrossRef]
59. Bean, C.P.; Livingston, J.D. Superparamagnetism. *J. Appl. Phys.* **1959**, *30*, S120–S129. [CrossRef]
60. Brown, W.F. Thermal Fluctuations of a Single-Domain Particle. *Phys. Rev.* **1963**, *130*, 1677–1686. [CrossRef]

 *mathematics*

Article

Understanding the Parameter Influence on Lesion Growth for a Mechanobiology Model of Atherosclerosis

Patricia Hernández-López [1,*], Miguel A. Martínez [1,2], Estefanía Peña [1,2] and Myriam Cilla [1,2,*]

1   Applied Mechanics and Bioengineering, Aragón Institute of Engineering Research (I3A), University of Zaragoza, 50018 Zaragoza, Spain
2   CIBER de Bioingeniería, Biomateriales y Nanomedicina (CIBER-BBN), 50018 Zaragoza, Spain
*   Correspondence: phernand@unizar.es (P.H.-L.); mcilla@unizar.es (M.C.)

Abstract: In this work, we analyse the influence of the parameters of a mathematical model, previously proposed by the authors, for reproducing atheroma plaque in arteries. The model uses Navier–Stokes equations to calculate the blood flow along the lumen in a transient mode. It also uses Darcy's law, Kedem–Katchalsky equations, and the three-pore model to simulate plasma and substance flows across the endothelium. The behaviours of all substances in the arterial wall are modelled with convection–diffusion–reaction equations, and finally, plaque growth is calculated. We consider a 2D geometry of a carotid artery, but the model can be extrapolated to other geometries or arteries, such as the coronaries or the aorta. A mono-variant sensitivity analysis of the model parameters was performed, with values of ±25% and ±10%, with respect to the values of the previous model. The results were analysed with respect to the volume in the plaque of foam cells (FC), synthetic smooth muscle cells (SSMC), and collagen fibre. It was observed that the volume in the plaque of the different substances (FC, SSMC, and collagen) has a strong influence on the results, so it could be used to analyse the vulnerability of plaque. The stenosis ratio of the plaque was also analysed, showing a strong influence on the results as well. Parameters that influence all the results considered when ranged ±10% are the rate of LDL degradation and the diffusion coefficients of LDL and monocytes in the arterial wall. Furthermore, it was observed that the change in the volume of foam cells in the plaque has a greater influence on the stenosis ratio than the change of synthetic smooth muscle cells or collagen fibre.

Keywords: atherosclerosis; mechanobiological model; parameter analysis; carotid artery

MSC: 92-08

Citation: Hernández-López, P.; Martínez, M.A.; Peña, E.; Cilla, M. Understanding the Parameter Influence on Lesion Growth for a Mechanobiology Model of Atherosclerosis. *Mathematics* **2023**, *11*, 829. https://doi.org/10.3390/math11040829

Academic Editor: Fernando Simões

Received: 22 December 2022
Revised: 31 January 2023
Accepted: 2 February 2023
Published: 6 February 2023

1. Introduction

Cardiovascular diseases, including atherosclerosis, are some of the main causes of mortality in developed countries [1]. They consist of the development of atheroma plaque in the arterial wall, which is caused by lipid deposition. This leads to an increase in the thickness of the arterial wall and, therefore, to a decrease in the area where blood flows, called the lumen. In addition, plaque can break and cause a blood clot that can travel through the circulatory system. For these reasons, it can lead to several consequences that are dependent on the artery affected, e.g., myocardial attacks, ischaemia, or ictus, among others.

The process of atheroma plaque formation begins with the deposition of low-density lipoproteins (LDLs) in the arterial wall and, once there, oxidise. Then, monocytes are deposited from the lumen into the arterial wall and differentiate into macrophages, which ingest oxidised LDL. Once they cannot ingest more oxidised LDL, they become foam cells. These foam cells compose the lipid core of the atheroma plaque. Concurrently, at the beginning of the inflammatory process, all muscular cells in the arterial wall are contractile, i.e., they cannot move or react with any other substance. However, due to the presence of cytokines in the arterial wall (segregated by macrophages), these muscle cells change their phenotype and become synthetic smooth muscle cells, which can move and interact with other substances in the arterial wall. Then synthetic smooth muscle cells surround foam cells, macrophages, and oxidised LDL, and segregate collagen fibre to isolate the lipid core forming the fibrous layer of the plaque [2–4].

Therefore, atheroma plaques are developed as a consequence of an increase in endothelial permeability. This variation in permeability was accepted to be due to a change in the shape of endothelial cells, depending on some mechanical stimuli caused by blood flow [5]. These mechanical stimuli could be, among others, the wall shear stress of blood with the endothelium (WSS), time-averaged wall shear stress ($TAWSS$), and oscillatory shear index (OSI) in a cardiac cycle [6,7], or a combination of some of them [8].

There are some studies in the literature related to the influence of blood hemodynamics on the location of atheroma plaque for patient-specific geometries. Malvè et al. [9] modelled a carotid artery bifurcation and analysed the influence of $TAWSS$ and OSI on the shape index (SI) of endothelial cells. Sáez et al. [6] analysed WSS on eight different geometries of coronary bifurcations. Alimohammadi et al. [7] analysed $TAWSS$, OSI, and a new index that they previously proposed [10], Highly Oscillatory Low Magnitude Shear (HOLMES), in the bifurcation of the abdominal aorta.

Some continuum mathematical models of the development of atherosclerosis have been developed. There is a huge quantity of these models developed as two-dimensional axisymmetric models. Olgac et al. [11] consider the LDL flow from the lumen to the arterial wall with the three-pore model, depending on the WSS, while Tomaso et al. [12] considered the flow of monocytes as well as LDL flow. Calvez et al. [13] used a two-dimensional model which, in addition to LDL and monocytes, also considered macrophages, cytokines, and foam cells. Finally, they modelled the fluid–structure interaction between the mesh displacement due to plaque formation and blood flow. Cilla et al. [4] modelled a two-dimensional axisymmetric geometry with LDL and monocyte flows into the arterial wall depending on WSS and the three-pore model, and the interactions between all the substances commented on before, as well as smooth muscle cells and collagen fibre. In addition, Shahzad et al. [14] studied the influence of hemodynamics in a two-dimensional geometry of an artery bifurcation with an obstacle plaque that disturbs blood flow. They used the fluid–structure interaction (FSI) with an elastic wall and computational fluid dynamics (CFD) considering a rigid wall.

There are also mechanobiological models with patient-specific geometries, such as by Siogkas et al. [15], who consider LDL that becomes oxidised, monocytes that differentiate into macrophages, and cytokines in the arterial wall. Filipovic et al. [16] modelled oxidised LDL, macrophages, and cytokines, depending on the WSS on coronary arteries, and adjusted the parameters of their model in order to obtain results based on experimental data. Hernández-López et al. [8] included also foam cells, smooth muscle cells, and collagen fibre, and analysed the effect of mechanical stimuli on the growth of atheroma plaque in carotid arteries. Kenjereš and De Loor [17] developed a computational model to simulate the flow of LDL through the arterial wall in a realistic geometry of a carotid bifurcation with a multilayered wall.

On the other hand, there are some agent-based models that study plaque development in the arteries [18–20]. The main advantage of continuum models against agent-based models is that they allow the simulation of plaque development in real geometries, whereas the advantage of agent-based models is that they can take into account the random behaviour of cells, which is not possible with continuum models. Olivares et al. [18] developed a model for early-stage atherosclerosis, in which they considered substances such as LDL, oxidised LDL, macrophage, foam cells, smooth muscle cells, endothelial cells, and autoantibodies. Bhui and Hayenga [19] developed a three-dimensional model of transendothelial migration during atherogenesis, in which they modelled endothelial cells, monocyte, macrophage, lymphocyte, neutrophil, and foam cells. In addition, Corti et al. [20] modelled atheroma plaque growth depending on *WSS* in three-dimensional geometry. They considered a multilayered wall, composed of the intima, media, and adventitia layers, modelling also the internal and external elastic laminae.

There is a large number of substances and parameters involved in the process of formation of atheroma plaque, so it is important to understand how each substance and parameter influences plaque growth.

There are also some studies focused on parameter influence in previous related models, such as Tomaso et al. [12], who analysed the influence of different LDL concentrations in blood to determine how it affects plaque growth. Olivares et al. [18] analysed, in 8 cases, the influence in the plaque of the oxidation rate of LDL, migration speed of agents, and auto-antibodies maximum concentration. Cilla et al. [21] analysed the effect of transmural transport properties on atheroma plaque formation, attending to LDL and oxidised LDL diffusion coefficient anisotropy. Finally, Corti et al. [20] made both, mono-parametric and multi-parametric sensitivity analyses to evaluate the changes in their results caused by changes in the parameters of an agent-based model of plaque growth.

The aim of this study is to analyse the influence of the parameters of a previous computational model on the formation of atheroma plaque in the arteries [8], to determine the effect of these parameters on the growth and composition of atheroma plaque.

The mathematical model used has a total of 52 parameters, which come from different studies and can be related to experimental or computational analysis. Among the experimental ones, there are some differences between the analysis conditions. For example, some of the parameters have been determined from in vitro tests, while others are from in vivo tests. Moreover, they come from studies of different species and arteries (coronary, carotid, or aorta). Finally, some of them have been estimated based on computational results. Therefore, the values of the parameters can have a large variation, so it is relevant to perform a sensitivity analysis of the parameters of the model.

Although the study of parameter variation was performed in a geometry of a carotid artery, the influence of the parameters in the model would be the same in the case of other arteries or geometries.

2. Geometry

Due to the high computational requirements of the model and because the objective of this study was to analyse the influences of the different parameters on plaque growth, a 2D axisymmetric approximation was considered enough for this study.

The geometry was developed based on that of Olgac et al. [11] as it reproduces the mechanical stimuli to which real patients are subjected, leading to the appearance of plaque (*TAWSS* and *OSI*). The geometry was adapted to carotid arteries by modifying the diameter and thickness of the vessel to the corresponding values for carotid arteries: 3.63 mm and 0.7 mm, respectively [22]. Figure 1 shows the geometry, which is composed of the lumen and the arterial wall, considered a monolayer. It also has an obstacle plaque that disturbs blood flow, causing low *TAWSS* and high *OSI* downstream and favouring the appearance of a new plaque in that area. It is a phenomenon that was observed in real patients [23,24]. This second plaque is the one that is going to be analysed.

The model was discretised using the finite element method. A sensitivity analysis of the length of the geometry and the mesh was developed, obtaining that the minimal length of the artery to ensure a fully developed blood flow is 90 mm. The domains of the model were meshed using quadrilateral elements, with three boundary layers in the lumen near the endothelium. A sensitivity analysis of the mesh was performed in both lumen and arterial wall meshes, in order to determine the most suitable mesh to solve the model with minimal computational costs, but without affecting the results. The total number of elements is equal to 36,731 for the lumen and 36,036 for the arterial wall, resulting in a total number of elements in the geometry of 72,767. A discretisation with P1-P1 elements was made for the calculation of blood flow. Therefore, linear interpolation was considered for both the velocity and pressure of blood flow in the lumen of the artery. To perform the sensitivity analysis of the mesh, a coarse mesh was selected in the first place and progressively refined, until the results obtained did not vary more than 5%. The same procedure was done with the number of boundary layers of the lumen mesh. In addition, linear and quadratic serendipity elements were used for the inflammatory process in the arterial wall and for plasma flow through the endothelium and the growth of the plaque, respectively. Two details of the selected mesh are shown in Figure 1. The solution to the transient problems was obtained using the Backwards differentiation formula (BDF) method, which is implicit, following Newton's method for nonlinear problems.

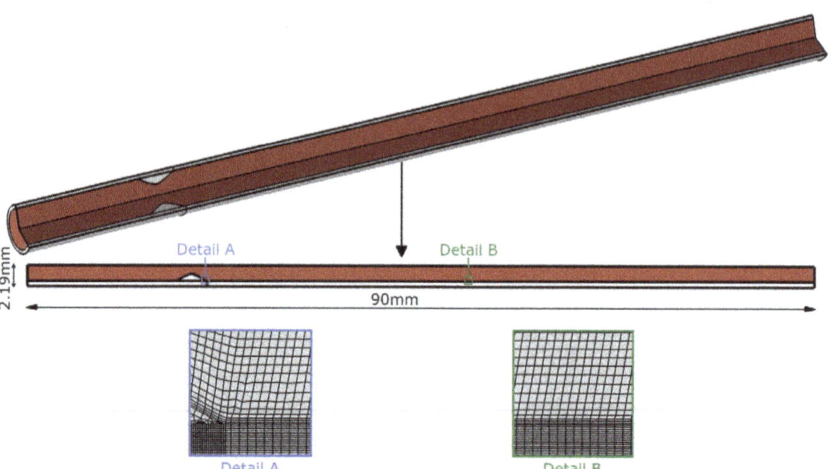

Figure 1. Two-dimensional axisymmetric geometry composed of lumen (in red), the arterial wall, and obstacle plaque (both in white color). Details (**A**,**B**) show the mesh near the endothelium at both sides (lumen and arterial wall) for the area near the obstacle plaque (**A**) and the area of development of the new plaque (**B**).

The total number of degrees of freedom of the lumen can be determined by multiplying the components of the velocity of blood and the number of elements in the lumen mesh. For the case of the arterial wall, the number of degrees of freedom would be equal to the multiplication of the components of the plasma flow velocity, the number of substance concentrations that we calculate and the number of elements of the arterial wall mesh.

3. Mathematical Model

In this section, the mathematical model is described, showing the equations in different subsections, referring to blood flow in the lumen, plasma flow across the endothelium, the inflammatory process in the arterial wall, and plaque growth.

3.1. Blood Flow in the Lumen

Blood flow along the arterial lumen is considered laminar due to the Reynolds number in the arteries under physiological conditions ($Re \approx 950$ for the mean diameter of the considered artery) [25]. Furthermore, blood is modelled as a Newtonian incompressible fluid, since the diameter of the lumen in the arteries is greater than 0.5 mm [26,27]. Moreover, it is modelled as a homogeneous fluid due to the small size of the particles with respect to the lumen diameter of the arteries [25].

The Navier–Stokes and continuity equations govern blood flow along the lumen:

$$\rho_b(u_l \cdot \nabla)u_l = \nabla \cdot [-P_l I + \mu_b(\nabla u_l + (\nabla u_l)^T)] + F_l \tag{1}$$

$$\rho_b \nabla \cdot u_l = 0, \tag{2}$$

where subscripts b and l refer to blood and lumen, respectively. The parameters ρ_b and μ_b are, respectively, the density and dynamic viscosity of blood, while u_l and P_l are the velocity and pressure of blood flow. F_l is the internal force of blood, which is negligible compared to the friction of blood flow with the arterial wall. All parameters referring to blood flow along the lumen are in Table 1.

Table 1. Parameters to calculate blood flow.

Blood Flow Parameters			
Parameter	**Description**	**Value**	**Reference**
ρ_b	Blood density	$1050\ \frac{kg}{m^3}$	Milnor [28]
μ_b	Blood viscosity	$0.0035\ Pa \cdot s$	Milnor [28]
T	Cardiac cycle period	0.85 s	Malvè et al. [25]

Blood flow is modelled in a transient step. An analysis of the number of cardiac cycles needed to model blood flow was performed, determining that three cardiac cycles are enough to develop blood flow in the lumen.

Transient mass flow and pressure are imposed, respectively, at the inlet and outlet of the geometry as boundary conditions [25], as can be seen in Figure 2. These boundary conditions influence the location and size of the generated plaque as they directly affect the recirculation that the obstacle plaque produces downstream. Therefore, they directly affect the mechanical stimuli that initiate plaque growth. For example, when increasing the inlet mass flow, the recirculation is bigger and, therefore, the areas with the mechanical stimuli that initiate plaque growth move to the end of the geometry. Finally, a no-slip condition is imposed at the endothelium.

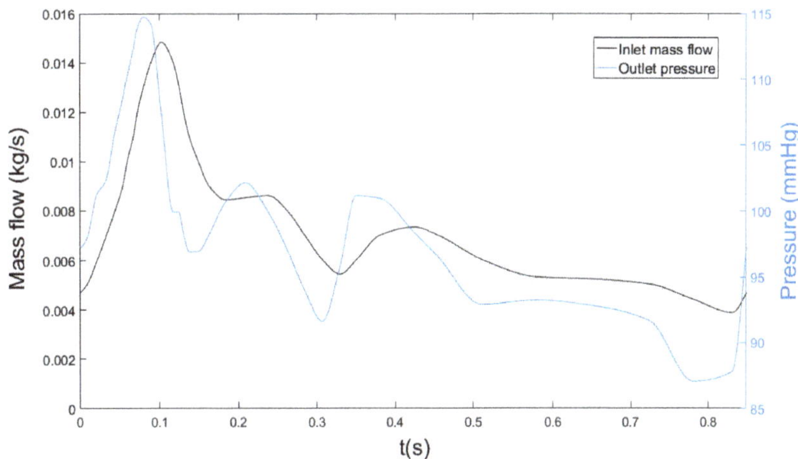

Figure 2. Mass flow (black line) and pressure (blue line) imposed as boundary conditions at the inlet and outlet of the geometry, respectively [25].

3.2. Plasma Flow across the Endothelium

As a result of the porous nature of the arterial wall, plasma can enter the bloodstream, causing a plasma flow through the endothelium. Plasma flow across the arterial wall is modelled with Darcy's law and continuity equations [11,29–33]. Darcy's law relates the velocity of the plasma flow with the pressure drop in the arterial wall and considers some parameters related to the porous nature of the arterial wall (permeability and porosity).

$$u_w = -\frac{k_w}{\mu_p} \nabla p_w, \tag{3}$$

$$\frac{\partial(\rho_p \epsilon_w)}{\partial t} + \rho_p \nabla \cdot u_w = J_v \tag{4}$$

Parameters related to the arterial wall are u_w, k_w, ϵ_w, and ∇p_w, and represent the velocity of the plasma flow, the permeability and porosity of the arterial wall, and the pressure gradient, respectively. ρ_p and μ_p are the dynamic viscosity and density of the plasma. Finally, J_v is the plasma flow across the endothelium, which is calculated with the Kedem–Katchalsky equations [34] and the three-pore model [11]:

$$J_v = Jv_{nj} + Jv_{lj} + Jv_{vp} \tag{5}$$

As can be seen in Equation (5), the three-pore model considers three different types of plasma flow through the endothelium: plasma flow through normal junctions (Jv_{nj}), leaky junctions (Jv_{lj}), and vesicular pathways (Jv_{vp}). Nevertheless, vesicular transport is intended for molecular transport, so plasma flow is negligible, relative to that across normal and leaky junctions.

Plasma flow through normal and vesicular pathways can be defined as:

$$Jv_{nj} = Lp_{nj} \cdot \Delta P_{End} \cdot (1 - \Phi_{lj}) \tag{6}$$

$$Jv_{lj} = Lp_{lj} \cdot \Delta P_{End} \tag{7}$$

Lp_{nj} and Lp_{lj} depend on the considered artery and are the hydraulic conductivity of normal and leaky junctions [35], while ΔP is the pressure drop across the endothelium, which depends on the intraluminal pressure [35]. Finally, Φ_{lj} is the endothelial fraction occupied by leaky cells [36–38].

The hydraulic conductivity of leaky junctions can be calculated as [38,39]:

$$Lp_{lj} = \frac{Ap}{S} \cdot Lp_{slj}, \tag{8}$$

with $\frac{Ap}{S}$ being the fraction of the total area occupied by leaky junctions, and Lp_{slj} being the hydraulic conductivity of a single leaky junction. To determine these parameters, we considered that leaky cells are enclosed by leaky junctions, which have ring shapes and are randomly distributed with a distance equal to ϵ_{lj}, being $2\epsilon_{lj}$ of the permeability of a leaky junction. Moreover, the spaces between endothelial cells are cylindrical [38,39]. Therefore, $\frac{Ap}{S}$ can be calculated as:

$$\frac{Ap}{S} = \frac{A_{slj}}{\pi \cdot \epsilon_{lj}{}^2}, \tag{9}$$

where $\pi \cdot \epsilon_{lj}{}^2$ is the total area occupied by leaky junctions, and A_{slj} is the area of a single leaky junction that can be calculated, assuming reduced thickness, as:

$$A_{slj} = 2\pi \cdot R_{cell} \cdot 2w_l, \tag{10}$$

where R_{cell} is the radius of endothelial cells and w_l is the half-width of a leaky junction [36,38,39].

On the other hand, Φ_{lj} can be defined as [36,38]:

$$\Phi_{lj} = \frac{R_{cell}{}^2}{\epsilon_{lj}{}^2} \tag{11}$$

Combining Equations (9)–(11), the fraction of area occupied by leaky junctions can be rewritten as (12):

$$\frac{A_p}{S} = \frac{4w_l}{R_{cell}} \cdot \Phi_{lj} \tag{12}$$

The ratio Φ_{lj} is known to be a function of some mechanical stimuli, such as the wall shear stress (WSS), the time-averaged wall shear stress ($TAWSS$), and the oscillation shear index (OSI) [40]. As blood flow is modelled in transient mode, WSS was not considered due to its changes according to the cardiac cycle. Instead of WSS, $TAWSS$ was defined. This variable takes into account the tangential stresses that blood produces in the endothelium during a cardiac cycle and that causes changes in the shape index of the endothelial cells. Therefore, in areas with low $TAWSS$, endothelial cells will be rounder, allowing plasma and substances to flow between their pores [16,40–42]:

$$TAWSS = \frac{1}{T} \int_o^T |\tau(t)| \cdot dt, \tag{13}$$

where T is the period of a cardiac cycle and $|\tau(t)|$ the magnitude of WSS dependent on time. On the other hand, high values of OSI imply that blood is highly oscillatory there. Thus, the tangential stresses of blood flow in the endothelium are also oscillatory and, due to their variation, they cause a change in the shape index of the endothelial cells, making them rounder and allowing the flow of plasma and substances through the endothelium, being areas of high OSI considered as atheroprones [8,40]. OSI is defined as:

$$OSI = 0.5 \cdot \left(1 - \frac{|\frac{1}{T}\int_o^T \tau(t) \cdot dt|}{TAWSS} \right) \tag{14}$$

Moreover, the ratio Φ_{lj}, $TAWSS$, and OSI are related by experimental correlations [6–8,11,40]. The first correlation is a parameter named shape index (SI), which can be calculated as:

$$SI = \frac{4\pi \cdot Area}{Perimeter^2}, \tag{15}$$

where $Area$ and $Perimeter$ correspond to a single endothelial cell.

To determine the behaviour of endothelial cells with $TAWSS$ and OSI, we proposed in a previous work [8] a numerical correlation based on the experimental results of Levesque et al. [40].

$$SI = 0.0264 \cdot e^{5.647 \cdot OSI} + 0.5513 \cdot e^{-0.1815 \cdot (TAWSS)^2} \tag{16}$$

The total number of mitotic cells (MC) in the endothelium depends on the SI [43], so the next experimental correlation was developed, with a unit area of 0.64 mm^2:

$$MC = 0.003797 \cdot e^{(14.75 \cdot SI)} \tag{17}$$

Finally, the number of leaky cells (LC) and mitotic cells (MC) is correlated by the following empirical equation [11,44]:

$$LC = 0.307 + \frac{0.805 \cdot MC}{0.453} \tag{18}$$

On the other hand, Φ_{lj} can be defined as the ratio between the area of leaky cells and the total area of cells. It can be calculated as:

$$\Phi_{lj} = \frac{LC \cdot \pi \cdot R_{cell}^2}{A_{unit}}, \tag{19}$$

considering A_{unit} equal to the unit area in all the experimental correlations before (0.64 mm^2). By calculating blood flow in the model, it is now possible to determine Φ_{lj} and therefore $\frac{A_p}{S}$.

Finally, the hydraulic conductivity of a single leaky junction, Lp_{slj}, is defined according to Olgac et al. [11]:

$$Lp_{slj} = \frac{w_l^2}{3 \cdot \mu_p \cdot l_{lj}},$$ (20)

where w_l and l_{lj} are, respectively, the width and the length of a leaky junction.

So, the plasma flow across the endothelium is completely defined, and the parameters needed to calculate it are shown in Table 2.

The boundary conditions of the plasma flow across the endothelium are the normal velocity of the plasma flow (J_v), which has already been defined, and the pressure in the adventitia, which [11] must be defined. Both an increase in the normal velocity of the plasma flow and a decrease in pressure at the adventitia would produce an increase in plasma flow, which will also cause an increase in substances flow, with the consequent increase in plaque size [8].

Table 2. Parameters to model plasma flow across the endothelium.

Plasma Flow Parameters			
Param.	**Description**	**Value**	**Reference**
R_{cell}	Endothelial cell radius	15 µm	Weinbaum et al. [38]
w_l	Half-width of a leaky junction	20 nm	Weinbaum et al. [38]
l_{lj}	Leaky junction length	2 µm	Weinbaum et al. [38]
ρ_p	Plasma density	1050 $\frac{kg}{m^3}$	Milnor [28]
μ_p	Plasma viscosity	0.001 $Pa \cdot s$	Milnor [28]
k_w	Darcian artery permeability	$1.2 \cdot 10^{-18}$ m^2	Huang et al. [36], Prosi et al. [45]
ϵ_p	Intima porosity	0.96	Ai and Vafai [46]
$L_{p,nj}$	Normal junction conductivity	$1.984 \cdot 10^{-12}$ $\frac{m}{s \cdot Pa}$	Tedgui and Lever [35]
ΔP_{End}	Endothelial pressure difference	20.727 mmHg	Tedgui and Lever [35]
A_{unit}	Unit area for the exp. correlations	0.64 mm^2	Chien [43]
P_{adv}	Pressure of the adventitia	17.5 mmHg	Olgac et al. [11]

3.3. Inflammatory Process in the Arterial Wall

There are many substances involved in the inflammatory process of the arterial wall. In the model, we consider molecules, such as low-density lipoproteins (LDL) and oxidised LDL (oxLDL). We also consider cells, such as monocytes (m), macrophages (M), foam cells (FC), and contractile and synthetic smooth muscle cells (CSMC and SSMC). We also model cytokines (C) and collagen fibre (Cg).

All of these substances can have convection and diffusion and interact with the others, so the inflammatory process of the arterial wall can be described with convection–diffusion–reaction equations.

In addition, the flow of substances across the arterial wall can be defined as:

$$N = -D_{X_i} \nabla X_i + u_w X_i,$$ (21)

where D_{X_i}, X_i, and u_w are the diffusion coefficient of the considered substance, its concentration and its convection velocity in the arterial wall. Due to the structure of the arterial wall, the diffusion is anisotropic, as the longitudinal diffusion is 3 times higher than radial diffusion [39].

At the beginning of the process, in the healthy artery, all muscle cells in the arterial wall are contractile. Thus, the only substance that has an initial concentration in the arterial wall is CSMC.

A high level of LDL concentration in blood is considered, according to a pathological model ($2.7 \frac{mg}{ml}$ [47], which corresponds to $6.98 \frac{mol}{m^3}$). Moreover, we impose a physiological monocyte inlet concentration in the lumen of $550 \cdot 10^{-9} \frac{cells}{m^3}$ [48]. An increase in both inlet concentrations would cause an increase in the growth of the resultant plaque.

3.3.1. LDL

Once the LDL molecules cross the endothelial barrier, they can suffer both convection and diffusion, because of their small sizes. When an LDL molecule enters the arterial wall, it is oxidised. Therefore, the reactive term of LDL molecules represents their oxidation.

$$\frac{\partial C_{LDL,w}}{\partial t} + \nabla \cdot (-D_{C_{LDL,w}} \nabla C_{LDL,w}) + K_{lag} \cdot u_w \cdot \nabla C_{LDL,w} = -d_{LDL} C_{LDL,w} \qquad (22)$$

d_{LDL} is the degradation ratio of LDL in the arterial wall and $C_{LDL,w}$ is its concentration at each time.

It is necessary to define as a boundary condition the LDL concentration in the adventitia ($C_{LDL,adv}$) [49]. The Kedem–Katchalsky equation is used to determine the LDL flow across the endothelium [11]:

$$J_{S,LDL} = C_{LDL,l} \cdot LDL_{dep} \cdot P_{app}, \qquad (23)$$

where $C_{LDL,l}$ is the LDL concentration in blood and LDL_{dep} is the amount of LDL molecules that are deposited from the lumen to the arterial wall. Finally, P_{app} is the apparent permeability of the arterial wall. LDL can flow from the lumen to the arterial wall through normal and leaky junctions and vesicular pathways, so the apparent permeability of the arterial wall is composed of three types of permeabilities [50,51]:

$$P_{app} = P_{app,lj} + P_{app,nj} + P_{app,vp}, \qquad (24)$$

with $P_{app,nj}$, $P_{app,lj}$, and $P_{app,vp}$ being the permeabilities of normal and leaky junctions and vesicular pathways, respectively.

The transport of molecules through the endothelium is dependent on the size of the particles. For the case of LDL, which has a radius of 11 nm [50]), transport across normal junctions is not possible due to their small size. Therefore, LDL transport through the endothelium can only occur through leaky junctions and vesicular pathways [52].

In addition, LDL transport through vesicular pathways is 10% of the flux through leaky junctions [11]:

$$P_{app,vp} = 0.1 \cdot P_{app,lj} \qquad (25)$$

The apparent permeability of leaky junctions can be defined as:

$$P_{app,lj} = P_{lj} Z_{lj} + J_{v,lj} \cdot (1 - \sigma_{f,lj}), \qquad (26)$$

where P_{lj}, Z_{lj}, and $\sigma_{f,lj}$ are the diffusive permeabilities of the leaky junctions, factors of reduction of the LDL concentration gradient at the inlet of the flow and the solvent-drag coefficient of leaky junctions. So, the LDL flux across the endothelium can be written as:

$$J_{S,LDL} = 1.1 \cdot C_{LDL,l} \cdot LDL_{dep} \cdot (P_{lj} Z_{lj} + J_{v,lj}(1 - \sigma_{f,lj})) \qquad (27)$$

The diffusive permeability of leaky junctions is defined as:

$$P_{lj} = \frac{A_p}{S} \chi P_{slj}, \qquad (28)$$

where χ is the difference between the total area of endothelial cells and the area of cells separated by leaky junctions:

$$\chi = 1 - \alpha_{lj}, \qquad (29)$$

with α_{lj} being the ratio between the radius of an LDL molecule (a_m) and the half-width of a leaky junction (w_l):

$$\alpha_{lj} = \frac{a_m}{w_l} \tag{30}$$

P_{slj} is the permeability of a single leaky junction, and can be determined by knowing the LDL diffusion coefficient in a leaky junction (D_{lj}) and the length of a leaky junction (l_{lj}):

$$P_{slj} = \frac{D_{lj}}{l_{lj}} \tag{31}$$

LDL diffusion coefficient in a leaky junction is related to the LDL diffusion coefficient by [11]:

$$\frac{D_{lj}}{D_l} = F\left(\alpha_{lj}\right) = 1 - 1.004\alpha_{lj} + 0.418\alpha_{lj}^3 - 0.16\alpha_{lj}^5 \tag{32}$$

On the other hand, Z_{lj} depends on a modified Péclet number:

$$Z_{lj} = \frac{Pe_{lj}}{e^{(Pe_{lj})} - 1} \tag{33}$$

This modified Péclet number can be defined as:

$$Pe_{lj} = \frac{Jv, lj \cdot (1 - \sigma_{f,lj})}{P_{lj}} \tag{34}$$

Finally, the solvent-drag coefficient of leaky junctions is given by [11]:

$$\sigma_{f,lj} = 1 - \frac{2}{3}\alpha_{lj}^2(1 - \alpha_{lj}) \cdot F(\alpha_{lj}) - (1 - \alpha_{lj})\left(\frac{2}{3} + \frac{2\alpha_{lj}}{3} - \frac{7\alpha_{lj}^2}{12}\right) \tag{35}$$

3.3.2. Oxidised LDL

Once LDL is oxidised, it is considered to not experience convection [4]. However, due to its similar size to LDL, oxidised LDL shows diffusion in the arterial wall. Once LDL enters the arterial wall, it becomes oxidised, so one of its reactive terms refers to this oxidation. On the other hand, oxidised LDL is phagocytosed by macrophages, which corresponds to its second reactive term.

$$\frac{\partial C_{oxLDL,w}}{\partial t} + \nabla \cdot (-D_{C_{oxLDL,w}} \nabla C_{oxLDL,w}) = d_{LDL}C_{LDL,w} - LDL_{ox,r}C_{oxLDL_w}C_{M,w}, \tag{36}$$

where C_{oxLDL_w} and $C_{M,w}$ are the oxidised LDL and macrophage concentrations in the arterial wall. In addition, $LDL_{ox,r}$ is a ratio of the quantity of oxidised LDL that a macrophage can ingest.

3.3.3. Monocytes

Due to their sizes, we hypothesise that monocytes do not have convection, but they experience diffusion in the arterial wall. Once monocytes are in the arterial wall, they differentiate into macrophages, and also, they suffer apoptosis. These two phenomena are represented by their reactive terms.

$$\frac{\partial C_{m,w}}{\partial t} + \nabla \cdot (-D_{C_{m,w}} \nabla C_{m,w}) = -d_m C_{m,w} - m_d C_{m,w}, \tag{37}$$

where $C_{m,w}$ is the monocyte concentration in the arterial wall. d_m is the rate of monocytes that differentiate into macrophages and m_d is the apoptosis rate of monocytes.

In addition, a physiological monocyte inlet concentration in the lumen of $550 \cdot 10^{-9} \frac{Monocyte}{m^3}$ is imposed [48].

Monocyte flow through the endothelium is dependent on hemodynamics. Areas of low $TAWSS$ are known to be atheroprone; particularly for carotid arteries, areas of $TAWSS$ lower than 2 Pa can develop atheroma plaque [16,41,42]. For the case of OSI, according to our previous work [8], it can be determined that areas of OSI greater than 0.1910 are atheroprones. Monocyte flow across the endothelium can be defined as [8]:

$$J_{s,m}(TAWSS, OSI) = m_r \cdot (0.8588 \cdot e^{-0.6301 \cdot TAWSS} + 0.1295 \cdot e^{3.963 \cdot OSI}) \cdot C_{LDL,ox,w} C_{m,l}, \quad (38)$$

with m_r being the monocyte recruitments from the lumen.

3.3.4. Macrophages

Similar to monocytes, we hypothesise that macrophages experience diffusion but not convection. Once monocytes are in the arterial wall, they differentiate into macrophages. Furthermore, macrophage phagocytes oxidised LDL, and once they cannot ingest a greater amount of oxidised LDL, they suffer apoptosis and become foam cells. These two phenomena are represented in the macrophage reactive terms:

$$\frac{\partial C_{M,w}}{\partial t} + \nabla \cdot (-D_{C_{M,w}} \nabla C_{M,w}) = d_m C_{m,w} - \frac{LDL_{ox,r}}{n_{FC}} \cdot C_{oxLDL_w} C_{M,w}, \quad (39)$$

In Equation (39), $LDL_{ox,r}$ is the constant rate of oxidised LDL uptaken by macrophages, and n_{FC} is the maximum amount of oxidised LDL that a single macrophage must ingest to transform into a foam cell.

3.3.5. Cytokines

Due to their sizes, cytokines do not have convection. In addition, they are enclosed by macrophages, so they do not have diffusion either. Cytokines in the arterial wall are segregated by macrophages and also experience degradation, phenomena that can be seen in their reactive terms:

$$\frac{\partial C_{c,w}}{\partial t} = C_r C_{oxLDL_w} C_{M,w} - d_c C_{c,w}, \quad (40)$$

where $C_{c,w}$ is cytokine concentration in the arterial wall. C_r and d_c are, respectively, the ratios of production and degradation of cytokines.

3.3.6. Contractile Smooth Muscle Cells

They have neither convection nor diffusion because of their large size. At the beginning of the process, all smooth muscle cells of the arterial wall have a contractile phenotype. Then, due to the presence of cytokines, they experience a change of phenotype into synthetic smooth muscle cells, which is represented in their reactive term:

$$\frac{\partial C_{CSMC,w}}{\partial t} = -C_{CSMC,w} \cdot S_r \cdot \left(\frac{C_{c,w}}{k_c \cdot C_{c,w}^{th} + C_{c,w}} \right) \quad (41)$$

$C_{CSMC,w}$ is CSMC concentration in the arterial wall. S_r is the CSMC differentiation rate into SSMC and $C_{c,w}^{th}$ is the maximum cytokine concentration allowed in the arterial wall. Finally, k_c is a constant for the saturation equation.

3.3.7. Synthetic Smooth Muscle Cells

Equal to CSMC and because of their sizes, SSMCs neither have convection nor diffusion. CSMCs change their phenotypes into SSMCs due to the presence of cytokines in the arterial wall. They also experience proliferation and apoptosis. These three phenomena are represented in their reactive terms:

$$\frac{\partial C_{SSMC,w}}{\partial t} = C_{CSMC,w} \cdot S_r \cdot \left(\frac{C_{c,w}}{k_c \cdot C_{c,w}^{th} + C_{c,w}} \right) +$$

$$\left(\frac{p_{ss} C_{c,w}}{C_{c,w/2}^{th} + C_{c,w}} \right) C_{SSMC,w} - r_{Apop} \cdot C_{SSMC,w} \tag{42}$$

$C_{SSMC,w}$ represents the SSMC concentration in the arterial wall. In addition, p_{ss} is the SSMC proliferation rate and r_{Apop} is the SSMC apoptosis rate.

3.3.8. Foam Cells

Due to their large sizes, foam cells neither have convection nor diffusion. Once a macrophage cannot ingest a greater quantity of oxidised LDL, it becomes a foam cell, which is represented in the reactive term of foam cells:

$$\frac{\partial C_{FC,w}}{\partial t} = \frac{LDL_{ox,r}}{n_{FC}} \cdot C_{LDLox_w} C_{M,w} \tag{43}$$

3.3.9. Collagen Fibre

Collagen fibre do not experience convection or diffusion because they are composed of many molecules. Collagen fibre experience segregation by SSMC, which can be seen in its reactive term.

$$\frac{\partial C_{Cg,w}}{\partial t} = G_r \cdot C_{SSMC,w}, \tag{44}$$

where $C_{Cg,w}$ is collagen concentration in the arterial wall, and G_r is its secretion rate due to plaque formation. Natural segregation and degradation of collagen in the arterial wall were not considered, as they occur in healthy areas of the arterial wall, not related to plaque generation.

Table 3 contains all the parameters to calculate the inflammatory process in the arterial wall.

Table 3. Parameters to calculate the inflammatory process in the arterial wall.

Inflammatory Process Parameters			
Parameter	**Description**	**Value**	**Reference**
$Dr_{LDL,w}$	LDL radial diff. coeff.	$8 \cdot 10^{-13}\ \frac{m^2}{s}$	Prosi et al. [45]
$Dr_{m,w}$	Monocyte radial diff. coeff.	$8 \cdot 10^{-15}\ \frac{m^2}{s}$	Cilla et al. [4]
$Dr_{oxLDL,w}$	Ox. LDL radial diff. coeff.	$8 \cdot 10^{-13}\ \frac{m^2}{s}$	Prosi et al. [45]
$Dr_{M,w}$	Macroph. radial diff. coeff.	$8 \cdot 10^{-15}\ \frac{m^2}{s}$	Cilla et al. [4]
d_{LDL}	LDL oxidation	$2.85 \cdot 10^{-4}\ \mathrm{s}^{-1}$	Ai and Vafai [46]
d_m	Monocyte differentiation	$1.15 \cdot 10^{-6}\ \mathrm{s}^{-1}$	Bulelzai and Dubbeldam [53]
m_d	Monocyte natural death	$\frac{1}{60}\ \mathrm{d}^{-1}$	Krstic [54]
$LDL_{ox,r}$	OxLDL uptake	$2.45 \cdot 10^{-23}\ \frac{m^3}{Macrophage \cdot s}$	Zhao et al. [55]
n_{FC}	Max. oxLDL uptake	$2.72 \cdot 10^{-11}\ \frac{mol_{oxLDL}}{Macrophage}$	Hernández-López et al. [8]
C_r	Cytokine production	$3 \cdot 10^{-10}\ \frac{mol_C \cdot m^3}{mol_{oxLDL} \cdot Macrophage \cdot s}$	Cilla et al. [4]

Table 3. *Cont.*

Inflammatory Process Parameters			
Parameter	**Description**	**Value**	**Reference**
d_c	Cytokine degradation	$2.3148 \cdot 10^{-5}\,\mathrm{s}^{-1}$	Zhao et al. [56]
S_r	SMC's differentiation	$0.0036\,\mathrm{d}^{-1}$	Cilla et al. [4]
p_{ss}	SSMC's proliferation	$0.0202\,\mathrm{d}^{-1}$	Budu-Grajdeanu et al. [57]
G_r	Collagen production	$2.49 \cdot 10^{-21}\,\frac{kg_{Cg}}{SSMC \cdot s}$	Zahedmanesh et al. [58]
w_l	Half-width of a leaky junct.	$20\,\mathrm{nm}$	Weinbaum et al. [38]
l_{lj}	Leaky junction length	$2\,\mu\mathrm{m}$	Weinbaum et al. [38]
R_{LDL}	LDL radius	$11\,\mathrm{nm}$	Tarbell [50]
$C_{c,w}^{th}$	Cytokine threshold	$1.235 \cdot 10^{13}\,\frac{mol_C}{m^3}$	Hernández-López et al. [8]
$C_{0,LDL}$	LDL initial conc.	$6.98\,\frac{mol_{LDL}}{m^3}$	Cilla [47]
$C_{0,m}$	Monocyte initial conc.	$550 \cdot 10^9\,\frac{Monocytes}{m^3}$	Khan [48]
$C_{0,CSMC}$	CSMC initial conc.	$3.16 \cdot 10^{13}\,\frac{CSMC}{m^3}$	Escuer et al. [59]
$C_{LDL,adv}$	LDL conc. at adventitia	$11.6\text{‰} \cdot C_{LDL,l}$	Meyer et al. [49]
k_c	Cytokine threshold factor	0.65093	Hernández-López et al. [8]
LDL_{dep}	LDL at the endothelium	$10^{-2} \cdot C_{LDL,l}$	Meyer et al. [49]
m_r	Monocyte recruitment	$6.636 \cdot 10^{-4}\,\frac{m^4}{mol_{oxLDL} \cdot day}$	Steinberg et al. [60]
r_{apop}	SSMC apoptosis rate	$8.011 \cdot 10^{-8}\mathrm{s}^{-1}$	Kockx et al. [61]
ρ_{LDL}	LDL density	$1063\,\frac{kg}{m^3}$	Ivanova et al. [62]
Mw_{LDL}	LDL molecular weight	$386.65\,\frac{g}{mol_{LDL}}$	Cilla [47]
k_{lag}	Solute lag coefficient of LDL	0.893	Dabagh et al. [63]

3.4. Plaque Growth

Finally, with Equation (45), it is possible to calculate the growth of plaque in the arterial wall. We consider plaque to be composed of a lipid nucleus of foam cells and a fibrous layer of synthetic smooth muscle cells and collagen fibre. Therefore, considering the isotropic growth of plaque, it is possible to determine the change in volume in the arterial wall due to plaque appearance:

$$\nabla \cdot v = \frac{\partial C_{FC,w}}{\partial t} \cdot Vol_{FC} + \frac{\partial \Delta C_{SSMC,w}}{\partial t} \cdot Vol_{SSMC} + \frac{\partial C_{Cg,w}}{\partial t} \cdot \frac{1}{\rho_{Cg}}, \tag{45}$$

where $\frac{\partial C_{i,w}}{\partial t}$ is the variation of concentration with respect to the initial concentration of the considered substance. Vol_{FC} and Vol_{SSMC} are the volumes of a foam cell and a synthetic smooth muscle cell, respectively. Finally, ρ_{Cg} is the collagen density.

To calculate the volume of foam cells, they have been approximated as spherical geometries, while synthetic smooth muscle cells are modelled as ellipsoids, so their volumes can be calculated with Equations (46) and (47).

$$Vol_{FC} = \frac{4}{3}\pi R_{FC}^3 \tag{46}$$

$$Vol_{SSMC} = \frac{4}{3}\pi R_{SSMC}^2 \cdot l_{SSMC}, \tag{47}$$

with R_{FC} and R_{SSMC} being foam cells and the synthetic smooth muscle cell radius, and l_{SSMC} being the lengths of synthetic smooth muscle cells. Parameters for calculating plaque growth in the arterial wall are given in Table 4.

Table 4. Parameters to calculate plaque growth in the arterial wall.

Plaque Growth Parameters			
Parameter	**Description**	**Value**	**Reference**
R_{FC}	Foam cell radius	15.264 µm	Krombach et al. [64], Cannon and Swanson [65]
R_{SSMC}	SMC radius	3.75 µm	Cilla et al. [4]
l_{SSMC}	SMC length	115 µm	Cilla et al. [4]
ρ_{Cg}	Collagen density	$1000\ \frac{kg}{m^3}$	Sáez et al. [66]

4. Numerical Methods

The software COMSOL Multiphysics (COMSOL AB, Burlington, MA, USA) was used to computationally solve the model. It was modelled using four consecutive steps, which can be seen in Figure 3. In the first step, three cardiac cycles and hemodynamic stimuli were calculated in a transient step, determining the values of $TAWSS$ and OSI for the third cardiac cycle. Then, in the second step, which is stationary, the plasma flow through the endothelium is calculated. In the third step, the inflammatory process is calculated in a transient mode, determining the concentrations of all substances in the arterial wall for a total of 30 years. Finally, a last stationary step was developed to calculate the growth of the plaque, knowing the concentrations of all substances for a period of 30 years.

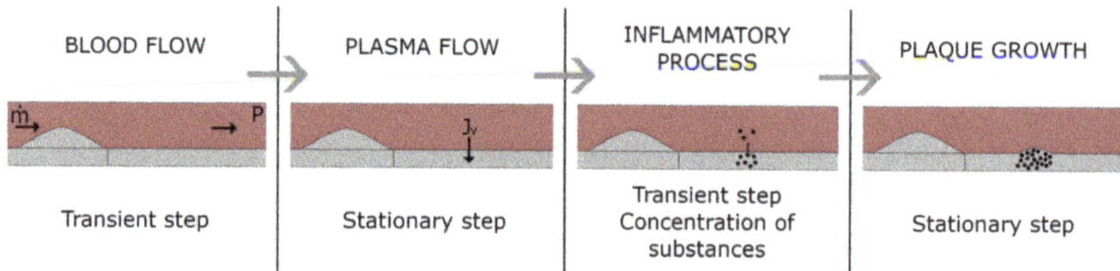

Figure 3. Pathline of the computational model.

A direct solver (multi-frontal massively parallel sparse direct solver, MUMPS) was used to calculate transient blood flow along the lumen, as well as plasma flow across the endothelium, the inflammatory process in the arterial wall, and plaque growth. The inflammatory process was calculated iteratively, using groups of segregated steps for the different substances.

5. Sensitivity Analysis

As can be seen in Section 3, the computational model has a large number of parameters that can affect the composition and growth of the plaque. There is a wide range of values of these parameters in the literature that also come from studies under different conditions, such as in vivo or in vitro experiments, or different species or arteries. Therefore, a sensitivity analysis can help to understand the role of every parameter in the generated plaque.

The objective of this study is to analyse plaque growth and its composition. To do so, a previous selection of the parameters to analyse was made. Therefore, geometric parameters, initial concentrations, material properties, flow properties, and other factors that are well known have not been considered in the analysis because their influence on the model seems to be clear (for example, an increase in the radius of foam cells would result in an increase in the stenosis ratio due to an increase in the volume of the plaque). Thus, in this study, only the parameters related to the reactive terms of the convection–diffusion–reaction equations of the inflammatory process are analysed, as well as the diffusion coefficients of substances in the arterial wall.

Table 5 contains all the parameters whose variation was analysed and their descriptions. The analysed parameters have been increased and reduced by 25 % and 10 % in 61 different simulations in a mono-variant sensitivity analysis. The values of the parameters for the case of $\pm\,10\%$ variation are included in Table 5, as an example:

Table 5. Analysed parameters and values.

Parameter	Description	Value -10%	Mean Value	Value $+10\%$
$D_{LDL,w} = D_{LDLox,w}\ (\frac{m^2}{s})$	LDL and oxLDL diff. coeff.	$7.2 \cdot 10^{-13}$	$8 \cdot 10^{-13}$	$8.8 \cdot 10^{-13}$
$D_{m,w} = D_{M,w}\ (\frac{m^2}{s})$	Monocyte and macrophage diff. coeff.	$7.2 \cdot 10^{-15}$	$8 \cdot 10^{-15}$	$8.8 \cdot 10^{-15}$
$d_{LDL}\ (s^{-1})$	LDL oxidation	$2.565 \cdot 10^{-4}$	$2.85 \cdot 10^{-4}$	$3.135 \cdot 10^{-4}$
$d_m\ (s^{-1})$	Monocyte differentiation	$1.035 \cdot 10^{-6}$	$1.15 \cdot 10^{-6}$	$1.265 \cdot 10^{-6}$
$m_d\ (d^{-1})$	Monocyte death	0.015	0.016	0.018
$LDL_{ox,r}\ (\frac{m^3}{Macrophage \cdot s})$	OxLDL uptake	$2.205 \cdot 10^{-23}$	$2.45 \cdot 10^{-23}$	$2.695 \cdot 10^{-23}$
$n_{FC}\ (\frac{mol_{oxLDL}}{Macrophage})$	Max. oxLDL uptake	$2.448 \cdot 10^{-11}$	$2.72 \cdot 10^{-11}$	$2.992 \cdot 10^{-11}$
$C_r\ (\frac{mol_C \cdot m^3}{mol_{oxLDL} \cdot Macrophage \cdot s})$	Cytokine production	$2.7 \cdot 10^{-10}$	$3 \cdot 10^{-10}$	$3.3 \cdot 10^{-10}$
$d_c\ (s^{-1})$	Cytokine degradation	$2.082 \cdot 10^{-5}$	$2.314 \cdot 10^{-5}$	$2.545 \cdot 10^{-5}$
$S_r\ (d^{-1})$	SMC differentiation	$3.24 \cdot 10^{-3}$	$3.6 \cdot 10^{-3}$	$3.96 \cdot 10^{-3}$
$p_{ss}\ (d^{-1})$	SSMC proliferation	0.01818	0.0202	0.0199
$G_r\ (\frac{kg_{Cg}}{SSMC \cdot s})$	Collagen production	$2.241 \cdot 10^{-21}$	$2.49 \cdot 10^{-21}$	$2.739 \cdot 10^{-21}$
$C_{c,w}^{th}\ (\frac{mol_C}{m^3})$	Cytokine threshold	$1.111 \cdot 10^{13}$	$1.235 \cdot 10^{13}$	$1.358 \cdot 10^{13}$
$m_r\ (\frac{m^4}{mol_{oxLDL} \cdot day})$	Monocyte recruitment	$5.972 \cdot 10^{-4}$	$6.636 \cdot 10^{-4}$	$7.299 \cdot 10^{-4}$
$r_{apop}\ (s^{-1})$	SSMC apoptosis rate	$7.209 \cdot 10^{-8}$	$8.011 \cdot 10^{-8}$	$8.812 \cdot 10^{-8}$

The percentage change in the volume of the plaque due to each of the substances involved in its growth (foam cells that compose the lipidic core of the plaque, and synthetic smooth muscle cells and collagen fibre, which correspond to the fibrous layer of the plaque) was analysed, as well as the variation in the stenosis ratio, which is defined as:

$$SR(\%) = \left(1 - \frac{Area\ of\ lumen\ with\ plaque}{Area\ of\ healthy\ lumen}\right) \cdot 100 \tag{48}$$

6. Results

Tables 6 and 7 show the results obtained by increasing and decreasing the selected parameters by 25% and 10%, respectively.

Table 6. Results of foam cells, synthetic smooth muscle cells, and collagen volume variations (second, third, and fourth columns, respectively), and stenosis ratio variation (fifth column), by reducing and increasing the values of the parameters of the first column by 25%.

Parameter	FC Volume Variation (%)		SSMC Volume Variation (%)		Cg Volume Variation (%)		SR Variation (%)	
	-25%	$+25\%$	-25%	$+25\%$	-25%	$+25\%$	-25%	$+25\%$
$D_{LDL,w} = D_{LDLox,w}$	7.35	-5.65	>100	-92.81	>100	-74.38	-	-36.61
$D_{m,w} = D_{M,w}$	4.82	-4.07	>100	-43.43	>100	-39.57	-	-19.71
d_{LDL}	-20.29	18.51	-99.89	>100	-93.98	>100	-82.68	-
d_m	-0.98	0.63	-42.43	21.00	-31.32	15.33	-7.28	4.69
m_d	0.23	-1.09	15.32	-34.09	12.95	-29.32	7.90	-8.16
$LDL_{ox,r}$	-3.73	2.90	>100	-99.89	>100	-93.27	-	-12.99
n_{FC}	48.43	-36.00	-0.21	0.13	-0.10	0.24	>100	-77.22
C_r	0	0	-99.89	>100	-95.52	>100	-22.05	-
d_c	0	0	>100	-99.89	>100	-94.14	-	-21.95
S_r	0	0	-5.50	6.46	-7.48	7.71	-0.64	0.032
p_{ss}	0	0	-99.89	>100	-96.88	>100	-22.17	-
G_r	0	0	0	0	-24.54	48.32	-1.08	2.09
$C_{c,w}^{th}$	0	0	>100	-99.89	>100	-94.14	-	-21.96
m_r	-3.77	3.75	-41.23	>100	-35.02	>100	-20.83	-
r_{apop}	0	0	>100	-99.91	>100	-96.72	-	-22.19

Table 7. Results of the foam cells, synthetic smooth muscle cells, and collagen volume variations (second, third, and fourth columns, respectively), and variation in the stenosis ratio (fifth column), reducing and increasing the values of the parameters of the first column by 10%.

Parameter	FC Volume Variation (%)		SSMC Volume Variation (%)		Cg Volume Variation (%)		SR Variation (%)	
	-10%	$+10\%$	-10%	$+10\%$	-10%	$+10\%$	-10%	$+10\%$
$D_{LDL,w} = D_{LDLox,w}$	2.69	-2.42	>100	-67.85	>100	-48.60	-	-21.31
$D_{m,w} = D_{M,w}$	2.32	-1.32	>100	-13.45	>100	-9.32	-	-11.23
d_{LDL}	-7.89	7.60	-97.67	>100	-81.81	>100	-45.91	-
d_m	-0.54	0.32	-25.54	12.55	-19.89	8.35	-2.67	2.10
m_d	0.05	-0.43	7.65	-25.43	4.56	-20.88	3.91	-3.06
$LDL_{ox,r}$	-1.36	1.23	>100	-97.85	>100	-82.55	–	-9.61
n_{FC}	15.50	-11.52	-0.07	0.04	-0.03	0.08	57.57	-26.82
C_r	0	0	-99.13	>100	-86.56	>100	-21.31	-
d_c	0	0	>100	-98.69	>100	-84.93	-	-21.16
S_r	0	0	-1.17	1.38	-1.59	1.64	-0.13	0.007
p_{ss}	0	0	-99.89	>100	-91.40	>100	-21.75	-
G_r	0	0	0	0	-4.50	8.77	-0.19	0.36
$C_{c,w}^{th}$	0	0	>100	-98.69	>100	-84.93	-	-21.15
m_r	-0.26	0.26	-2.89	>100	-2.45	67.59	-1.49	-
r_{apop}	0	0	>100	-99.90	>100	-91.14	-	-21.77

The first column of Tables 6 and 7 represents the parameter whose influence is being analysed. The second, third, and fourth double columns are, respectively, changes in the volume of foam cells, synthetic smooth muscle cells, and collagen fibre in the plaque, caused by the considered parameter variation. Finally, the last double column is the change in the stenosis ratio of the artery due to the change in the considered parameter.

As can be seen in Tables 6 and 7, the trend of the results is the same in the cases of variation of 25% and 10% parameters variation. Therefore, the results will be discussed only with reference to the 10% variation table (Table 7), and can be extrapolated for the 25% variation table (Table 6).

As can be seen in Tables 6 and 7, the variation of the substance was limited to a maximum of 100%. Therefore, in cases where the variation in the percentage of a substance was greater than 100%, the stenosis ratio was not calculated.

As can be seen in Table 7, an increase in the diffusion parameters, m_d, $LDL_{ox,r}$, n_{FC}, d_c, $c_{c,w}^{th}$, and r_{apop} causes a decrease in the stenosis ratio produced by the plaque. In contrast, an increase in d_{LDL}, d_m, C_r, S_r, p_{ss}, G_r, and m_r induces an increase in it. In addition, there are some parameters of the model that have more influence on the results than others, and their variation causes a change greater than 100% in the volume of any of the substances in the plaque.

When considering the parameters that influence the change in the volume of foam cells in the plaque (Table 7), these are, in order of influence, n_{FC}, d_{LDL}, $D_{LDL,w} = D_{LDLox,w}$ and $D_{m,w} = D_{M,w}$, which are related to foam cells, LDL, and the diffusion properties of substances in the arterial wall, respectively. As can be seen, none of the variations produces a change in the volume of foam cells greater than 100%.

The parameters that cause a higher change in the volume of synthetic smooth muscle cells are, in order of influence: d_{LDL}, C_r, p_{ss}, m_r, r_{apop}, d_c, $C_{c,w}^{th}$, and $LDL_{ox,r}$, when increased (the first four produce a change greater than 100%). When their values decrease, the most influential are, in order: $D_{LDL,w} = D_{LDLox,w}$, $D_{m,w} = D_{M,w}$, $LDL_{ox,r}$, d_c, $C_{c,w}^{th}$, r_{apop}, p_{ss}, C_r, and d_{LDL}, the first six of which cause changes greater than 100%. Therefore, for the change in the volume of synthetic smooth muscle cells volume in the plaque, the parameters d_{LDL}, C_r, p_{ss}, r_{apop}, d_c, $C_{c,w}^{th}$, and $LDL_{ox,r}$ have huge influences, regardless if their values are increased or decreased.

For the case of the influence on volume change due to SSMC, d_{LDL}, C_r, p_{ss}, r_{apop}, d_c, $C_{c,w}^{th}$ and $LDL_{ox,r}$ have a great influence on the results obtained.

For the variation of collagen volume in the plaque, the most influential parameters are, when increased: d_{LDL}, C_r, p_{ss}, r_{apop}, d_c, $C_{c,w}^{th}$ and $LDL_{ox,r}$, while when decreased: $D_{LDL,w} = D_{LDLox,w}$, $D_{m,w} = D_{M,w}$, $LDL_{ox,r}$, d_c, $C_{c,w}^{th}$, and r_{apop}.

In Figure 4, the variation in the volume of foam cells, synthetic smooth muscle cells, and collagen fibre is represented in a graph of parallel bars for variations of $\pm 10\%$. In cases of parameters that cause a volume variation higher than 100% in any of the considered substances, the bar of this substance is represented in red, i.e., the cases of the diffusion coefficients d_{LDL}, $LDL_{ox,r}$, C_r, d_c, p_{ss}, $c_{c,w}^{th}$, m_r, and r_{apop}.

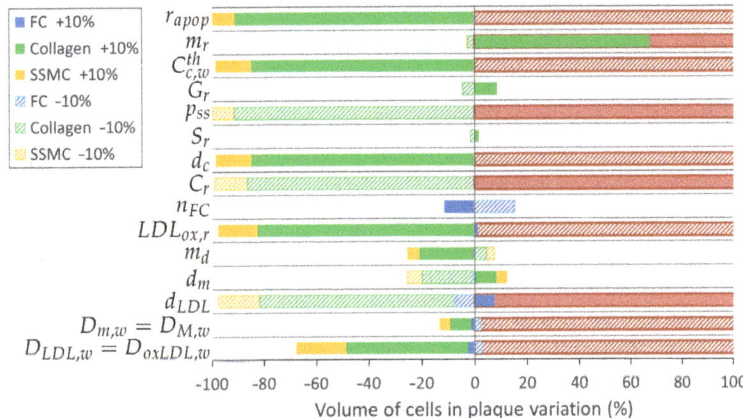

Figure 4. Variation of the volume of FC (blue color), SSMC (yellow color), and collagen (green color) when increasing and decreasing the parameters by 10% (solid and striped colors, respectively). Bars in red represent a variation of one of the substances higher than 100%.

Figure 5 represents the change in the stenosis ratio due to the variation of the considered parameters when varied $\pm 10\%$. The red bars refer to cases in which at least one of the substances that adds volume to the plaque has a volume variation greater than 100%.

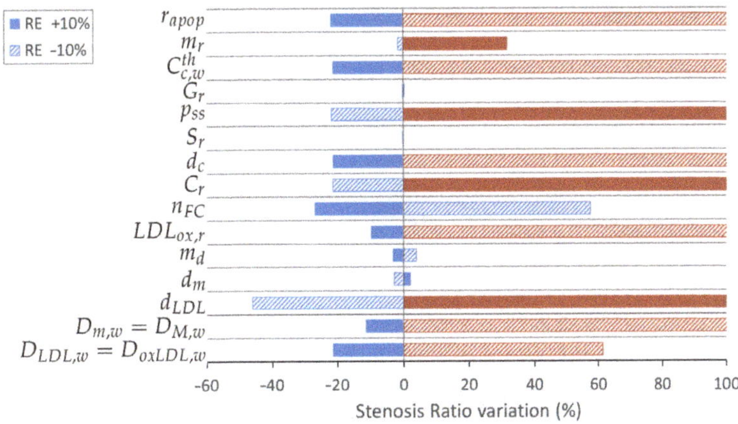

Figure 5. Variation of the stenosis ratio (blue color) when increasing and decreasing the parameters by 10% (solid and striped colors, respectively). Bars in red represent a variation of one of the substances higher than 100%.

7. Discussion

In this work, an analysis of the influence of some parameters of a previous mathematical model of atherosclerosis development in arteries was performed [8]. The mathematical model has a large number of parameters that can affect the growth of the plaque. However, some of them are considered well-known due to, for example, corresponding to geometrical properties of arteries or substances. Therefore, the parameters whose influence on plaque growth was analysed are related to the reactive terms of the equations referred to substances in the arterial wall. These parameters have been modified by increasing and decreasing their value in different cases by 10% and, to determine how they affect plaque growth, variations in the volume of substances that add volume to the plaque have been analysed (foam cells, synthetic smooth muscle cells, and collagen fibre). In addition, the variation in the plaque stenosis ratio was analysed.

As can be seen in the results, on the one hand, the parameters whose variations are directly proportional to the stenosis ratio are the degradation rate of LDL (d_{LDL}), the monocyte differentiation rate (d_m), the parameters referring to the production and degradation of cytokines (C_r and S_r), the proliferation rate of synthetic smooth muscle cells (p_{ss}), the segregation rate of collagen (G_r) and the parameter related to the amount of monocyte recruited by the endothelium (m_r). On the other hand, some of the analysed parameters are inversely proportional to the growth of the plaque, and an increase in their values will cause a decrease in the volume of the plaque and, therefore, in the stenosis ratio. It is the case of the diffusion parameters of substances on the arterial wall ($D_{LDL,w} = D_{LDLox,w}$ and $D_{m,w} = D_{M,w}$), the rate of death of monocytes (m_d), the rate of oxidised LDL that is uptaken by macrophages, and the maximum amount of oxidised LDL that a macrophage can ingest ($LDL_{ox,r}$ and n_{FC}), the cytokine degradation rate, its threshold in the arterial wall (d_c and $C_{c,w}^{th}$), and the rate of apoptosis of synthetic smooth muscle cells (r_{apop}).

As can be noticed, the parameters that influence the change in the volume of synthetic smooth muscle cells also influence the change in the volume of collagen. It is due to the segregation of collagen fibre by SSMC, so collagen depends directly on them.

The less influential parameters in the volume change of substances in the plaque are: d_m, m_d, S_r, and G_r. It should be noted that d_m and m_d are parameters referring to monocytes, which act at the beginning of the process. Therefore, a great influence on them could be expected. However, monocytes highly affect the results with the parameter m_r, which is the monocytes recruitment from the lumen, and its variation has a huge influence on the volume of FC in the plaque and, therefore, in the stenosis ratio.

As can be seen in Figure 4, r_{apop} has a large influence on the variation of volume of synthetic smooth muscle cells and collagen for both cases, when increasing and decreasing its value by 10%. The smaller the r_{apop} value, the more plaque is generated, as it is an apoptosis factor of synthetic smooth muscle cells. However, its influence on the variation of the stenosis ratio is greater in the case of increasing its value than in the case of decreasing it. As it is a parameter related to the apoptosis ratio of synthetic smooth muscle cells, its change does not cause variation in the results of foam cells (Figure 5).

m_r has more influence in both concentrations and stenosis ratio variation when increased (Figures 4 and 5). This is because, when its value is decreased, the amount of monocytes deposited in the arterial wall is reduced.

$C_{c,w}^{th}$ has a large influence in the variations of the results, having more influence when increased (the change of volume and stenosis ratio are greater than 100% in this case, as can be seen in Figures 4 and 5). It is a parameter involved in the differentiation of contractile smooth muscle cells into synthetic ones due to the presence of cytokines in the arterial wall. Thus, it does not influence the volume of foam cells.

As G_r is a parameter of collagen fibre segregation, it only has influence on the change of the volume of collagen in the plaque (Figure 4). Therefore, its influence on the stenosis ratio is limited (Figure 5).

p_{ss} is related to the proliferation of synthetic smooth muscle cells, so it does not influence the results of foam cells. On the contrary, as can be seen in Figure 4, increasing it by 10% produces a change greater than 100% in the variation of synthetic smooth muscle cells and collagen fibre.

S_r changes do not cause a large variation in the volume of any substance or the plaque stenosis ratio (Figure 5).

d_c is a parameter that also has a large influence on the volume variations (Figure 4). This parameter represents the cytokine degradation; thus, the higher it is, the more cytokines are degraded and, thus, the volume and the ratio of stenosis are lower (Figure 5). The same occurs with C_r, which represents the cytokine production.

n_{FC} represents the maximum amount of oxidised LDL that a macrophage can ingest before becoming a foam cell. Therefore, an increase in its value produces a reduction in the volume of foam cells and the stenosis ratio (Figures 4 and 5). Its influence on the variation of the volume of the substances is not very large, but it produces an important variation in the stenosis ratio. When these results are contrasted with those of a substance that produces a large variation in synthetic smooth muscle cells and collagen volumes (for example, d_c), it can be observed that a smaller change in the volume of foam cells produces a larger change in the stenosis ratio. It can be explained by attending to Equation (45): The volume of a foam cell is equal to $1.489 \cdot 10^{-14} m^3$, while the volume of a synthetic smooth muscle cell is equal to $6.774 \cdot 10^{-15} m^3$. Therefore, due to their size difference, less change in foam cell volume is needed to produce a similar stenosis ratio variation.

$LDL_{ox,r}$ is related to the oxidised LDL uptaken by macrophages, so it affects the volume of each of the considered substances. An increase in its value produces a reduction in the volume of substances (Figure 4) and therefore of the stenosis ratio (Figure 5).

d_m and m_d are both parameters referring to monocytes. The first one is related to their differentiation, while the second one refers to their apoptosis. Therefore, their influences are opposite. Their influence is more notable for synthetic smooth muscle cells and collagen volumes (Figure 4).

d_{LDL} is the degradation rate of LDL, so it has an influence on foam cells, synthetic smooth muscle cells, and collagen fibre and, therefore, in the stenosis ratio of the plaque (Figures 4 and 5). So, it is one of the most influential parameters of the model and the most influential in the stenosis ratio when it is reduced.

$D_{LDL,w} = D_{LDLox,w}$ and $D_{m,w} = D_{M,w}$ are related to the diffusion of substances in the arterial wall, so they affect all the processes. Therefore, they influence the results of all the substances, and are some of the most important parameters in the model (Figures 4 and 5).

With all of this information, knowing the influence of all the parameters, they could be adjusted to achieve more or less vulnerable atheroma plaque, according to the percentage volume of foam cells, synthetic smooth muscle cells, and collagen fibre [67,68]. The vulnerability of a plaque is dependent on multiple factors, such as its size or stresses caused by blood flow in it, but it is also dependent on its composition. There is a high risk of rupture of plaque with a large lipidic nucleus and a thin fibrous cap [67,69]. Therefore, a high quantity of foam cells and a small amount of synthetic smooth muscle cells and collagen fibre will be indicators of plaque with a high risk of rupture (and, therefore, less stable) than one with a large quantity of synthetic smooth muscle cells and collagen [67–70]. Therefore, reducing the maximum amount of oxidised LDL that a macrophage can ingest and the ratio of oxidation of LDL (n_{FC}) will cause plaque with bigger lipid nuclei, which can develop into more unstable plaque. However, as can be seen, it has no influence on SSMC and collagen volumes in the plaque. Conversely, increasing the apoptosis ratio of SSMCs, the cytokine threshold in the arterial wall and its degradation rate, and the rate of oxidised LDL uptaken by macrophages (r_{apop}, $c_{c,w}^{th}$, d_c and $LDL_{ox,r}$, respectively), and reducing SSMC proliferation, cytokine production, and the oxidation LDL ratio (p_{ss}, C_r, and d_LDL, respectively) will produce plaque with less fibrotic layer and, thus, a high risk of rupture.

Some studies in the literature focus on determining the most influential parameters in the development of atheroma plaque in different mathematical models. It is the case of Cilla et al. [21], in which the effect of the anisotropy of the arterial wall on the diffusion coefficients of LDL was analysed. Their results have been considered to implement the anisotropy of the diffusion coefficients in the present model. There are also some studies on parameter influence in agent-based models. It is the case of Olivares et al. [18], who focus on determining how the migration of agents, the velocity of oxidation of LDL, and the maximum amount of autoantibodies can affect the plaque. In addition Corti et al. [20] have mono-parametric and multi-parametric studies to determine the influence of the parameters on their model. However, it is not possible to contrast their results with the ones obtained in this article, as each of the models considers different substances in the process of atheroma plaque formation and, therefore, their parameters are referred to other substance values.

The findings of this study should be interpreted in the context of its limitations. For example, the study of the influence of parameters was done only in the geometry of the carotid artery. However, the behaviour of the mathematical model would be the same for other geometrical configurations and arteries, for example, coronary or aorta arteries, adapting the values of the corresponding parameters if necessary. Another limitation of the study is that it was done with a 2D-axisymmetric model instead of a real geometry. This produces a higher plaque growth than in real cases as diffusion in the circumferential direction is not allowed and therefore causes a higher accumulation of substances in the area of the plaque. However, it should not influence the development of the plaque according to the variations in the parameters that were analysed here. In addition, blood flow and the inflammatory process are not coupled, which could influence the shape and stenosis ratio of the developed plaque. In this study, we also do not consider the influence of the mechanics of the arterial wall in the development of the plaque (such as tortuosity or changes in the permeability of the arterial wall due to the thickness variation of it with the cardiac cycle).

8. Conclusions

The mathematical model has a large number of parameters and their values influence the plaque obtained. They can affect the volume of the substances that provide volume to the plaque and also its stenosis ratio. Therefore, the dependence of the model on the variation of its parameters was analysed. For that, a previous selection of the parameters to analyse was done, and those referred to geometrical parameters, initial concentrations, and material properties were discarded.

As can be seen in Section 6, the variation of the selected parameters carries important variations on the results, and in some cases, this variation can be greater than 100%.

d_{LDL}, d_m, C_r, S_r, p_{ss}, G_r, and m_r are directly proportional to the change of substances volume (FC, SSMC, and collagen) and to the stenosis ratio, while $D_{LDL,w} = D_{LDLox,w}$, $D_{m,w} = D_{M,w}$, m_d, $LDL_{ox,r}$, n_{FC}, d_c, $C_{c,w}^{th}$, and r_{apop} are inversely proportional.

In addition, it was noticed that a variation in foam cell volume results in more of a change in the plaque stenosis ratio than in the volume of synthetic smooth muscle cells or collagen, due to the larger volume.

For all of this, it could be interesting to study the vulnerability of plaque by changing the analysed parameters, knowing how each one of them affects the volume of foam cells, synthetic smooth muscle cells, and collagen fibre in the plaque.

Author Contributions: M.C., M.A.M. and E.P. conceived and designed the study; P.H.-L., M.C., M.A.M. and E.P. developed the mathematical model; P.H.-L. and M.C. conducted the computational implementation of the model and post-processing of the results; P.H.-L., M.C., M.A.M. and E.P. wrote, reviewed, and edited the manuscript. All authors have read and agreed to the published version of the manuscript.

Funding: This research was funded by the Spanish Ministry of Science and Technology through research project PID2019-107517RB-I00; financial support was given to P. Hernández-López via grant BES-2017-080239 and the regional Government of Aragón via research project T24-20R.

Data Availability Statement: Not applicable.

Acknowledgments: The authors gratefully acknowledge the research support from CIBER– Bioengineering, Biomaterials & Nanomedicne (CIBER-BBN at the University of Zaragoza). CIBER Actions are financed by Instituto de Salud Carlos III with assistance from the European Regional Development Fund.

Conflicts of Interest: The authors declare no conflict of interest.

Abbreviations

The following abbreviations are used in this manuscript:

LDL	Low-density lipoproteins
FC	Foam cells
CSMC	Contractile smooth muscle cells
SSMC	Synthetic smooth muscle cells
WSS	Wall shear stress
$TAWSS$	time-averaged wall shear stress
OSI	Oscillatory shear index
SI	Shape index
FSI	Fluid–structure interaction
CFD	Computational fluid dynamics
BDF	Backwards differentiation formula

Subscripts:

b	Blood
l	Lumen
p	Plasma
w	Arterial wall
nj	Normal junctions
lj	Leaky junctions
vp	Vesicular pathways

Nomenclature list:

ρ	Density
μ	Dynamic viscosity
ϵ	Porosity
k	Permeability
u	Velocity
P	Pressure
v	Volume
R	Radius
l	Length
F_l	Internal forces of blood
T	Cardiac cycle period
J_v	Plasma flow across the endothelium
ΔP_{End}	Pressure drop in the endothelium
L_p	Hydraulic conductivity
Φ_{lj}	Endothelial fraction of leaky cells
$\frac{A_p}{S}$	Endothelial area occupied by leaky junctions
ϵ_{lj}	Leaky junction permeability
A_{slj}	Area of a single leaky junction
R_{cell}	Endothelial cell radius
w_l	Half width of a leaky junction
$\tau(t)$	Blood flow tangential stresses

A_{unit}	Unit area
Lp_{slj}	Hydraulic conductivity of a single leaky junction
l_{lj}	Length of a leaky junction
N	Substances flow
D_{X_i}	Diffusion coefficient of the substance X_i in the arterial wall
C_i	i substance concentration
K_{lag}	Solute lag coefficient of LDL
$J_{S,i}$	i substance flow across the endothelium
LDL_{dep}	LDL deposited into the arterial wall
P_{app}	Apparent permeability
P_{lj}	Diffusive permeability of leaky junctions
P_{slj}	Diffusive permeability of a single leaky junction
Z_{lj}	Reduction factor of leaky junctions
$\sigma_{f,lj}$	Solvent-drag coefficient of leaky junctions
χ	Fraction of endothelial cells separated by leaky junctions
α_{lj}	Geometric ratio
a_m	LDL molecule radius
Pe_{lj}	Modifies Péclet number
d_{LDL}	Degradation rate of LDL
$LDL_{ox,r}$	LDL that a macrophage can ingest
d_m	Monocyte differentiation rate
m_d	Monocyte natural death
m_r	Monocyte recruitment from lumen
n_{FC}	Quantity of oxLDL that a macrophage has to ingest to turn into a FC
C_r	Cytokine production rate
d_c	Cytokine degradation rate
S_r	CSMC differentiation rate
k_c	Saturation constant
$C_{c,w}^{th}$	Cytokine threshold
p_{ss}	SSMC proliferation rate
r_{Apop}	SSMC apoptosis rate
G_r	Collagen secretion rate
LDL	Low-density lipoproteins
oxLDL	Oxidised LDL
m	Monocyte
M	Macrophage
c	Cytokine
CSMC	Contractile Smooth Muscle Cells
SSMC	Synthetic Smooth Muscle Cells
FC	Foam Cells
Cg	Collagen

References

1. Gaziano, T.; Gaziano, J.M. Chapter 1: Global burden of cardiovascular disease. In *Brunwald's Heart Disease: A Textbook of Cardiovascular Medicine*, 9th ed.; Bonow, R., Mann, D., Zupes, D.P.L., Eds.; Elsevier: Philadlephia, PA, USA, 2012.
2. Libby, P.; Ridker, P.M.; Hansson, G.K. Progress and challenges in translating the biology of atherosclerosis. *Nature* **2011**, *473*, 317–325. [CrossRef] [PubMed]
3. Libby, P. Inflammation during the life cycle of the atherosclerotic plaque. *Cardiovasc. Res.* **2021**, *117*, 2525–2536. [CrossRef] [PubMed]
4. Cilla, M.; Peña, E.; Martínez, M.A. Mathematical modelling of atheroma plaque formation and development in coronary arteries. *J. R. Soc. Interface* **2014**, *11*, 20130866. [CrossRef] [PubMed]
5. Dai, G.H.; Kaazempur-Mofrad, M.R.; Natarajan, S.; Zhang, Y.Z.; Vaughn, S.; Blackman, B.R.; Kamm, R.D.; Garcia-Gardena, G.; Gimbrone, M.A. Distinct endothelial phenotypes evoked by arterial waveforms derived from atherosclerosis-susceptible and -resistant regions of human vasculature. *Proc. Natl. Acad. Sci. USA* **2004**, *101*, 14871–14876. [CrossRef] [PubMed]
6. Sáez, P.; Malvè, M.; Martínez, M.A. A theoretical model of the endothelial cell morphology due to different waveforms. *J. Theor. Biol.* **2015**, *379*, 16–23. [CrossRef]

7.   Alimohammadi, M.; Pichardo-Almarza, C.; Agu, O.; Díaz-Zuccarini, V. A multiscale modelling approach to understand atherosclerosis formation: A patient-specific case study in the aortic bifurcation. *Proc. Inst. Mech. Eng. Part H J. Eng. Med.* **2017**, *231*, 378–390. [CrossRef] [PubMed]

8.   Hernández-López, P.; Cilla, M.; Martínez, M.A.; Peña, E. Effects of the haemodynamic stimulus on the location of carotid plaque based on a patient-specific mechanobiological plaque atheroma formation model. *Front. Bioeng. Biotechnol.* **2021**, *9*, 690685. [CrossRef] [PubMed]

9.   Malvè, M.; Gharib, A.M.; Yazdani, S.K.; Finet, G.; Martínez, M.A.; Pettigrew, R.; Ohayon, J. Tortuosity of Coronary Bifurcation as a Potential Local Risk Factor for Atherosclerosis: CFD Steady State Study Based on In Vivo Dynamic CT Measurements. *Ann. Biomed. Eng.* **2014**, *43*, 82–93. [CrossRef] [PubMed]

10.   Alimohammadi, M.; Pichardo-Almarza, C.; Agu, O.; Díaz-Zuccarini, V. Development of a patient-specific multi-scale model to understand atherosclerosis and calcification locations: Comparison with in vivo data in an aortic dissection. *Front. Physiol.* **2016**, *7*, 238. [CrossRef]

11.   Olgac, U.; Kurtcuoglu, V.; Poulikakos, D. Computational modeling of coupled blood-wall mass transport of LDL: Effects of local wall shear stress. *Am. J. Physiol. Heart Circ. Physiol.* **2008**, *294*, 909–919. [CrossRef]

12.   Tomaso, G.D.; Díaz-Zuccarini, V.; Pichardo-Almarza, C. A multiscale model of atherosclerotic plaque formation at its early stage. *IEEE Trans. Biomed. Eng.* **2011**, *58*, 3460–3463. [CrossRef] [PubMed]

13.   Calvez, V.; Ebde, A.; Meunier, N.; Raoult, A. Mathematical modelling of the atherosclerotic plaque formation. *ESAIM Proc.* **2009**, *28*, 1–12. [CrossRef]

14.   Shahzad, O.; Wang, X.; Ghaffari, A.; Iqbal, K.; Hafeez, M.B.; Krawczuk, M.; Wojnicz, W. Fluid structure interaction study of non-Newtonian Casson fluid in a bifurcated channel having stenosis with elastic walls. *Nat. Sci. Rep.* **2022**, *12*, 12219. [CrossRef] [PubMed]

15.   Siogkas, P.; Sakellarios, A.; Exarchos, T.P.; Athanasiou, L.; Karvounis, E.; Stefanou, K.; Fotiou, E.; Fotiadis, D.I.; Naka, K.K.; Michalis, L.K.; et al. Multiscale—Patient-specific artery and atherogenesis models. *IEEE Trans. Biomed. Eng.* **2011**, *58*, 3464–3468. [CrossRef] [PubMed]

16.   Filipovic, N.; Teng, Z.; Radovic, M.; Saveljic, I.; Fotiadis, D.; Parodi, O. Computer simulation of three-dimensional plaque formation and progression in the carotid artery. *Med. Biol. Eng. Comput.* **2013**, *51*, 607–616. [CrossRef] [PubMed]

17.   Kenjereš, S.; De Loor, A. Modelling and simulation of low-density lipoprotein transport through multi-layered wall of an anatomically realistic carotid artery bifurcation. *J. R. Soc. Interface* **2014**, *11*, 20130941. [CrossRef]

18.   Olivares, A.L.; González Ballester, M.A.; Noailly, J. Virtual exploration of early stage atherosclerosis. *Bioinformatics* **2016**, *32*, 3798–3806. [CrossRef]

19.   Bhui, R.; Hayenga, H.N. An agent-based model of leukocyte transendothelial migration during atherogenesis. *PLoS Comput. Biol.* **2017**, *13*, e1005523. [CrossRef]

20.   Corti, A.; Chiastra, C.; Colombo, M.; Garbey, M.; Migliavaca, F.; Casarin, S. A fully coupled computational fluid dynamics—Agent based model of atherosclerotic plaque development: Multiscale modeling framework and parameter sensitivity analysis. *Comput. Biol. Med.* **2020**, *118*, 103623. [CrossRef]

21.   Cilla, M.; Martínez, M.A.; Peña, E. Effect of Transmural Transport Properties on Atheroma Plaque Formation and Development. *Ann. Biomed. Eng.* **2015**, *43*, 1516–1530. [CrossRef]

22.   Sommer, G.; Regitnig, P.; Költringer, L.; Holzapfel, G.A. Biaxial mechanical properties of intact and layer-dissected human carotid arteries at physiological and supraphysiological loadings. *Am. J. Physiol. Heart Circ. Physiol.* **2010**, *298*, 898–912. [CrossRef] [PubMed]

23.   De Bruyne, B.; Pijls, N.H.J.; Heyndrickx, G.R.; Hodeige, D.; Kirkeeide, R.L.; Gould, K.L. Pressure-derived fractional flow reserve to assess serial epicardial stenoses: Theoretical basis and animal validation. *Circulation* **2000**, *101*, 1840–1847. [CrossRef] [PubMed]

24.   Pijls, N.H.J.; De Bruyne, B.; Bech, G.J.W.; Liistro, F.; Heyndrickx, G.R.; Bonnier, J.J.R.M.; Koolen, J.J. Coronary pressure measurement to assess the hemodynamic significance of serial stenosis within one coronary-artery: Validation in humans. *Circulation* **2000**, *102*, 2371–2377. [CrossRef] [PubMed]

25.   Malvè, M.; Chandra, S.; García, A.; Mena, A.; Martínez, M.A.; Finol, E.A.; Doblaré, M. Impedance-based outflow boundary conditions for human carotid haemodynamics. *Comput. Methods Biomech. Biomed. Eng.* **2014**, *17*, 1248–1260. [CrossRef] [PubMed]

26.   Caro, C.G.; Pedley, T.J.; Schroter, R.C.; Seed, W.A.; Fung, Y.C. *The Mechanics of the Circulation*; Oxford University Press: Oxford, UK, 1978.

27.   Perktold, K.; Resch, M.; Florian, H. Pulsatile non-newtonian flow characteristics in a three-dimensional human carotid bifurcation model. *J. Biomech. Eng.* **1991**, *113*, 464–475. [CrossRef]

28.   Milnor, W.R. *Hemodynamics*, 2nd ed.; Lippincott Williams & Wilkins: Baltimore, MD, USA, 1989.

29.   Khaled, A.R.; Vafai, K. The role of porous media in modeling flow and heat transfer in biological tissues. *Int. J. Heat Mass Transf.* **2003**, *46*, 4989–5003. [CrossRef]

30.   Huang, Y.; Rumschitzki, D.; Chien, S.; Weinbaum, S. A fiber matrix model for the filtration through fenestral pores in a compressible arterial intima. *Am. J. Physiol.* **1997**, *272*, H2023–H2039. [CrossRef]

31.   Sun, N.; Wood, N.B.; Hughes, A.D.; Thom, S.A.; Xu, X.Y. Effects of transmural pressure and wall shear stress on LDL accumulation in the arterial wall: A numerical study using a multilayered model. *Am. J. Physiol. Heart Circ. Physiol.* **2007**, *292*, 3148–3157. [CrossRef]

32. Sun, N.; Wood, N.B.; Hughes, A.D.; Thom, S.A.M.; Xu, X.Y. Fluid-wall modelling of mass transfer in an axisymmetric Stenosis: Effects of shear-dependent transport properties. *Ann. Biomed. Eng.* **2006**, *34*, 1119–1128. [CrossRef] [PubMed]
33. Karner, G.; Perktold, K.; Zehentner, H.P. Computational modeling of macromolecule transport in the arterial wall. *Comput. Methods Biomech. Biomed. Eng.* **2001**, *4*, 491–504. [CrossRef]
34. Kedem, O.; Katchalsky, A. Thermodynamic analysis of the permeability of biological membranes to non-electrolytes. *Biochim. Biophys. Acta* **1958**, *27*, 229–246. [CrossRef]
35. Tedgui, A.; Lever, M.J. Filtration through damaged and undamaged rabbit thoracic aorta. *Am. J. Physiol. Heart Circ. Physiol.* **1984**, *247*, 784–791. [CrossRef] [PubMed]
36. Huang, Y.; Rumschitzki, D.; Chien, S.; Weinbaum, S. A fiber matrix model for the growth of macromolecular leakage spots in the arterial intima. *J. Biomech. Eng.* **1994**, *116*, 430–445. [CrossRef] [PubMed]
37. Huang, Z.J.; Tarbell, J.M. Numerical simulation of mass transfer in porous media of blood vessel walls. *Am. J. Physiol. Heart Circ. Physiol.* **1997**, *273*, H464–H477. [CrossRef] [PubMed]
38. Weinbaum, S.; Tzeghai, G.; Ganatos, P.; Pfeffer, R.; Chien, S. Effect of cell turnover and leaky junctions on arterial macromolecular transport. *Am. Physiol. Soc.* **1985**, *248*, 945–960. [CrossRef] [PubMed]
39. Yuan, F.; Chien, S.; Weinbaum, S. A new view of convective-diffusive transport processes in the arterial intima. *J. Biomech. Eng.* **1991**, *113*, 314–329. [CrossRef] [PubMed]
40. Levesque, M.J.; Liepsch, D.; Moravec, S.; Nerem, R.M. Correlation of Endothelial Cell Shape and Wall Shear Stress in a Stenosed Dog Aorta. *Am. Heart Assoc. J.* **1986**, *6*, 220–229. [CrossRef] [PubMed]
41. Zhao, S.Z.; Ariff, B.; Long, Q.; Hughes, A.D.; Thom, S.A.; Stanton, A.V.; Xu, X.Y. Inter-individual variations in wall shear stress and mechanical stress distributions at the carotid artery bifurcation of healthy humans. *J. Biomech.* **2002**, *35*, 1367–1377. [CrossRef]
42. Younis, H.F.; Kaazempur-Mofrad, M.R.; Chan, R.C.; Isasi, A.G.; Hinton, D.P.; Chau, A.H.; Kim, L.A.; Kamm, R.D. Hemodynamics and wall mechanics in human carotid bifurcation and its consequences for atherogenesis: Investigation of inter-individual variation. *Biomech. Model. Mechanobiol.* **2004**, *3*, 17–32. [CrossRef]
43. Chien, S. Molecular and mechanical bases of focal lipid accumulation in arterial wall. *Prog. Biophys. Mol. Biol.* **2003**, *83*, 131–151. [CrossRef]
44. Lin, S.J.; Jan, K.M.; Weinbaum, S.; Chien, S. Transendothelial transport of low density lipoprotein in association with cell mitosis in rat aorta. *Arteriosclerosis* **1989**, *9*, 230–236. [CrossRef] [PubMed]
45. Prosi, M.; Zunino, P.; Perktold, K.; Quarteroni, A. Mathematical and numerical models for transfer of low-density lipoproteins through the arterial walls: A new methodology for the model set up with applications to the study of disturbed lumenal flow. *J. Biomech.* **2005**, *38*, 903–917. [CrossRef] [PubMed]
46. Ai, L.; Vafai, K. A coupling model for macromolecule transport in a stenosed arterial wall. *Int. J. Heat Mass Transf.* **2006**, *49*, 1568–1591. [CrossRef]
47. Cilla, M. Mechanical Effects on the Atheroma Plaque Appearance, Growth and Vulnerability. Ph.D. Thesis, University of Zaragoza, Zaragoza, Spain, 2012.
48. Khan, F.H. *The Elements of Immunology*; Pearson Education: Delhi, India, 2009.
49. Meyer, G.; Merval, R.; Tedgui, A. Effects of pressure-induced stretch and convection on low-density lipoprotein and albumin uptake in the rabbit aortic wall. *Circ. Res.* **1996**, *79*, 532–540. [CrossRef] [PubMed]
50. Tarbell, J.M. Mass Transport in Arteries and the Localization of Atherosclerosis. *Annu. Rev. Biomed. Eng.* **2003**, *5*, 79–118. [CrossRef]
51. Hu, X.; Weinbaum, S. A new view of Starling's hypothesis at the microstructural level. *Microvasc. Res.* **1999**, *58*, 281–304. [CrossRef]
52. Ogunrinade, O.; Kameya, G.T.; Truskey, G.A. Effect of Fluid Shear Stress on the Permeability of the Arterial Entohelium. *Ann. Biomed. Eng.* **2002**, *30*, 430–446. [CrossRef]
53. Bulelzai, M.A.K.; Dubbeldam, J.L.A. Long time evolution of atherosclerotic plaque. *J. Theor. Biol.* **2012**, *297*, 1–10. [CrossRef]
54. Krstic, R.V. *Human Microscopic Anatomy: An Atlas for Students of Medicine and Biology*; Springer: Berlin/Heidelberg, Germany, 1997.
55. Zhao, B.; Li, Y.; Buono, C.; Waldo, S.W.; Jones, N.L.; Mori, M.; Kruth, H.S. Constitutive receptor-independent low density lipoprotein uptake and cholesterol accumulation by macrophages differentiated from human monocytes with macrophage-colony-stimulating factor (M-CSF). *J. Biol. Chem.* **2006**, *281*, 15757–15762. [CrossRef]
56. Zhao, W.; Oskeritzian, C.A.; Pozez, A.L.; Schwartz, L.B. Cytokine Production by Skin-Derived Mast Cells: Endogenous Proteases Are Responsible for Degradation of Cytokines. *J. Immunol.* **2005**, *175*, 2635–2642. [CrossRef]
57. Budu-Grajdeanu, P.; Schugart, R.C.; Friedman, A.; Valentine, C.; Agarwal, A.K.; Rovin, B.H. A mathematical model of venous neointimal hyperplasia formation. *Theor. Biol. Med. Model.* **2008**, *5*, 1–9. [CrossRef] [PubMed]
58. Zahedmanesh, H.; Van Oosterwyck, H.; Lally, C. A multi-scale mechanobiological model of in-stent restenosis: Deciphering the role of matrix metalloproteinase and extracellular matrix changes. *Comput. Methods Biomech. Biomed. Eng.* **2014**, *17*, 813–828. [CrossRef]
59. Escuer, J.; Martínez, M.A.; McGinty, S.; Peña, E. Mathematical modelling of the restenosis process after stent implantation. *J. R. Soc. Interface* **2019**, *16*, 20190313. [CrossRef]

50. Steinberg, D.; Khoo, J.C.; Glass, C.K.; Palinski, W.; Almazan, F. A new approach to determining the rates of recruitment of circulating leukocytes into tissues: Application to the measurement of leukocyte recruitment into atherosclerotic lesions. *Proc. Natl. Acad. Sci. USA* **1997**, *94*, 4040–4044. [CrossRef] [PubMed]

51. Kockx, M.M.; Muhring, J.; Bortier, H.; De Meyer, G.R.Y.; Jacob, W. Biotin- or digoxigenin-conjugated nucleotides bind to matrix vesicles in atherosclerotic plaque. *Am. J. Pathol.* **1996**, *148*, 1771–1777. [PubMed]

52. Ivanova, E.A.; Myasoedova, V.A.; Melnichenko, A.A.; Grechko, A.V.; Orekhov, A.N. Small Dense Low-Density Lipoprotein as Biomarker for Atherosclerotic Diseases. *Oxidative Med. Cell. Longev.* **2017**, *2017*, 1273042. [CrossRef]

53. Dabagh, M.; Jalali, P.; Konttinen, Y.T. The study of wall deformation and flow distribution with transmural pressure by three-dimensional model of thoracic aorta wall. *Med. Eng. Phys.* **2009**, *31*, 816–824. [CrossRef]

54. Krombach, F.; Münzing, S.; Allmeling, A.M.; Gerlach, J.T.; Behr, J.; Dörger, M. Cell size of alveolar macrophagues: an interspecies comparison. *Environ. Health Perspect.* **1997**, *105*, 1261–1263.

55. Cannon, G.J.; Swanson, J.A. The macrophage capacity for phagocytosis. *J. Cell Sci.* **1992**, *101*, 907–913. [CrossRef]

56. Sáez, P.; Peña, E.; Ángel Martínez, M.; Kuhl, E. Mathematical modeling of collagen turnover in biological tissue. *J. Math. Biol.* **2013**, *67*, 1765–1793. [CrossRef]

57. Le Floc'h, S.; Ohayon, J.; Tracqui, P.; Finet, G.; Gharib, A.M.; Maurice, R.L.; Cloutier, G.; Pettigrew, R.I. Vulnerable Atherosclerotic Plaque Elasticity Reconstruction Based on a Segmentation-Driven Optimization Procedure Using Strain Measurements: Theoretical Framework. *IEEE Trans. Med. Imaging* **2009**, *28*, 1126–1137. [CrossRef] [PubMed]

58. Pan, J.; Cai, Y.; Wang, L.; Maehara, A.; Mintz, G.; Tang, D.; Li, Z. A prediction tool for plaque progression based on patient-specific multi-physical modeling. *PLoS Comput. Biol.* **2021**, *17*, e1008344. [CrossRef]

59. Fuster, V.; Moreno, P.R.; Fayad, Z.A.; Corti, R.; Badimon, J.J. Atherothrombosis and High-Risk Plaque: Part I: Evolving Concepts. *J. Am. Coll. Cardiol.* **2005**, *46*, 937–954. [CrossRef] [PubMed]

60. Virmani, R.; Burke, A.P.; Farb, A.; Kolodgie, F.D. Pathology of the Vulnerable Plaque. *J. Am. Coll. Cardiol.* **2006**, *47*, C13–C18. [CrossRef] [PubMed]

 *mathematics*

Article

Hamilton–Jacobi Inequality Adaptive Robust Learning Tracking Controller of Wearable Robotic Knee System

Houssem Jerbi [1,*], Izzat Al-Darraji [2], Georgios Tsaramirsis [3], Lotfi Ladhar [4] and Mohamed Omri [5]

[1] Department of Industrial Engineering, College of Engineering, University of Ha'il, Ha'il 81451, Saudi Arabia
[2] Automated Manufacturing Department, Al-Khwarizmi College of Engineering, University of Baghdad, Baghdad 10081, Iraq; izzat.a@kecbu.uobaghdad.edu.iq
[3] Abu Dhabi Women's Campus, Higher Colleges of Technology, Abu Dhabi 25026, United Arab Emirates; gtsaramirsis@hct.ac.ae
[4] Department of Electrical and Computer Engineering, Faculty of Engineering, King Abdul Aziz University, Jeddah 21589, Saudi Arabia; lladhar@kau.edu.sa
[5] Deanship of Scientific Research (DSR), King Abdulaziz University, Jeddah 21589, Saudi Arabia; mnomri@kau.edu.sa
* Correspondence: h.jerbi@uoh.edu.sa

Abstract: A Wearable Robotic Knee (WRK) is a mobile device designed to assist disabled individuals in moving freely in undefined environments without external support. An advanced controller is required to track the output trajectory of a WRK device in order to resolve uncertainties that are caused by modeling errors and external disturbances. During the performance of a task, disturbances are caused by changes in the external load and dynamic work conditions, such as by holding weights while performing the task. The aim of this study is to address these issues and enhance the performance of the output trajectory tracking goal using an adaptive robust controller based on the Radial Basis Function (RBF) Neural Network (NN) system and Hamilton–Jacobi Inequality (HJI) approach. WRK dynamics are established using the Lagrange approach at the outset of the analysis. Afterwards, the L_2 gain technique is applied to enhance the control motion solutions and provide the main features of the designed WRK control systems. To prove the stability of the controlled system, the HJI approach is investigated next using optimization techniques. The synthesized RBF NN algorithm supports the easy implementation of the adaptive controller, as well as ensuring the stability of the WRK system. An analysis of the numerical simulation results is performed in order to demonstrate the robustness and effectiveness of the proposed tracking control algorithm. The results showed the ability of the suggested controller of this study to find a solution to uncertainties.

Keywords: wearable robotic knee; tracking controller; radial basis function neural network; L_2 gain; Hamilton–Jacobi Inequality; robust control; adaptive control

MSC: 93A30; 93C40; 93C95; 93D05; 93D09

Citation: Jerbi, H.; Al-Darraji, I.; Tsaramirsis, G.; Ladhar, L.; Omri, M. Hamilton–Jacobi Inequality Adaptive Robust Learning Tracking Controller of Wearable Robotic Knee System. *Mathematics* **2023**, *11*, 1351. https://doi.org/10.3390/math11061351

Academic Editor: Mauro Malvè

Received: 3 January 2023
Revised: 3 March 2023
Accepted: 7 March 2023
Published: 10 March 2023

1. Introduction

After middle-age, the possibility of knee weakness might increase as a result of knee fragility or weak muscles [1]. For this reason, devices that assist people in walking are essential to reduce the pressure on the knee. People with physical disabilities can significantly enhance their quality of life with assistive technologies such as knee prostheses. To make these technologies more effective, however, main scientific research requirements must be met. This study is motivated by the need to establish and validate an advanced controller for simulation-assisted prosthesis design, which covers typical assumptions about the appropriate form and functionality of knee disabilities. In more specific terms, this can be divided into two specific goals: (i) Design a control strategy that analyzes real knee dynamics and behavior and then translates the outcomes into an appropriate real prosthesis. It is imperative that the control system be stable, flexible, and safe. (ii) Develop a set of analytical and smart tools that will enable the enhancement of design techniques by incorporating realistic testing constraints as a basis for validating simulated trends. Various control simulation scenarios based on a Lagrangian dynamics model will be investigated in this work to validate the overall design process. The scope of the present study is restricted to creating the framework necessary to enable this significant design process—fundamentally, the pipeline from theory to experimentation. The Wearable Robotic Knee (WRK) is an assist system that relieves the knee by offering support in performing various types of movements [2]. Generally, WRK interacts with the motion of the elderly or injured person during activity by actuated orthosis [3]. In [4], a WRK was designed using elastic actuators via a special technique that includes kinematic analysis, topology options, and optimization of structure. The developed WRK demonstrated low inertia on the person wearing the device, back-drive, and the ability to overcome the issues of misalignments. In [5], a WRK exoskeleton with variable stiffness actuators [6,7] was designed for lower limbs. Its structure was flexible, and it was equipped with six joints with variable stiffness actuators. By virtue of its structure, it showed the ability of assisting persons in walking by applying a simple torque. In [8], a functional and general gait assist robot was designed that could carry weight using a suggested mechanical design. The structure was dependent on the model of the human knee and the analysis of body movement. The designed gait-assisting robot could be worn without any restrictions on persons. Furthermore, it could achieve various movements for different human leg length and body weight. In [9], a prototype of a wearable robot was designed to assist people in walking by applying eight electric motors and inertial measurement unit sensors. A special controller was designed for this prototype to control the mechanical structure of the WRK. The designed prototype showed its ability to carry and move the person who wore it while keeping human body balance. In [10], the path planning of WRK was simplified by applying a Probabilistic Fam technique. This technique provides safe motion via an active knee orthosis. In [11], a WRK device was integrated within electromyogram sensors to assist the muscle of the person's knee. The human knee extension and flexion movements were controlled by manipulating the electromyogram sensor signals corresponding to rectus femoris and biceps femoris. The developed prototype showed functionality in practical application. A clinical control technique for an exoskeleton WRK is presented in [2]. The presented method applied the action of the human joint synergistically, using a gait help stick, to predict and supply the required assistance for injured limbs. The result of injured persons' walking with the proposed control technique achieves the aims that they can walk with the developed system. In [12], one of the major benefits of the proposed method is that it does not cancel the non-linear properties of the process, but compensates for these features by dampening them instead. The designed controller achieves the system's optimal performance in terms of robustness, rapidity and steady-state accuracy. However, a nonlinear observer reconstructs the unmeasurable signals to ensure that the process output trajectory closely follows the desired path. In the work of Belkhier et al. [13], the authors propose a passivity-based controller, designed to cope with nonlinear systems exhibiting ODEs with a high degree of coupling and a wide range of modelling uncertainties.

This study contributes to the development of an adaptive robust stable tracking controller using the Radial Basis Function (RBF) Neural Network (NN) and Hamilton–Jacobi Inequality (HJI) method. The following steps are implemented:

1. By considering model errors and external disturbances, the Lagrange theory is used to derive the dynamics equations of the WRK system.
2. A control input is developed for the WRK system from feedback action, which determines the difference between the current knee angle value and the required angle value.
3. The HJI-based L_2 gain technique is established to prove the stability of the WRK's control system and to enable control motion solutions.

4. By using RBF NN, the control strategy is designed to overcome problems associated with random variations and uncertain operating conditions.

In Table 1, we provide the standard nomenclature used throughout this study in order to introduce the various symbols that appear in the equations presented.

Table 1. Standard nomenclature.

Notation	
T	Height of the person
m_{body}	Weight of the person
K_{WRK}	Kinetic energy of the WRK system
K_{thigh}	Kinetic energy of the thigh
K_{calf}	Kinetic energy of the calf
P_{WRK}	Potential energy of the WRK system
P_{thigh}	Potential energy of the thigh
P_{calf}	Potential energy of the calf
l_{thigh}	Length of the thigh
l_{calf}	Length of the calf
x_{thigh}	Position of the center of thigh in x-axis
y_{thigh}	Position of the center of thigh in y-axis
v_{thigh}	Velocity of the thigh
x_{calf}	Position of the center of calf in x-axis
y_{calf}	Position of the center of calf in y-axis
v_{calf}	Velocity of the calf
θ_{hip}	Angle of the hip
θ_{knee}	Angle of the knee
$\tau_{WRK,1}$	Input torque at hip joint
$\tau_{WRK,2}$	Input torque at knee joint
m_{thigh}	Mass of the thigh
m_1	Mass of WRK device upper arm
I_{thight}	Moment of inertia of thigh
I_1	Moment of inertia of WRK device upper arm
m_{calf}	Mass of the calf
m_2	Mass of WRK device lower arm
I_{calf}	Moment of inertia of calf
I_2	Moment of inertia of WRK device lower arm
M_{WRK}	Matrix of WRK system inertia
C_{WRK}	Matrix of Coriolis centrifugal forces
G_{WRK}	Matrix of gravity
ε_{WRK}	Model errors
D_{WRK}	External disturbances
δ_{WRK}	Uncertainties
$u_{i/p}$	Control input

Table 1. *Cont.*

Notation	
c_i	Inputs of RBF NN vector
d_j	Gaussian function
d^T_{WRK}	Gaussian function vector
$w_{WRK,hj}$	RBF NN weight matrix
$\dot{\hat{W}}_{WRK}$	Derivative of the estimated RBF NN weight matrix
E_{WRK}	Error vector
Q	Function of positive value
E	Input energy
$\beta, \gamma, \varepsilon, \beth$	Constant of positive value

The rest of the paper is organized as follows: In Section 2, the related works and contributions of this study are presented. In Section 3, dynamics equations of the WRK system are derived. In Section 4, the design of the tracking error controller is outlined. In Section 5, the HJI-based L_2 gain technique is introduced. In Section 6, the RBF NN–based Adaptive Robust Controller is detailed. In Section 7, the simulation study and results are presented and discussed. The conclusions of this study and future work suggestions are included in Section 8.

2. Related Works

Developing a suitable tracking controller for a WRK is important to ensure the operation of the designed device in an independent and intelligent manner. This section presents the existing research on WRK control motion planning design. In [14], a WRK is designed using actuator-type elastomeric muscles and a body of soft fabric material. The knee movements are improved by the produced prototype. The process of producing this WRK was simple by virtue of the selected components. Moreover, producing the platform of the WRK from soft fabric material gives the prototype the essential property of light weight, making it a device that can be worn with ease. In [15], a control technique is proposed for an exoskeletal WRK in order to help with a specific motion, taking in consideration the weight and balance of the person. By equipping pressure sensors on the shoes of the person wear exoskeletal WRK, the center of the downward walking pressure by the person can be measured. Next, a suggested control algorithm is developed to provide balance. The required torque for improving the gait is supplied by actuators of the elastic type, which are installed at the wearer's hip and knee joints. In [16], a new WRK motion planning technique is introduced for helping humans to walk. The presented algorithm is based on the principles of the dynamic mechanism of a specifically developed lower limb wearable robot. The process of helping humans to walk is developed via applying a learning algorithm into the mechanism of the robot. Obtaining the path of the knee joint requires the value of the ankle joint. The uncertainties in joints are treated by a reinforcement learning algorithm. In [17], virtual force is used to obtain a technique for path control planning for both the hip joint and knee joint. The designed technique addresses the issue of uncertainty. In [18], an on-line learning algorithm is presented for a lower limb wearable robot to assist in walking. The objective of the suggested technique is to minimize the effort of the person who wears the exoskeleton robot and the consumed energy of the actuator. The results showed the functionality of the introduced on-line learning algorithm in reducing the effort of the person during walking and the consumed power by actuator. In [19], a control algorithm of the human–robot cooperation type is introduced to aid the wearer of the wearable robot in locomotion during stair climbing. The trajectory is divided into two control patterns. The first one is called the human control space, where the person does not need help from the wearable device. The second one is the robot control space, where the person needs help from the wearable device. These two control spaces are managed through a new function. This new function constrains the motion of the person to achieve the required trajectory. An adaptive controller is designed to guarantee the smooth transmission between the human and robot motions. The main features of the controller algorithms of the mentioned works are listed in Table 2.

Table 2. Main motion control features of the related work.

Benchmark Paper	Compatibility of All WRK Mechanisms	Uncertainties	Disturbances	Adaptivity	Learning Ability	Independence of Knee Joint, Hip Joint	Walking Motion	Self-Working Ability	Safety
Park Y.-L. et al. (2014) [14]	No	Yes	No	No	Yes	No	Yes	Yes	No
Eong M. et al. (2020) [15]	Yes	Yes	No	No	No	No	No	Yes	No
Yuan Y. et al. (2020) [16]	Yes	No	No	No	Yes	Yes	Yes	No	No
Bian Y. et al. (2021) [17]	Yes	No	No	Yes	No	Yes	Yes	Yes	No
Kagawa T. et al. (2017) [18]	Yes	No	Yes	No	No	Yes	Yes	Yes	Yes
Li Z. et al. (2020) [19]	Yes	No	No	No	No	Yes	Yes	Yes	Yes
Current work	Yes	Yes	Yes	Yes	Yes	Yes	Yes	Yes	Yes

In view of the literature review, in this paper we primarily address issues related to modeling uncertainties, robustness, and random external disturbances by satisfying the above criteria. Furthermore, since the WRK is a human medical prosthesis, it should be expected that operating conditions and model parameters will vary widely. Thus, maintaining WRK stability is a key performance factor. Equally significant, the controller running time must be reduced to allow online tuning and real-time update of the control input. As a means of overcoming the above-mentioned problems, this paper examines the combination of three conventional and advanced artificial intelligence techniques: the Hamilton–Jacobi Inequality method, the gain technique, and the Radial Basis Function Neural Network system. Generally, a WRK device is a nonlinear system of high uncertainty. The path planning of a WRK system should take in consideration path accuracy, consumed energy, and smooth functionality. Moreover, it is essential to consider the issues of external disturbances, modeling errors, unknown human weight, and other uncertainties. The other most important working condition that needs to be considered is changing of the control inputs during operation. For example, the person who wears the WRK may hold weight during movement or may feel tired and exert lower torque by the knee. All of these issues are treated in this recent research via the design an adaptive robust learned controller that aims to achieve the following contributions: (1) enhancing the control of the WRK system, (2) achieving better accuracy of knee joint trajectory tracking, (3) manipulating nonlinearities in WRK dynamics, (4) treating external disturbances, (5) achieving path tracking that does not depend on the weight of the user, and (6) most importantly, ensuring the designed controller can learn during operation to adapt to changing conditions.

3. Dynamics Analysis

The exoskeleton WRK is a mechatronics device that integrates a person with a machine to form a human–machine approach. The WRK model within the human body is depicted in Figure 1. The thigh and calf are rotated by hip angle θ_{hip} and knee angle θ_{knee}, respectively. The WRK orthosis actuators are link-1 and link-2, fixed at the wearer's knee, thigh, and calf. Note that link-1 is fixed to thigh and knee, and link-2 is fixed to calf and knee. Consider the height of the person is T, then, the lengths of the thigh and calf, WRK link-1, and WRK link-2 are obtained from Equations (1)–(4), respectively [20]:

$$l_{thigh} = 0.245T \tag{1}$$

$$l_{calf} = 0.246T \tag{2}$$

$$l_1 = 0.12T \tag{3}$$

$$l_2 = 0.12T \tag{4}$$

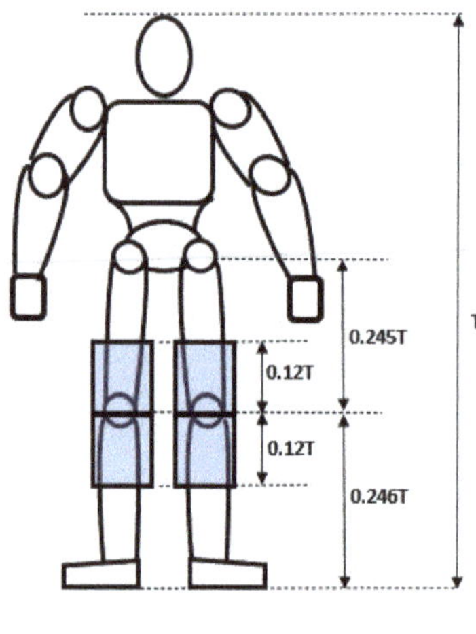

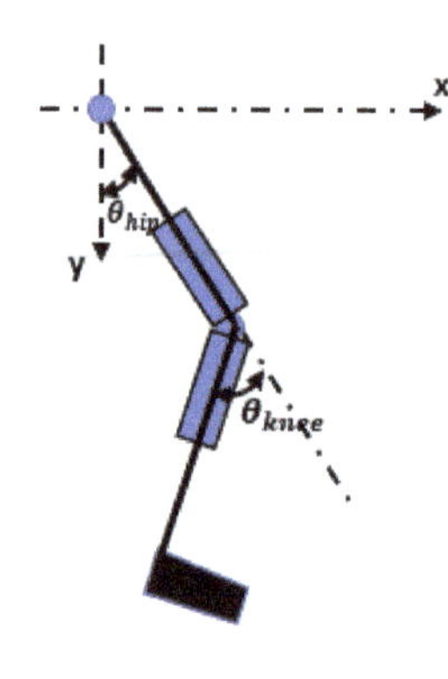

(a) (b)

Figure 1. WRK device model. (**a**) Front view of human body. (**b**) Side view of human leg.

The mass of the thigh and calf of a female and a male is calculated from total weight of the body m_{body} [21] as follows.

For a female:

$$m_{thigh} = 0.118m_{body} \tag{5}$$

$$m_{calf} = 0.0535m_{body} \tag{6}$$

For a male:

$$m_{thigh} = 0.105m_{body} \tag{7}$$

$$m_{calf} = 0.0475m_{body} \tag{8}$$

It is assumed that each part, including the thigh, the calf, the WRK link-1, and the WRK link-2, can be considered as a thin rod that rotates around an axis at its end. Hence, the moment of inertia of a part of mass M_{part} and length L_{part} is calculated according to Equation (9):

$$I_{part} = \frac{1}{3}M_{part}L^2_{part} \tag{9}$$

where part refers to the thigh, calf, WRK link-1, or WRK link-2.

The WRK controller design is based on the dynamics of the WRK manipulator. There are various techniques that can be implemented to obtain the dynamics of manipulators [22]. In this study, the Lagrange approach is applied to obtain WRK equations of motion; the Lagrangian formula is as follows:

$$L_{WRK} = K_{WRK} - P_{WRK} \tag{10}$$

where K_{WRK} and P_{WRK} denote the kinetic and potential energy, respectively. Human thigh and calf length are denoted as l_{thigh} and l_{calf}, respectively. The position of center of thigh is obtained analytically from Figure 1b, as below:

$$x_{thigh} = 0.5l_{thigh}\sin\theta_{hip} \tag{11}$$

$$y_{thigh} = 0.5l_{thigh}\cos\theta_{hip} \tag{12}$$

According to Equations (11) and (12), the velocity is obtained as:

$$\dot{x}_{thigh} = 0.5l_{thigh}\cos\theta_{hip}\dot{\theta}_{hip} \tag{13}$$

$$\dot{y}_{thigh} = -0.5l_{thigh}\sin\theta_{hip}\dot{\theta}_{hip} \tag{14}$$

$$v_{thigh}^2 = \dot{x}_{thigh}^2 + \dot{y}_{thigh}^2 = 0.25l_{thigh}^2\dot{\theta}_{hip}^2 \tag{15}$$

In the same way, the position of center of calf is obtained as:

$$x_{calf} = l_{thigh}\sin\theta_{hip} - 0.5l_{calf}\sin\left(\theta_{knee} - \theta_{hip}\right) \tag{16}$$

$$y_{calf} = l_{thigh}\cos\theta_{hip} + 0.5l_{calf}\cos\left(\theta_{knee} - \theta_{hip}\right) \tag{17}$$

The velocity of calf is obtained from Equations (16) and (17) as:

$$\dot{x}_{calf} = \left[l_{thigh}\cos\theta_{hip} + 0.5l_{calf}\cos\left(\theta_{knee} - \theta_{hip}\right)\right]\dot{\theta}_{hip} - 0.5l_{calf}\cos\left(\theta_{knee} - \theta_{hip}\right)\dot{\theta}_{knee} \tag{18}$$

$$\dot{y}_{calf} = \left[-l_{thigh}\sin\theta_{hip} + 0.5l_{calf}\sin\left(\theta_{knee} - \theta_{hip}\right)\right]\dot{\theta}_{hip} - 0.5l_{calf}\sin\left(\theta_{knee} - \theta_{hip}\right)\dot{\theta}_{knee} \tag{19}$$

$$v_{calf}^2 = \dot{x}_{calf}^2 + \dot{y}_{calf}^2$$

$$v_{calf}^2 = l_{thigh}^2\dot{\theta}_{hip}^2 + 0.25l_{calf}^2\left(\dot{\theta}_{hip} - \dot{\theta}_{knee}\right)^2 + l_{thigh}l_{calf}\dot{\theta}_{hip}\left(\dot{\theta}_{hip} - \dot{\theta}_{knee}\right)\cos\theta_{knee} \tag{20}$$

The WRK system is represented by the following Lagrangian equation:

$$\tau_{WRK,i} = \frac{d}{dt}\left(\frac{\partial L_{WRK}}{\partial\dot{\theta}_i}\right) - \frac{\partial L_{WRK}}{\partial\theta_i}, i = hip, knee \tag{21}$$

The input torque $\tau_{WRK,hip}$ and $\tau_{WRK,knee}$ represent the hip joint torque and knee joint torque, respectively. The total kinetic energy is:

$$K_{WRK} = K_{thigh} + K_{Calf}$$

$$K_{WRK} = \tfrac{1}{2}\left(m_{thigh} + m_1\right)v_{thigh}^2 + \tfrac{1}{2}\left(I_{thigh} + I_1\right)\dot{\theta}_{hip}^2 + \tfrac{1}{2}\left(m_{calf} + m_2\right)v_{calf}^2 + \tfrac{1}{2}\left(I_{calf} + I_2\right)\dot{\theta}_{knee}^2 \tag{22}$$

where m_1, m_2, I_{thigh}, I_1, I_{calf}, and I_2 are the mass of WRK upper arm, mass of WRK lower arm, moment of inertia of thigh, moment of inertia of WRK upper arm, moment of inertia of calf, and moment of inertia of WRK lower arm, respectively. Assuming the center of mass of the thigh and calf are at the geometrical center of the thigh and calf, then the potential energy is obtained as:

$$P_{WRK} = P_{thigh} + P_{Calf}$$

$$P_{WRK} = -0.5\left(m_{thigh} + m_1\right)gl_{thigh}\cos\theta_{hip} - \left(m_{calf} + m_2\right)gl_{thigh}\cos\theta_{hip}$$
$$-0.5\left(m_{calf} + m_2\right)gl_{calf}\cos\left(\theta_{hip} - \theta_{knee}\right) \tag{23}$$

After substituting the values of v_{thigh}^2 and v_{calf}^2 from Equations (15) and (20), respectively, into Equation (22), then Equations (22) and (23) are inserted into Equation (10) to obtain the Lagrangian formula L_{WRK}, shown in Equation (24):

$$
\begin{aligned}
L_{WRK} &= 0.125\left(m_{thigh} + m_1\right)l_{thigh}^2\dot{\theta}_{hip}^2 + 0.5\left(I_{thigh} + I_1\right)\dot{\theta}_{hip}^2 \\
&+ 0.5\left(I_{calf} + I_2\right)\dot{\theta}_{knee}^2 + 0.5\left(m_{calf} + m_2\right)l_{thigh}^2\dot{\theta}_{hip}^2 \\
&+ 0.125\left(m_{calf} + m_2\right)l_{calf}^2\left(\dot{\theta}_{hip} - \dot{\theta}_{knee}\right)^2 \\
&+ 0.5\left(m_{calf} + m_2\right)l_{thigh}l_{calf}\dot{\theta}_{hip}\left(\dot{\theta}_{hip} - \dot{\theta}_{knee}\right)\cos\theta_{knee} \\
&+ 0.5\left(m_{thigh} + m_1\right)gl_{thigh}\cos\theta_{hip} \\
&+ \left(m_{calf} + m_2\right)gl_{thigh}\cos\theta_{hip} \\
&+ 0.5\left(m_{calf} + m_2\right)gl_{calf}\cos\left(\theta_{hip} - \theta_{knee}\right)
\end{aligned}
\tag{24}
$$

Considering Equation (21), the torque at hip and knee are obtained according to Equations (25) and (26), respectively:

$$
\begin{aligned}
\tau_{WRK,hip} &= \left[\left(m_{thigh} + m_1\right)\left(\tfrac{l_{thigh}}{2}\right)^2 + \left(m_{calf} + m_2\right)l^2_{thigh} + \left(m_{calf} + m_2\right)\left(\tfrac{l_{calf}}{2}\right)^2\right. \\
&+ \left.\left(m_{calf} + m_2\right)l_{thigh}l_{calf}\cos\theta_{knee} + \left(I_{calf} + I_2\right)\right]\ddot{\theta}_{hip} \\
&- \left[\left(m_{calf} + m_2\right)\left(\tfrac{l_{calf}}{2}\right)^2\right. \\
&+ 0.5\left.\left(m_{calf} + m_2\right)l_{thigh}l_{calf}\cos\theta_{knee}\right]\ddot{\theta}_{knee} \\
&- \left(m_{calf} + m_2\right)l_{thigh}l_{calf}\sin\theta_{knee}\dot{\theta}_{hip}\dot{\theta}_{knee} \\
&+ 0.5\left(m_{calf} + m_2\right)l_{thigh}l_{calf}\sin\theta_{knee}\dot{\theta}_{knee}^2 \\
&+ \left[\left(m_{thigh} + m_1\right)g\tfrac{l_{thigh}}{2} + \left(m_{calf} + m_2\right)gl_{thigh}\right]\sin\theta_{hip} \\
&+ \left(m_{calf} + m_2\right)\tfrac{l_{calf}}{2}\sin\left(\theta_{hip} - \theta_{knee}\right)
\end{aligned}
\tag{25}
$$

$$
\begin{aligned}
\tau_{WRK,knee} &= -\left[\left(m_{calf} + m_2\right)\left(\tfrac{l_{calf}}{2}\right)^2 + 0.5\left(m_{calf} + m_2\right)l_{thigh}l_{calf}\cos\theta_{knee}\right]\ddot{\theta}_{hip} \\
&+ \left[\left(I_{calf} + I_2\right) + \left(m_{calf} + m_2\right)\left(\tfrac{l_{calf}}{2}\right)^2\right]\ddot{\theta}_{knee} \\
&+ 0.5\left(m_{calf} + m_2\right)l_{thigh}l_{calf}\sin\theta_{knee}\dot{\theta}_{hip}^2 \\
&- \left(m_{calf} + m_2\right)g\tfrac{l_{calf}}{2}\sin\left(\theta_{hip} - \theta_{knee}\right)
\end{aligned}
\tag{26}
$$

By rearranging the resulting above two equations into matrix forms, the WRK dynamics Equation (27) is obtained, as below:

$$
\begin{aligned}
\left[\tau_{WRK,hip}\,\tau_{WRK,knee}\right]^T &= M_{WRK}\left(\theta_{hip}, \theta_{knee}\right)\left[\ddot{\theta}_{hip}\ddot{\theta}_{knee}\right]^T + \\
C_{WRK}\left(\theta_{hip}, \theta_{knee}, \dot{\theta}_{hip}, \dot{\theta}_{knee}\right)&\left[\dot{\theta}_{hip}\dot{\theta}_{knee}\right]^T + G_{WRK}\left(\theta_{hip}, \theta_{knee}\right)
\end{aligned}
\tag{27}
$$

where $M_{WRK}\left(\theta_{hip},\theta_{knee}\right)$, $C_{WRK}\left(\theta_{hip},\theta_{knee},\dot{\theta}_{hip},\dot{\theta}_{knee}\right)$, and $G_{WRK}\left(\theta_{hip},\theta_{knee}\right)$ are the matrix of WRK system inertia, Coriolis centrifugal forces, and gravity, which are obtained as follows:

$$M_{WRK}\left(\theta_{hip},\theta_{knee}\right) = \begin{bmatrix} M_{WRK,11} & M_{WRK,12} \\ M_{WRK,21} & M_{WRK,22} \end{bmatrix}$$

where

$$\begin{cases} M_{WRK,11} = \left(I_{thigh} + I_1\right) + \left(m_{thigh} + m_1\right)\left(\frac{l_{thigh}}{2}\right)^2 + \left(m_{calf} + m_2\right)l_{thigh}^2 + \left(m_{calf} + m_2\right)\left(\frac{l_{calf}}{2}\right)^2 + \left(m_{calf} + m_2\right)l_{thigh}l_{calf}\cos\theta_{knee} \\ M_{WRK,12} = \left(m_{calf} + m_2\right)\left(\frac{l_{calf}}{2}\right)^2 + 0.5\left(m_{calf} + m_2\right)l_{thigh}l_{calf}\cos\theta_{knee} \\ M_{WRK,21} = \left(m_{calf} + m_2\right)\left(\frac{l_{calf}}{2}\right)^2 + \left(m_{calf} + m_2\right)l_{thigh}l_{calf}\cos\theta_{knee} \\ M_{WRK,22} = \left(I_{calf} + I_2\right) + \left(m_{calf} + m_2\right)\left(\frac{l_{calf}}{2}\right)^2 \end{cases}$$

$$C_{WRK}\left(\theta_{hip},\theta_{knee},\dot{\theta}_{hip},\dot{\theta}_{knee}\right) = \begin{bmatrix} C_{WRK,11} & C_{WRK,12} \\ C_{WRK,21} & C_{WRK,22} \end{bmatrix}$$

where

$$\begin{cases} C_{WRK,11} = -\left(m_{calf} + m_2\right)l_{thigh}l_{calf}\sin\theta_{knee}\dot{\theta}_{knee} \\ C_{WRK,12} = 0.5\left(m_{calf} + m_2\right)l_{thigh}l_{calf}\sin\theta_{knee}\dot{\theta}_{knee} \\ C_{WRK,21} = 0.5\left(m_{calf} + m_2\right)l_{thigh}l_{calf}\sin\theta_{knee}\dot{\theta}_{hip} \\ C_{WRK,22} = 0 \end{cases}$$

and

$$G_{WRK}\left(\theta_{hip},\theta_{knee}\right) = \begin{bmatrix} G_{WRK,1} \\ G_{WRK,2} \end{bmatrix}$$

where

$$\begin{cases} G_{WRK,1} = \left[\left(m_{thigh} + m_1\right)g\frac{l_{thigh}}{2} + \left(m_{calf} + m_2\right)gl_{thigh}\right]\sin\theta_{hip} + \left(m_{calf} + m_2\right)\frac{l_{calf}}{2}\sin\left(\theta_{hip} - \theta_{knee}\right) \\ G_{WRK,2} = -\left(m_{calf} + m_2\right)g\frac{l_{calf}}{2}\sin\left(\theta_{hip} - \theta_{knee}\right) \end{cases}$$

The dynamics equation of motion, Equation (27), of WRK system with uncertainties $\delta_{WRK}\left(\theta_{hip},\theta_{knee},\dot{\theta}_{hip,r},\dot{\theta}_{knee,r}\right)$ can be written as:

$$\begin{aligned} \left[\tau_{WRK,hip}\,\tau_{WRK,knee}\right]^T &= M_{WRK}\left(\theta_{hip},\theta_{knee}\right)\left[\ddot{\theta}_{hip}\ddot{\theta}_{knee}\right]^T \\ &+ C_{WRK}\left(\theta_{hip},\theta_{knee},\dot{\theta}_{hip},\dot{\theta}_{knee}\right)\left[\dot{\theta}_{hip}\dot{\theta}_{knee}\right]^T \\ &+ G_{WRK}\left(\theta_{hip},\theta_{knee}\right) + \delta_{WRK}\left(\theta_{hip},\theta_{knee},\dot{\theta}_{hip,r},\dot{\theta}_{knee,r}\right) \end{aligned} \tag{28}$$

In term of model errors $\varepsilon_{WRK}\left(\theta_{hip},\theta_{knee},\dot{\theta}_{hip,r},\dot{\theta}_{knee,r}\right)$ and external disturbances D_{WRK}, the uncertainty value is obtained as: $\delta_{WRK}\left(\theta_{hip},\theta_{knee},\dot{\theta}_{hip,r},\dot{\theta}_{knee,r}\right) = D_{WRK} + \varepsilon_{WRK}\left(\theta_{hip},\theta_{knee},\dot{\theta}_{hip,r},\dot{\theta}_{knee,r}\right)$.

4. Tracking Error

In this section, the control input and the trajectory of a hip and knee joint are presented to address the tracking error of the WRK system. The control input is based on the difference between the current and the desired value, of the hip and knee joints, in the feedback section. When there are no uncertainties and no errors in the initial values of states, the laws of control will move the WRK system on a desired path in the feedforward control scheme.

The unknown weight of the person is included in the outer disturbances, while all the uncertainties are involved in the error of the dynamics model. The uncertainties, including model errors and disturbance errors, are applied to train the RBF NN. This will be necessary to continue the WRK device on the required track, i.e., to achieve robustness. On the other hand, hip and knee reference trajectories for a healthy person are derived to simulate the values of hip and knee joints during the gait cycle.

4.1. Control Input

The control input $u_{i/p}$ affecting the WRK is developed from feedback action that is applied to provide the difference between the current value of the hip angle θ_{hip} and knee angle θ_{knee} and the required value of hip angle $\theta_{hip,r}$ and knee angle $\theta_{knee,r}$, respectively. Hence, the error vector of tracking will be $E_{WRK} = \begin{bmatrix} \theta_{hip} - \theta_{hip,r} & \theta_{knee} - \theta_{knee,r} \end{bmatrix}^T$. On the other hand, the feedforward control design is applied to the WRK system. The laws of control will move the WRK system on a desired path when there are no uncertainties and no errors in the initial values of states. Consequently, Equation (29) is used to determine the feedforward control loop:

$$
\begin{aligned}
\begin{bmatrix} \tau_{WRK,hip} & \tau_{WRK,knee} \end{bmatrix}^T \\
= u_{i/p} + M_{WRK}\left(\theta_{hip}, \theta_{knee}\right)\begin{bmatrix} \ddot{\theta}_{hip,r} & \ddot{\theta}_{knee,r} \end{bmatrix}^T \\
+ C_{WRK}\left(\theta_{hip}, \theta_{knee}, \dot{\theta}_{hip}, \dot{\theta}_{knee}\right)\begin{bmatrix} \dot{\theta}_{hip,r} & \dot{\theta}_{knee,r} \end{bmatrix}^T \\
+ G_{WRK}\left(\theta_{hip}, \theta_{knee}\right)
\end{aligned}
\tag{29}
$$

By considering the uncertainties, the WRK robust closed loop system is obtained by inserting Equation (29) in the WRK dynamics Equation (28), as below:

$$
\begin{aligned}
u_{i/p} = M_{WRK}\left(\theta_{hip}, \theta_{knee}\right)\ddot{E}_{WRK} + C_{WRK}\left(\theta_{hip}, \theta_{knee}, \dot{\theta}_{hip}, \dot{\theta}_{knee}\right)\dot{E}_{WRK} \\
+ \delta_{WRK}\left(\theta_{hip}, \theta_{knee}, \dot{\theta}_{hip,r}, \dot{\theta}_{knee,r}\right)
\end{aligned}
\tag{30}
$$

The unknown weight of the person is included in the outer disturbances D_{WRK}, while all the uncertainties involving the error of the dynamics model are represented as $\varepsilon_{WRK}\left(\theta_{hip}, \theta_{knee}, \dot{\theta}_{hip,r}, \dot{\theta}_{knee,r}\right)$. In the model-based control system, the tracking error of the WRK device moves toward a linear trajectory when the model errors and disturbances errors are close to zero. Thus, a lack of congruence between the tracking error of the WRK control-based model and the actual WRK device causes unacceptable control. This issue will result in the inability to support appropriate actuator torque by the WRK device to reach the required trajectory. Thus, the WRK system will be unrobust. To solve this issue, the uncertainties, including model errors and disturbance errors, are applied to train the RBF NN. This will be necessary for the WRK device to continue on the required track, i.e., to achieve robustness. The applied RBF NN structure includes the following:

(1) Inputs of vector $c = \begin{bmatrix} c_1 & c_2 & \ldots & c_n \end{bmatrix}^T$ where n is the number of inputs of the RBF NN.

(2) For neural net j in the hidden layer, the Gaussian function value is obtained as:

$$
d_j = \frac{1}{\left[1 + e^{-c_j^2}\right]}, j = 1, 2, \ldots, k
$$

where k is the number of neurons in the hidden layer, which is supposed to be equal to the number of the inputs of the RBF NN.

(3)   Output, as calculated from the following equation:

$$\varepsilon_{WRK}\left(\theta_{hip}, \theta_{knee}, \dot{\theta}_{hip,r}, \dot{\theta}_{knee,r}\right) = \sum_{j=1}^{k} w_{WRK,hj} d_j, \, h = 1, 2, \ldots, m \tag{31}$$

where m and $w_{WRK,hj}$ denote the number of outputs and the RBF NN weight, respectively. Hence,

$$\delta_{WRK}\left(\theta_{hip}, \theta_{knee}, \dot{\theta}_{hip,r}, \dot{\theta}_{knee,r}\right) = \sum_{j=1}^{k} w_{WRK,hj} d_j + D_{WRK} \tag{32}$$

Inserting Equation (32) into (30) yields

$$u_{i/p} = M_{WRK}\left(\theta_{hip}, \theta_{knee}\right) \ddot{E}_{WRK} + C_{WRK}\left(\theta_{hip}, \theta_{knee}, \dot{\theta}_{hip}, \dot{\theta}_{knee}\right) \dot{E}_{WRK} \\ + \sum_{j=1}^{k} w_{WRK,hj} d_j + D_{WRK} \tag{33}$$

The training algorithm for RBF NN weight $w_{WRK,hj}$ dominates the associated computation for the suggested technique. In what follows, an examination of the adaptive computational complexity involved is provided [23]. Suppose that the input data $x(t)$ and output data $y(t)$ of the WRK system are sampled in the following interval $\{t = 1, \cdots, N\}$, where N denotes the number of samples. To evaluate how the RBF NN performs, a metric known as the normalized prediction error (NPE_{WRK}) is implemented, which is defined as

$$NPE_{WRK} = \sqrt{\frac{\sum_{t=1}^{N} \left(\breve{y}(t) - y(t)\right)^2}{\sum_{t=1}^{N} y^2(t)}} \times 100\% \tag{34}$$

where $\breve{y}(t)$ represents the calculated values of the WRK output system.

4.2. Hip and Knee Joint Trajectory

The categorization of walking features allows us to separate a movement cycle into a stance period and a swing period. As shown in Figure 2, the percentage of gait cycle can be separated into 60% and 40% for the stance period and swing period, respectively. The leg, considering the right leg in this analysis, normally moves according to the following stages: heel strike, foot flat, midstance, heel off, toe off, midswing, heel strike [24]. During typical walking on level ground, a person's knee bends at an approximate extension angle of 2° to 15° during the standing phase and bends at an approximately extension angle of 2° to 60° during the swing phase. To be considered healthy, a hip joint must be able to bend and extend across an approximate angle of −15° to 22°. For a gait cycle, hip and knee reference swing trajectories are fitted for typical walking on level ground, as presented in Equations (35) and (36) and detailed in [25]:

$$\theta_{hip} = \sum_{j=1}^{6} f_{1j} \sin\left(f_{2j}\omega t + f_{3j}\right) \tag{35}$$

$$\theta_{knee} = \sum_{j=1}^{6} g_{1j} \sin\left(g_{2j}\omega t + g_{3j}\right) \tag{36}$$

where

ω denotes the frequency of motion;

f_{1j} denotes the amplitude of $\sin\left(f_{2j}\omega t + f_{3j}\right)$ component of θ_{hip};

g_{1j} denotes the amplitude of $\sin\left(g_{2j}\omega t + g_{3j}\right)$ component of θ_{knee};
f_{2j} and g_{2j} denote the harmonic numbers for θ_{hip} and θ_{knee}, respectively;
f_{3j} and g_{3j} denote the phase shift for θ_{hip} and θ_{knee}, respectively.

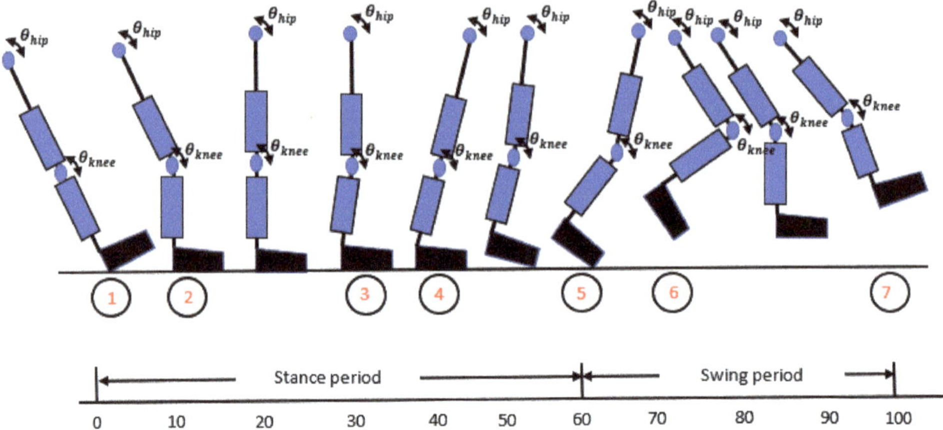

Figure 2. Hip and knee joints in a complete gait cycle of walking, with the right leg moving in the following stages: 1—heel strike, 2—foot flat, 3—midstance, 4—heel off, 5—toe off, 6—midswing, 7—heel strike.

The sum of sines model in the curve fitting toolbox of MATLAB is implemented to fit the periodic functions in Equations (35) and (36) to obtain the fitting parameters presented in Table 3.

Table 3. Fitting parameters of hip and knee reference trajectories.

j	f_{1j}	f_{2j}	f_{3j}	g_{1j}	g_{2j}	g_{3j}
1	9.17	23.9	0.6	2.8	0.5	−2.3
2	9.5	14.2	0.04	12.5	1.5	2.5
3	23.5	37.1	3.7	5.18	2.4	1.2
4	10.4	3.6	9.8	19.4	−6.7	−0.36
5	−13.07	3.12	5.4	24.8	0.2	−2.36
6	1.8	1.9	19.1	27.3	−6.4	−0.36

5. Separation System and L_2 Gain

This study is only a first step towards developing a comprehensive theory of nonlinear-state feedback H_∞ control; and important issue is the more complex dynamics measurement feedback problem, particularly involving disturbances that corrupt the process states variables measurements. In fact, the assumption that only affected measurements are available for feedback was an essential part of the main purpose behind linear H_∞ analysis design (see, e.g., [26,27] and the references cited therein). As a matter of terminology, the H_∞ norm is defined as a norm for transfer matrices and therefore cannot be directly generalized to nonlinear systems. The H_∞ norm, however, is nothing more than the L_2 induced norm converted into the time domain. The application of L_2 gain in the adaptive robust dynamics RBF NN is developed in this section in order to facilitate the control motion solutions and provide the feature of intelligent WRK control techniques. In the control algorithm, the characteristics of separation and style L_2 gain for the WRK system are crucial due to the status of relations between the input and output. The separation of system is implemented to establish the WRK dynamic system, as follows:

$$\dot{x} = f_{WRK}(x) + v_{WRK}(x)D_{WRK}y = q_{WRK}(x) \tag{37}$$

where f_{WRK} is a smooth function; y represents the output of the WRK system implemented to monitor the error; $v_{WRK}(x)$ is an $n \times m$ matrix, where n represents the number of local coordinates, $x = (x_1, \ldots, x_n)$ of the smooth function f_{WRK}, and m is the number of disturbances $D_{WRK} = (d_1, \ldots, d_m)$; $q_{WRK}(x)$ is the output that results from D_{WRK} corresponding to initial states [28,29]. However, the WRK system will be dissipative in the case that all of its accumulated energy will be consumed during operation to the moment of reaching the equilibrium position. Consequently, we obtained:

$$Q(x(t_n)) - Q(x(t_0)) \leq \int_{t_0}^{t_n} E(D_{WRK}, y(t))dt \tag{38}$$

where Q is a function that should have a value greater than or equal to zero, and $E(d_{WRK}, y(t))$ denotes the input energy. On the other hand, the dynamic WRK system is stable when:

$$\|y(t)\|_{L_2} \leq \beta\|D_{WRK}(t)\|_{L_2} \tag{39}$$

where L_2 gain is less than or equal to β and $\beta \geq 0$. The inequality of Equation (38) is solved next based on the optimization technique to establish that function $Q(x)$ has a minimum value, as follows:

$$\frac{\partial Q}{\partial x}\left(f_{WRK}(x) + v_{WRK}(x)D_{WRK}\right) \leq 0.5\left(\beta^2\|D_{WRK}\|^2 - \|y\|^2\right) \tag{40}$$

The achievement of this equation is the solution of Q as greater than or equal to zero for all disturbances D_{WRK} and $y(t) = q_{WRK}(x)$. The WRK system L_2 gain is developed by the following equation to determine the robustness of the WRK system against disturbance:

$$I_{WRK} = \sup_{D_{WRK} \neq 0} \frac{\|y\|_{L_2}}{\|D_{WRK}\|_{L_2}} \tag{41}$$

The operation of the WRK device is more robust against disturbance for a lower I_{WRK} value. When the inequality presented in Equation (40) is fulfilled, the following inequality is obtained as:

$$\dot{V} \leq 0.5\left(\beta^2\|D_{WRK}\|^2 - \|y\|^2\right) \tag{42}$$

where $I_{WRK} \leq \beta$. The inequality mentioned in Equation (42) will be used in the next section to find the solution of WRK motion for the adaptive RBF NN.

6. RBF NN–Based Adaptive Robust Controller

RBF NNs are reported to be efficient in designing control techniques for complex uncertain nonlinear systems in a significant number of theoretical studies and practical applications [30,31]. Although there have been notable achievements in the field of NN, due to their capabilities in analyzing nonlinear system dynamics, the current research on adaptive RBF NN controllers remains primarily focused on investigating fundamental approaches. Assuming that discontinuous RBF NN functions are defined properly, it is possible to simulate models with nonlinear dynamics. Thus, adaptive RBF NNs are most effective in practical applications where system dynamics are inherently nonlinear, vary significantly, and are not fully analyzed (see, e.g., Ref. [32] and the references cited therein). In this section, a robust design technique is discussed for stable adaptive-control WRK systems using RBF NN approximation-based methods.

Adaptive robust controllers involve a tracking error control algorithm that can be applied to enhance the performance of manipulators in trajectory planning [33–35]. First, the control input $u_{i/p}$ in Equation (29) is used to compensate the WRK system in Equation (28). Then, the closed loop of the WRK system is developed as explained in Equation (33). By assuming the external disturbances D_{WRK} and presenting a new signal S_{WRK}, the index signal of Equation (41) can be rewritten as follows:

$$I_{WRK} = \frac{Sup}{\|D_{WRK}\| \neq 0} \|S_{WRK}\|_2 / \|D_{WRK}\|_2 \tag{43}$$

The aim of the RBF NN-based robust controller is to obtain control signal $u_{i/p}$ and RBF NN learning $\dot{W}_{WRK}$, where $I_{WRK} < \epsilon$. The variable ϵ represents a predefined level. As explained in Section 2, the proposed WRK model is derived using the Lagrange approach. Hence, the obtained WRK dynamics model shown in Equation (27) generally contains the acceleration that is essential to be measured for the design of the WRK controller. However, measuring the acceleration signal is not an easy process due to its noise. In turn, assuming $\varepsilon > 0$, the following two variables are defined to avoid using the acceleration signal:

$$z_1 = E_{WRK}$$

$$z_2 = \dot{E}_{WRK} + \varepsilon E_{WRK}$$

Hence, Equation (33) is rewritten as

$$\dot{z}_1 = z_2 - \varepsilon z_1 \tag{44}$$

$$\begin{aligned}
M_{WRK}\left(\theta_{hip}, \theta_{knee}\right)\dot{z}_2 &= M_{WRK}\left(\theta_{hip}, \theta_{knee}\right)\varepsilon\dot{E}_{WRK} \\
&\quad + C_{WRK}\left(\theta_{hip}, \theta_{knee}, \dot{\theta}_{hip}, \dot{\theta}_{knee}\right)\varepsilon E_{WRK} \\
&\quad - C_{WRK}\left(\theta_{hip}, \theta_{knee}, \dot{\theta}_{hip}, \dot{\theta}_{knee}\right)z_2 - D_{WRK} \\
&\quad - \sum_{j=1}^{k} w_{WRK,hj}d_j + u_{i/p}
\end{aligned} \tag{45}$$

To track the position and speed of the WRK system, an adaptive law is designed based on RBF NN, as follows:

$$\dot{\hat{W}}_{WRK} = -\gamma z_2 d_{WRK}^T \tag{46}$$

where the constant $\gamma > 0$. On the other hand, $\dot{\hat{W}}_{WRK}$ and d_{WRK}^T denote the derivative of estimated weight and Gaussian function vector of RBF NN, respectively.

$$w_{WRK} = \begin{bmatrix} w_{WRK,11} & w_{WRK,12} & \cdots & w_{WRK,1k} \\ w_{WRK,21} & w_{WRK,22} & \cdots & w_{WRK,2k} \\ \vdots & \vdots & \vdots & \vdots \\ w_{WRK,n1} & w_{WRK,n2} & \cdots & w_{WRK,nk} \end{bmatrix}, d_{WRK}^T = \begin{bmatrix} d_1 & d_2 & \cdots & d_k \end{bmatrix}$$

The control law is now designed with feedback based on Equations (44) and (45), as below:

$$\begin{aligned}
u_{\frac{i}{p}} &= -M_{WRK}\left(\theta_{hip}, \theta_{knee}\right)\varepsilon\dot{E}_{WRK} \\
&\quad - C_{WRK}\left(\theta_{hip}, \theta_{knee}, \dot{\theta}_{hip}, \dot{\theta}_{knee}\right)\varepsilon E_{WRK} - \frac{1}{2\varepsilon^2}z_2 \\
&\quad + \hat{W}_{WRK}d_{WRK} - \frac{1}{2}z_2,
\end{aligned} \tag{47}$$

where the index signal of Equation (43) is such that $I_{WRK} < \beth$. System stability can be described by two major concepts: system stability, which indicates its trajectories depend on initial conditions at a nearby equilibrium point, or asymptotical stability. A system's asymptotic stability refers to its ability to achieve equilibrium under relatively minor variations for a reasonable period of time. Robustly asymptotically stable systems may still be stable despite being subjected to random disturbances and modeling uncertainties [36]. Considering the WRK closed system, the stability is ensured by introducing the following Lyapunov function:

$$L_{WRK} = \frac{1}{2} z_2^T M_{WRK}\left(\theta_{hip}, \theta_{knee}\right) z_2 + \frac{1}{2\gamma}\left(\tilde{W}_{WRK}^T \tilde{W}_{WRK}\right) \tag{48}$$

$$\tilde{W}_{WRK} = \hat{W}_{WRK} - W_{WRK}$$

Now, considering Equations (44), (45) and (47) and assuming the WRK dynamics of Equation (27), we obtain:

$$\dot{L}_{WRK} = z_2^T M_{WRK}\left(\theta_{hip}, \theta_{knee}\right)\dot{z}_2 + \frac{1}{2} z_2^T \dot{M}_{WRK}\left(\theta_{hip}, \theta_{knee}\right)\dot{z}_2$$
$$+ \frac{1}{\gamma} tr\left(\dot{\tilde{W}}_{WRK}^T \tilde{W}_{WRK}\right) \tag{49}$$

Inserting $M_{WRK}\left(\theta_{hip}, \theta_{knee}\right)\dot{z}_2$ from Equation (45) into above equation yields:

$$\begin{aligned}
\dot{L}_{WRK} &= z_2^T\Big[M_{WRK}\left(\theta_{hip}, \theta_{knee}\right)\varepsilon\dot{E}_{WRK} \\
&\quad + C_{WRK}\left(\theta_{hip}, \theta_{knee}, \dot{\theta}_{hip}, \dot{\theta}_{knee}\right)\varepsilon E_{WRK} \\
&\quad - C_{WRK}\left(\theta_{hip}, \theta_{knee}, \dot{\theta}_{hip}, \dot{\theta}_{knee}\right) z_2 - D_{WRK} \\
&\quad - \sum_{j=1}^{k} w_{WRK,hj} d_j + u_{i/p}\Big] + \frac{1}{2} z_2^T \dot{M}_{WRK}\left(\theta_{hip}, \theta_{knee}\right)\dot{z}_2 \\
&\quad + \frac{1}{\gamma}\left(\dot{\tilde{W}}_{WRK}^T \tilde{W}_{WRK}\right)
\end{aligned}$$

$$\dot{L}_{WRK} = -z_2^T D_{WRK} - \frac{1}{2\beth^2} z_2^T z_2 + z_2^T \tilde{W}_{WRK} d_{WRK}^T - \frac{1}{2} z_2^T z_2$$
$$+ \frac{1}{\gamma} tr\left(\dot{\tilde{W}}_{WRK}^T \tilde{W}_{WRK}\right) \tag{50}$$

Considering the external disturbance D_{WRK}, assume:

$$H_{WRK} = \dot{L}_{WRK} - \frac{1}{2}\beth^2 \|D_{WRK}\|^2 + \frac{1}{2}\|S_{WRK}\|^2 \tag{51}$$

Inserting Equation (44) into above equation gives

$$\begin{aligned}
H_{WRK} &= -z_2^T D_{WRK} - \frac{1}{2\beth^2} z_2^T z_2 - \frac{1}{2}\beth^2 \|D_{WRK}\|^2 \\
&\quad + z_2^T \tilde{W}_{WRK} d_{WRK}^T \\
&\quad + \frac{1}{\gamma} tr\left(\dot{\tilde{W}}_{WRK}^T \tilde{W}_{WRK}\right) - \frac{1}{2} z_2^T z_2 \\
&\quad + \frac{1}{2}\|S_{WRK}\|^2
\end{aligned} \tag{52}$$

The components of Equation (52) are considered as follows:

1. The first three components involve the following:

$$-z_2^T D_{WRK} - \frac{1}{2\beth^2} z_2^T z_2 - \frac{1}{2}\beth^2 \|D_{WRK}\|^2 = -\frac{1}{2}\left\|\frac{1}{\beth} z_2 + \beth D_{WRK}\right\|^2 \leq 0$$

2. Regarding the second two components, since $z_2^T \widetilde{W}_{WRK} d_{WRK}^T = tr\left(\widetilde{W}_{WRK} d_{WRK}^T z_2^T \right)$ and using the adaptive law, i.e., $\dot{\hat{W}}_{WRK} = -\gamma z_2 d_{WRK}^T$", introduced in Equation (46), we obtain

$$tr\left(\widetilde{W}_{WRK} d_{WRK}^T z_2^T \right) = -\frac{1}{\gamma} tr\left(\dot{\widetilde{W}}_{WRK}^T \widetilde{W}_{WRK} \right)$$

Hence,

$$z_2^T \widetilde{W}_{WRK} d_{WRK}^T + \frac{1}{\gamma} tr\left(\dot{\widetilde{W}}_{WRK}^T \widetilde{W}_{WRK} \right) = 0$$

3. For the last two components, considering the approximation error as the disturbance D_{WRK} and defining $S_{WRK} = z_2 = \dot{E}_{WRK} + \varepsilon E_{WRK}$, we obtain:

$$-\frac{1}{2} z_2^T z_2 + \frac{1}{2} \|z_2\|^2 = 0$$

Hence, $H_{WRK} \leq 0$. According to Equation (45), we obtain:

$$\dot{L}_{WRK} \leq \frac{1}{2} \beth^2 \|D_{WRK}\|^2 - \frac{1}{2} \|S_{WRK}\|^2 \tag{53}$$

Equation (53) represents the derived HJI in Equation (42). Consequently, based on Section 5, it can be concluded that the system is stable for $I_{WRK} < \beth$. This finalizes the proof of the following theorem:

Theorem 1. *Considering the implemented nonlinear system of a wearable robotic knee, as defined in Equation (37), and the Hamiltonian–Jacobi Inequality. which is defined in (42), the RBF NN–based adaptive control law given by (47) ensures the asymptotic convergence to 0 for the trajectory tracking error if, and only if:*

$$I_{WRK} < \beth$$

where $\beth$ is a positive constant number. The designed control algorithm of this system is depicted in Figure 3.

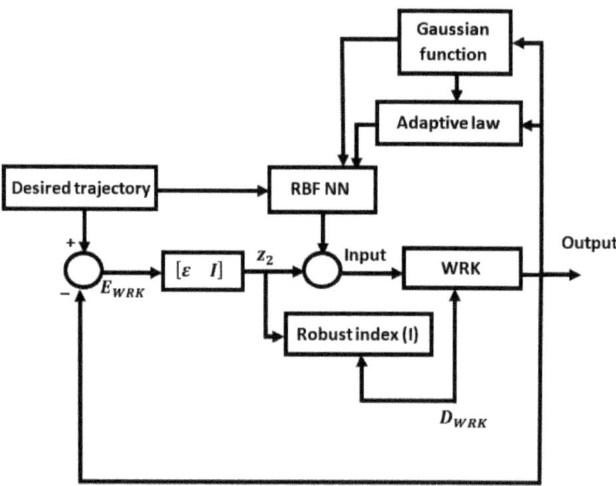

Figure 3. Block diagram of the WRK tracking control system.

7. Numerical Simulation Validation

The designed control algorithm of this study is tested in this section on the tracking motion of a WRK system by simulation experiments. Five tests were performed as part of the numerical simulation analysis for validating the controller developed for this study. Test 1 was designed as a means of assessing the overall performance of the controller without taking into account the weight and the height of the user. Tests 2 and 3 were conducted using real users' [19,20] (one male and one female) data in order to evaluate the performance of the designed controller in terms of position and speed tracking errors. Tests 4 and 5 evaluated the torques at each joint of the hip and knee using data from two real human users [19,20] (one male and one female), allowing the torque applied to the calf and thigh to be measured in a realistic context. All tests, except Test 1, used authentic human parameters to test the controller during realistic human walking, based on the hip and knee joint trajectories derived from Section 4.2 and reference [24]. Four additional tests were implemented based on different combinations of human characteristics (such as male, female, height, weight) as presented in Table 4, widening the potential user base for the device. Table 4 contains the parameters for the simulation experiments of the four persons. The parameters of person 1, person 2, person 3, and person 4 are used in Test 2, Test 3, Test 4, and Test 5, respectively. In this table, the tests are considered for four persons of four different heights T and weights m_{body}. The length of thigh, calf, WRK link-1, and WRK link-2 are obtained according to Equations (1)–(4), respectively. The weights of the thigh and calf are calculated as a percentage of total weight of the body according to Equations (5)–(8). The inertia of the thigh, calf, WRK link-1, and WRK link-2 are calculated referring to Equation (9). The weights of each of the WRK link are selected as 0.25 kg. The main features of the simulation study methodology are presented in Figure 4.

Table 4. Parameters of the simulation tests.

Person #	Test #	Thigh	Calf	WRK Link-1	WRK Link-2
Person 1 Gender: female Hight = 170 cm Weight = 74 kg	Test 2	$l_{thigh} = 41.65$ cm $m_{thigh} = 8.732$ kg $I_{thigh} = 0.505$ kgm^2	$l_{calf} = 41.82$ cm $m_{calf} = 3.959$ kg $I_{calf} = 0.23$ kgm^2	$l_1 = 20.4$ cm $m_1 = 0.25$ kg $I_1 = 0.00346$ kgm^2	$l_2 = 20.4$ cm $m_2 = 0.25$ kg $I_2 = 0.003468$ kgm^2
Person 2 Gender: Male Hight = 180 cm Weight = 88 kg	Test 3	$l_{thigh} = 44.1$ cm $m_{thigh} = 9.24$ kg $I_{thigh} = 0.599$ kgm^2	$l_{calf} = 44.28$ cm $m_{calf} = 4.18$ kg $I_{calf} = 0.273$ kgm^2	$l_1 = 21.6$ cm $m_1 = 0.25$ kg $I_1 = 0.00388$ kgm^2	$l_2 = 21.6$ cm $m_2 = 0.25$ kg $I_2 = 0.00388$ kgm^2
Person 3 Gender: female Hight = 185 cm Weight = 80 kg	Test 4	$l_{thigh} = 45.325$ cm $m_{thigh} = 9.44$ kg $I_{thigh} = 0.6372$ kgm^2	$l_{calf} = 45.51$ cm $m_{calf} = 4.28$ kg $I_{calf} = 0.295$ kgm^2	$l_1 = 22.2$ cm $m_1 = 0.25$ kg $I_1 = 0.0041$ kgm^2	$l_2 = 22.2$ cm $m_2 = 0.25$ kg $I_2 = 0.0041$ kgm^2
Person 4 Gender: Male Hight = 175 cm Weight = 72 kg	Test 5	$l_{thigh} = 42.875$ cm $m_{thigh} = 7.56$ kg $I_{thigh} = 0.4632$ kgm^2	$l_{calf} = 43.05$ cm $m_{calf} = 3.42$ kg $I_{calf} = 0.211$ kgm^2	$l_1 = 21$ cm $m_1 = 0.25$ kg $I_1 = 0.0036$ kgm^2	$l_2 = 21$ cm $m_2 = 0.25$ kg $I_2 = 0.0036$ kgm^2

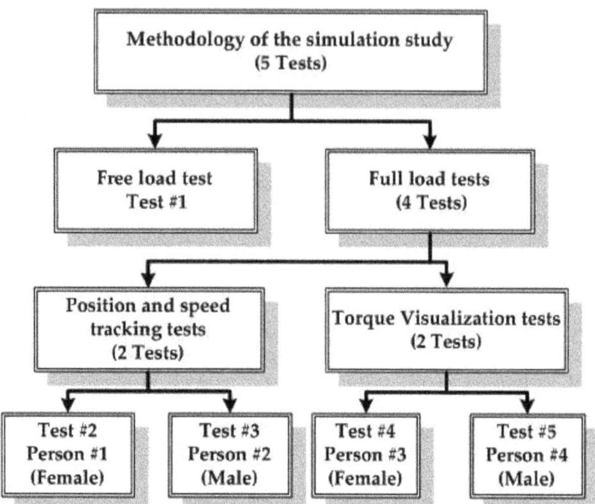

Figure 4. Features of the simulation study methodology.

7.1. Simulation Results

The computing simulation analysis is conducted using MATLAB programming software on a computer characterized by the following configurations: (i) Windows 10 Enterprise for 64 bits; (ii) 16 Gigabytes of RAM; and (iii) Intel(R) Core(TM) i7-4790T CPU @ 2.70 GHz. The uncertainty due to model errors is considered as:

$$\varepsilon_{WRK} = \begin{bmatrix} 0.7sgn\left(\dot{\theta}_{hip,r} - \dot{\theta}_{hip}\right) + 0.7sgn\left(\dot{\theta}_{hip,r} - \dot{\theta}_{hip}\right)exp\left(-\left|\dot{\theta}_{hip,r} - \dot{\theta}_{hip}\right|\right) \\ 0.6sgn\left(\dot{\theta}_{knee,r} - \dot{\theta}_{knee}\right) + 0.3sgn\left(\dot{\theta}_{knee,r} - \dot{\theta}_{knee}\right)exp\left(-\left|\dot{\theta}_{knee,r} - \dot{\theta}_{knee}\right|\right) \end{bmatrix} \text{N.m} \tag{54}$$

The second part of the uncertainty due to external disturbances is assumed as

$$D_{WRK} = \begin{bmatrix} 7random() \\ 6random() \end{bmatrix} + D_{weight} \tag{55}$$

$$D_{weight} = \begin{bmatrix} 8sgn\left(\dot{\theta}_{hip,r} - \dot{\theta}_{hip}\right)exp\left(-\left|\dot{\theta}_{hip,r} - \dot{\theta}_{hip}\right|\right) \\ 7sgn\left(\dot{\theta}_{knee,r} - \dot{\theta}_{knee}\right)exp\left(-\left|\dot{\theta}_{knee,r} - \dot{\theta}_{knee}\right|\right) \end{bmatrix}$$

where D_{weight} represents the disturbances due to the extra weight that the person may hold during walking, and "random()" represents a random number between 0 and 1. The disturbances that are represented by Equation (55) are selected to be changed in a random way to simulate the unknown external disturbances. Hence, for each test, the disturbance signal is not the same. In addition, the exponential function is included in the disturbance signal of Equation (55) to simulate the nonlinearity. The sign of the difference between the current and the required value of tracking are taken into consideration in the "*sgn*" function. The number of inputs, hidden layers, and outputs of the NN are set as 4, 7, and 1, respectively. By applying the adaptive control law of Equation (46) and the feedback control law in Equations (29) and (47) for all the five tests, the tracking controller is designed with the following constant values: $\gamma = 1000$, $\varepsilon = 15$, $\beth = 0.06$.

In the first test, the parameters of the WRK system are set as $m_1 = 0.25$ kg, $m_2 = 0.25$ kg, $I_1 = 0.012$ kgm^2, $I_2 = 0.012$ kgm^2. The free load test is conducted under the assumption that WRK has not been worn by any individual (i.e., $m_{thigh} = m_{calf} = 0$). The initial states are $\theta_{hip} = 0.2$ rad, $\theta_{knee} = 0.2$ rad, $\dot{\theta}_{hip} = 0$ rad/s, and $\dot{\theta}_{knee} = 0$ rad/s. It is assumed that the length of the WRK-thigh-link and the WRK-calf-link will be 25 cm so as to be suitable for adults, who usually have thigh and calf lengths that exceed 35 cm [20,37]. The ideal trajectory of the hip and knee joint is assumed as a sinusoid function of $\theta_{hip,r} = \sin 4\pi t$ and $\theta_{knee,r} = \sin 4\pi t$, respectively. The tracking errors results are shown in Figures 5 and 6. As can been seen from these figures, there are no steady errors, since the actual trajectory matches the desired trajectory.

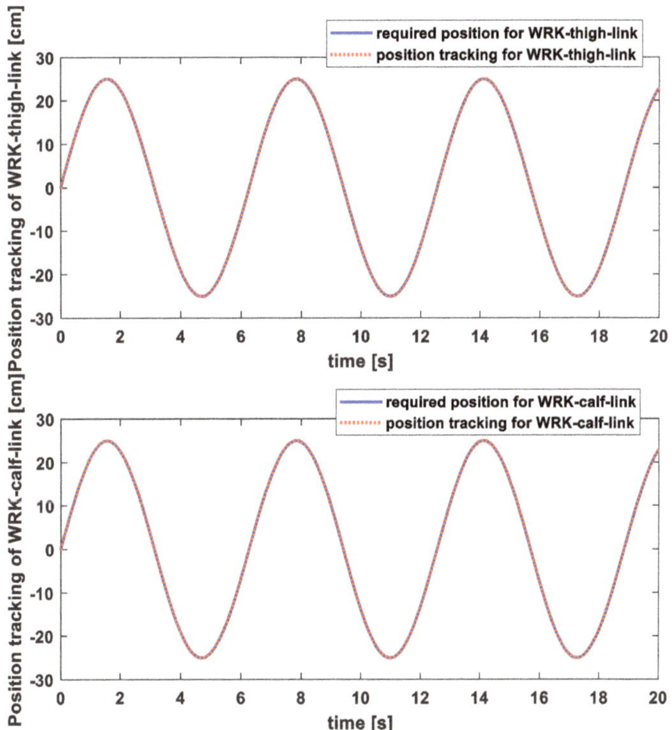

Figure 5. Position tracking of WRK thigh and calf links.

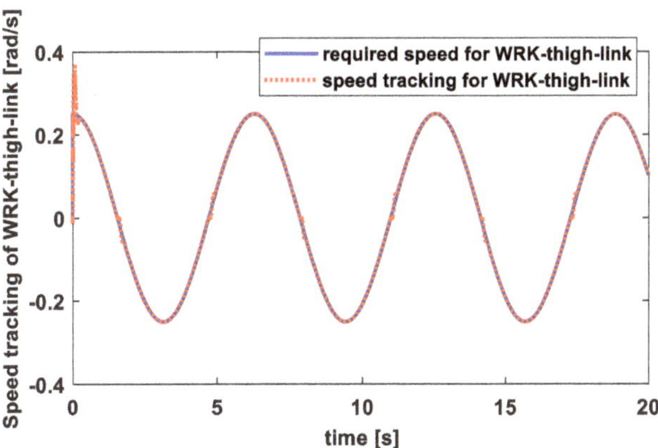

Figure 6. *Cont.*

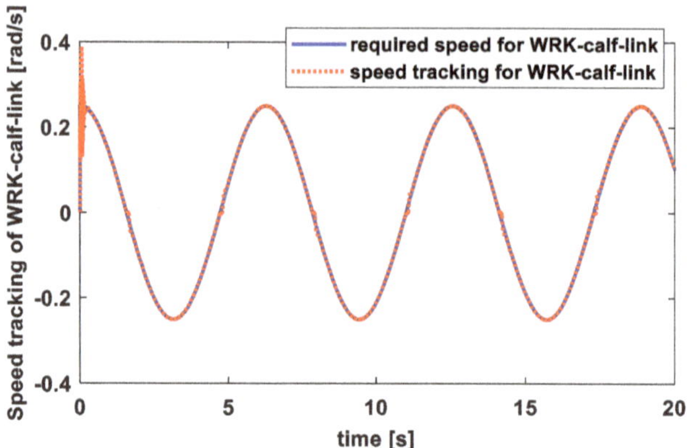

Figure 6. Speed tracking of WRK thigh and calf links.

As seen in Figure 7, the ideal trajectory for each hip joint ($\theta_{hip,r}$) and knee joint ($\theta_{knee,r}$) during a swing period simulates a realistic human path during walking (Tests 2 and 3). Specifically, such a trajectory is based on the hip and knee joint trajectories derived from Section 4.2 and reference [24]. As explained in Equations (35) and (36), this is normal movement for a normal individual. The initial states here are assumed to be $\theta_{hip} = -0.158$ degree, $\theta_{knee} = -23.6$degree, $\dot{\theta}_{hip} = 0$ rad/s, and $\dot{\theta}_{knee} = 0$ rad/s.

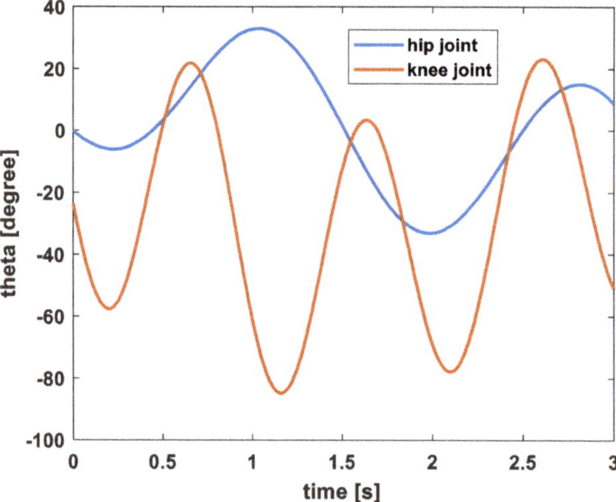

Figure 7. Swing angle of hip and knee joints.

Considering the trajectories of hip and knee joints of Figure 7 and the parameters of the simulation tests of Table 4 corresponding to person 1 female and person 2 male, the required positions of the end of thigh and calf are calculated according to Equations (1) and (2), respectively, as depicted in Figure 8. The end of the thigh and calf is located at knee and ankle joint, respectively. These positions, as specified in Figure 8, will be implemented in the simulation of Test 2 and Test 3 as the required position of the controller of person 1 female and person 2 male, respectively.

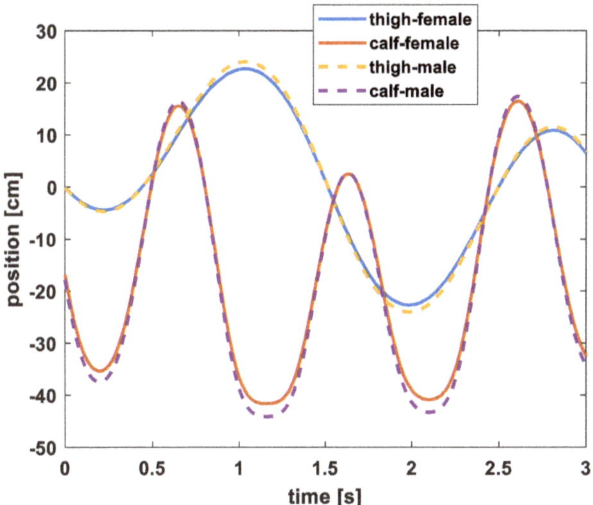

Figure 8. Position of the end of the thigh and calf of person 1 female and person 2 male.

The number of inputs, hidden layers, and outputs of the NN, γ, ε, and $\sqsupset$, are set as the values of the first test. Again, by applying the adaptive control law of Equation (46) and the feedback control law in Equations (29) and (47), the tracking error results for person 1 female and person 2 male are shown in Figures 9–12.

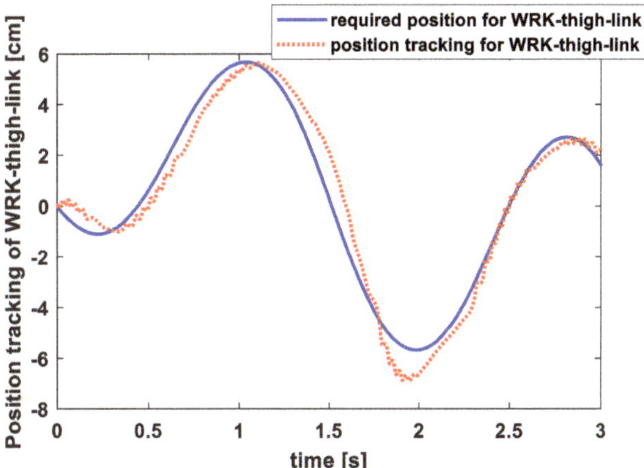

Figure 9. *Cont.*

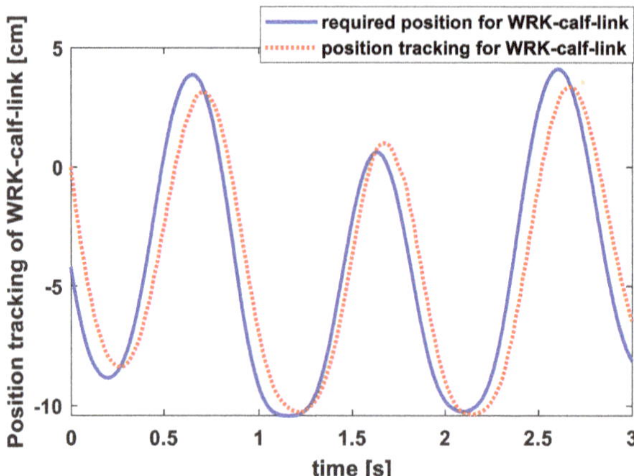

Figure 9. Position tracking of WRK thigh and calf links for person 1 female.

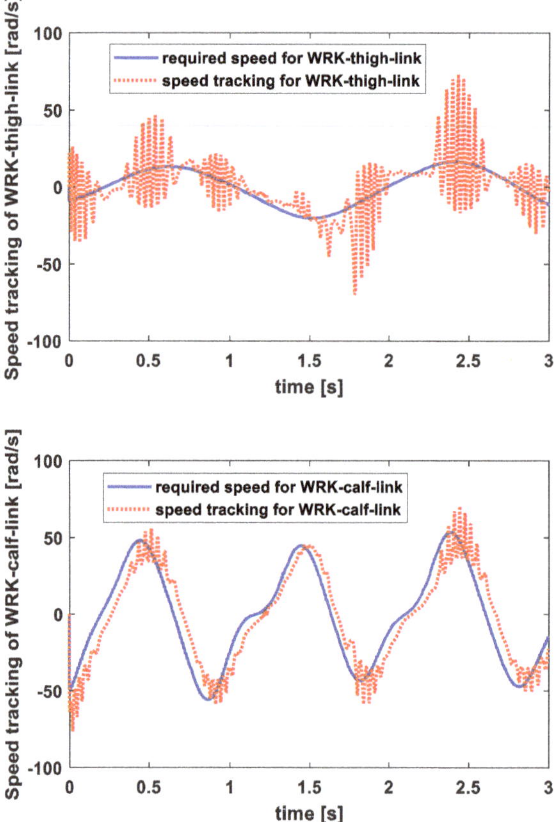

Figure 10. Speed tracking of WRK thigh and calf links for person 1 female.

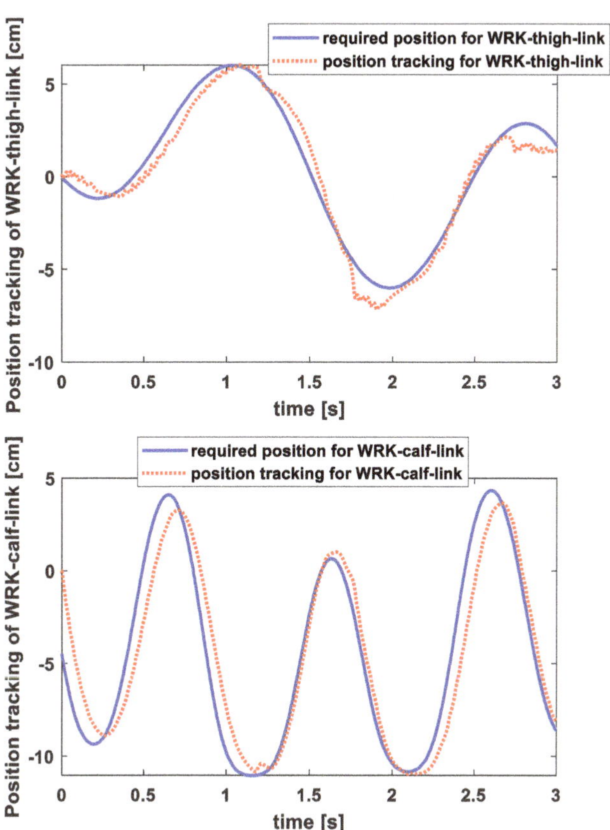

Figure 11. Position tracking of WRK thigh and calf links for person 2 male.

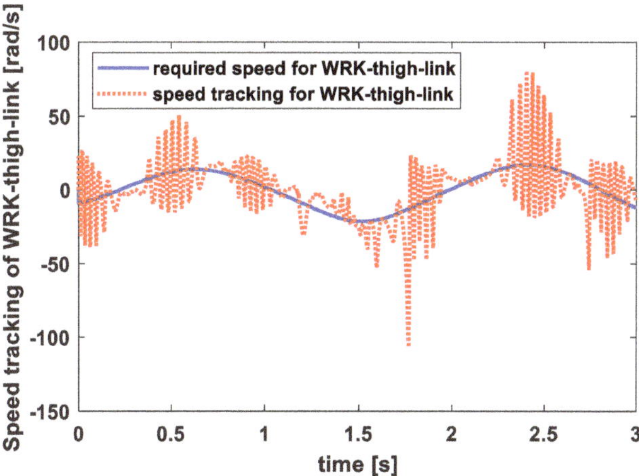

Figure 12. *Cont.*

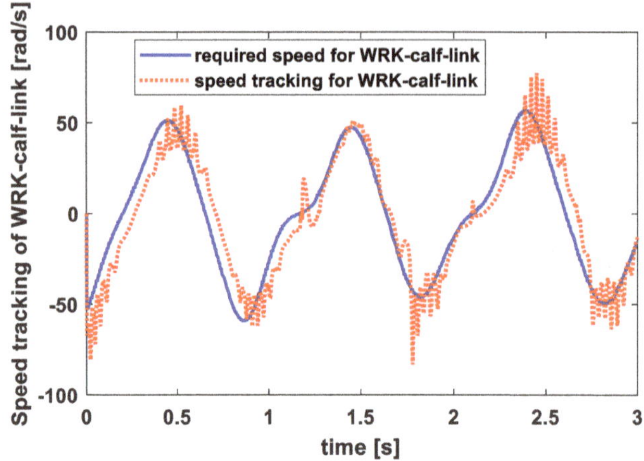

Figure 12. Speed tracking of WRK thigh and calf links for person 2 male.

Based on the results shown in Figures 9–12, the controller was able to track the desired position and speed of the thigh and calf. There was, however, a small delay between the actual position and the required position in position tracking. It was expected that this would occur since the delay was been taken into account in the current study; it will be examined in a future study. As a result of the calf link, the controller was able to track the required speed with acceptable results. The tracking speed of the thigh link was slightly inaccurate due to the fact that in this study we exclusively focused on the robustness and adaptiveness of the system to random external disturbances, as shown by Equations (54) and (55). Hence the controller responded with the same plot when the experiment was repeated, which was predictable.

The fourth and fifth tests were implemented to depict the control input torque on thigh and calf. The parameters of person 3 and person 4 presented in Table 4 were used in Test 4 and Test 5, respectively. Considering the values of hip and knee joints of Figure 7 and the parameters of person 3 and person 4, the required position of the end of thigh and calf were calculated according to Equations (1) and (2), respectively, as depicted in Figure 13.

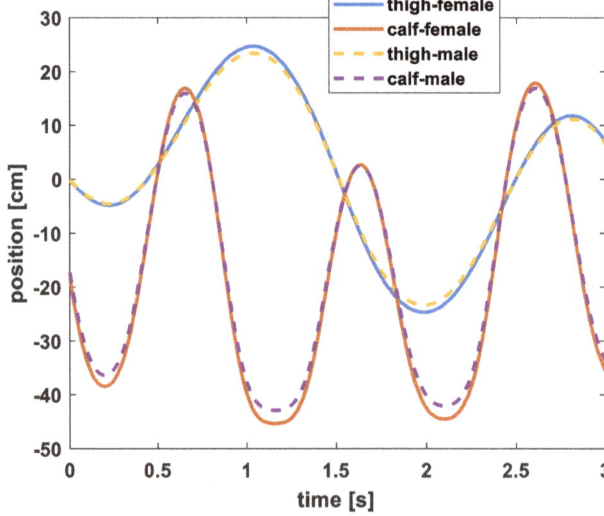

Figure 13. Position of the end of thigh and calf of person 3 female and person 4 male.

The tracking error results for person 3 in Test 4 are shown in Figures 14 and 15, while the torque of the thigh and calf are shown in Figure 16. Regarding person 4, the tracking error results from Test 5 are shown in Figures 17 and 18, while the torque of the thigh and calf are shown in Figure 19.

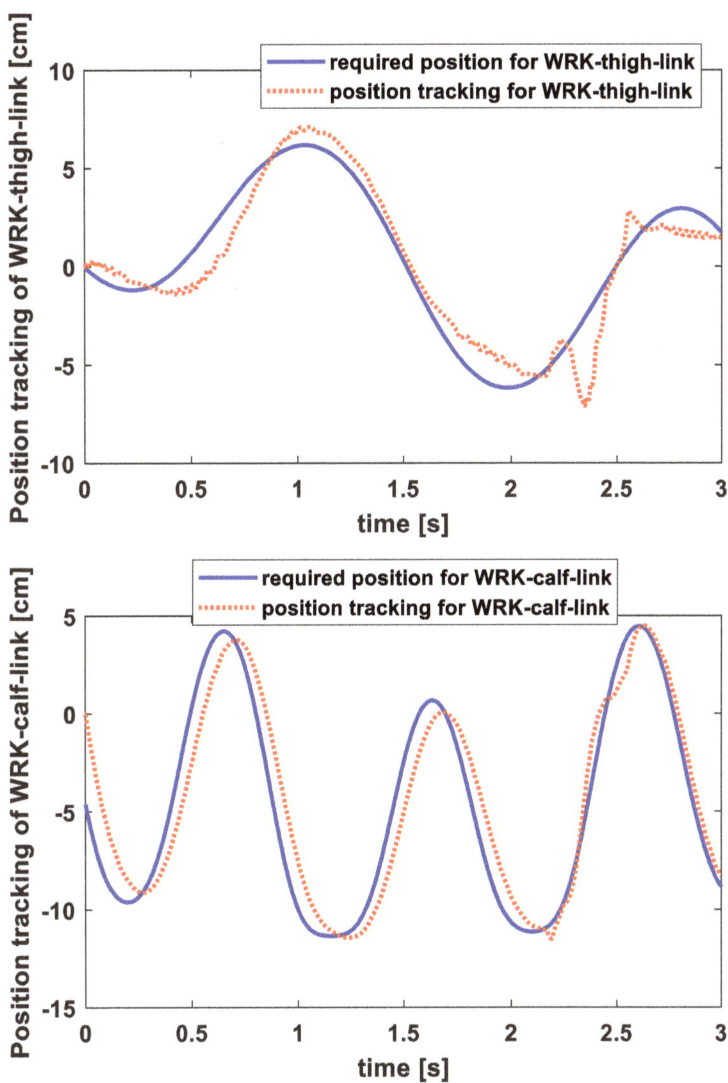

Figure 14. Position tracking of WRK thigh and calf links for person 3 female.

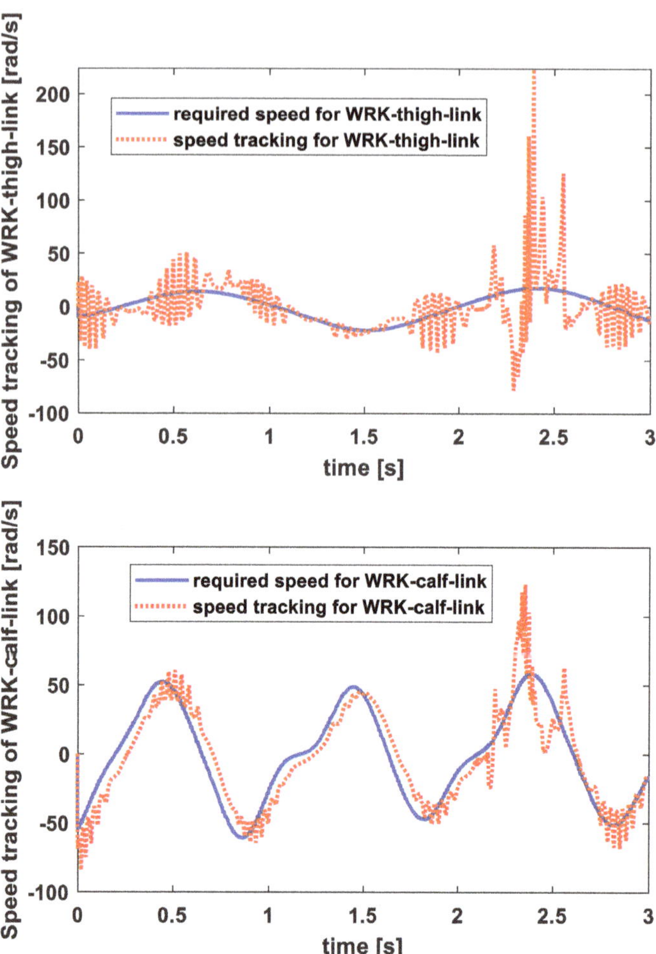

Figure 15. Speed tracking of WRK thigh and calf links for person 3 female.

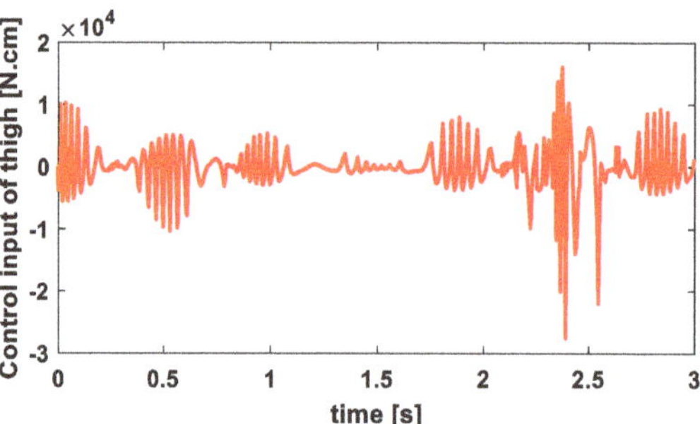

Figure 16. *Cont.*

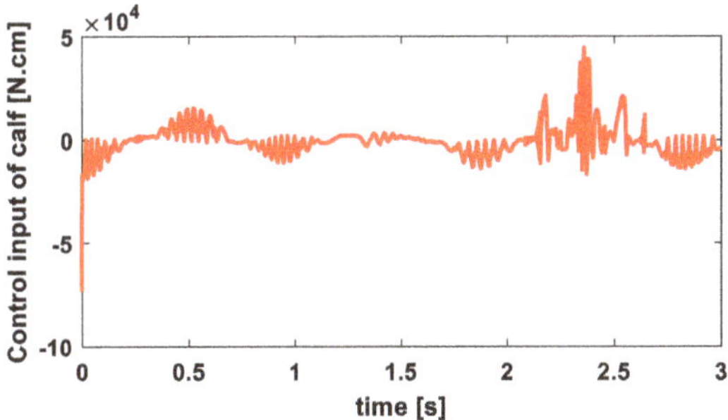

Figure 16. Torque input of thigh and calf of person 3 female.

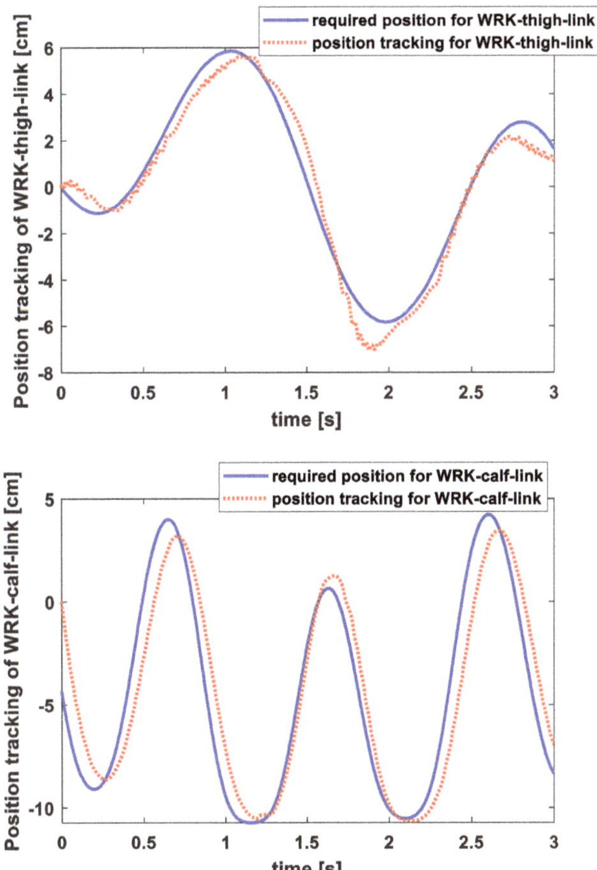

Figure 17. Position tracking of WRK thigh and calf links for person 4 male.

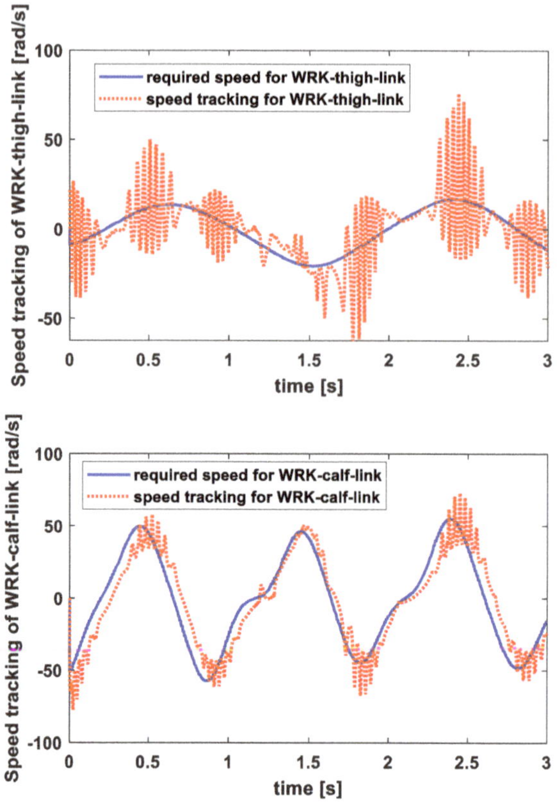

Figure 18. Speed tracking of WRK thigh and calf links for person 4 male.

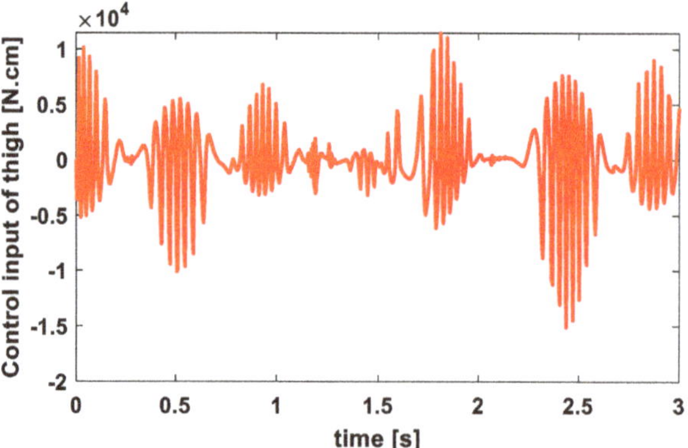

Figure 19. *Cont.*

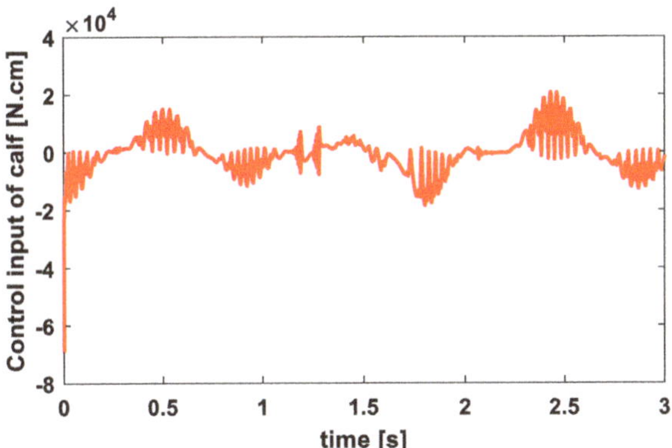

Figure 19. Torque input of thigh and calf of person 4 male.

Furthermore, as previously discussed, the controller's adaptivity and robustness to a combination of user parameters in Tests 4 and 5 resulted in the same performance as in Tests 2 and 3. It was observed that there is a delay in response when tracking the positions of calf and thigh links (Figures 14 and 17), as well as errors in terms in tracking thigh link speed (Figures 15 and 18) as a result of these tests. As shown in Figures 16 and 19, the thigh and calf input torques for persons 3 and 4, respectively, are depicted.

7.2. Discussion

The numerical simulation analysis, regarding Test 1, obtained from Figure 5 confirms that the designed control technique of this study is able to track a defined position of both WRK links (thigh and calf) with zero steady-state error in the free load test. On the other hand, in Figure 6, the speeds of the WRK links (thigh and calf) are tracked with zero steady-state error. In both the position and velocity tracking, the robustness is adaptively achieved by the proposed controller against model errors and external disturbances.

According to the findings shown in Figures 9–12, the controller was able to successfully monitor the appropriate position and speed of the calf and the thigh. When it came to position monitoring, however, there was a small delay between the real location and the position. It is reasonable to anticipate that this takes place, given that the delay was not taken into account. As a consequence of the calf link, the controller was in a position to track the necessary speed with satisfactory outcomes. The thigh link's tracking speed had a error term due to the fact that in this study we exclusively focused on the robustness and adaptiveness of the system to random external disturbances, as shown by Equations (54) and (55). Upon retesting, the controller consistently provided the same performance results. The adaptability and robustness of the controller to different user parameter combinations in Tests 4 and 5 resulted in the same performance as in tests 2 and 3. As a consequence of these tests, it has been noticed that there is a delay in response when tracking the positions of calf and thigh links (Figures 14 and 17), as well as errors in tracking thigh link speed (Figures 15 and 18). Additionally, it has been observed that there is a delay in response when tracking the angle of rotation of the thigh link (Figures 14 and 17). The controller designed in this paper achieves asymptotic tracking even in the presence of unstructured disturbances. It should however be noted that high control gains may lead to inappropriate responses, including inadmissible overshooting. Thus, in order to limit such high controller gains, an adaptive controller is included that reduces tracking error and overshoots. The input torques for the thighs and calves of persons 3 and 4 are represented in Figures 16 and 19, respectively. The hard motional dynamics constraints considered in the output trajectories show that the stability of the controlled closed loop system was achieved and the dynamical behavior is quite satisfactory. The RBF NN proved its efficiency in compensating for all model uncertainties and random external disturbances.

The remaining three tests confirm the ability to apply the designed controller to users with different physical parameters (height, weight and gender). The proposed controller was designed for both males and females with different user parameters (thigh length and mass, calf length and mass) that differ slightly according to biological features, as described in [20,21]. Thus, the controller was adaptive to the potential physical parameters of users and robust to external unstructured disturbances.

8. Conclusions

In this study, a motion controller was developed and investigated for a WRK system to obtain an appropriate output trajectory tracking performance. The designed controller showed its ability to perform without errors against model uncertainties and external disturbances. The designed control algorithm includes the adaptive law, a RBF NN compensator, and a robust feedback part. The adaptive law is applied to supply the feedback controller with the estimation of RBF NN weights by updating the weights of the RBF NN compensator. The robust feedback controller tracks the desired trajectory in the presence of uncertainties. All the uncertainties are considered in the WRK dynamics using the Lagrange approach. The L_2 gain technique facilitates the control motion solutions and provides the feature of intelligent WRK control. The stability is proved by applying an HJI approach based on the optimization technique and Lyapunov stability theory. The presented robust RBF NN algorithm ensures the ability to implement the adaptive controller easily as well as the stability of the WRK system. The results demonstrate that the designed WRK controller can be successfully implemented when the parameter values are modified to accommodate the potential user of the device. As a future work, it is recommended to apply the obtained control of this study practically to a real WRK device. A deep learning approach is recommended as a way to enhance the performance of RBF NNs. Furthermore, the implementation of advanced meta-heuristic optimization techniques is a promising future area of study.

Author Contributions: Conceptualization, H.J., I.A.-D., G.T., L.L. and M.O.; Methodology, H.J., I.A.-D., G.T., L.L. and M.O.; Software, H.J., I.A.-D. and G.T.; Validation, I.A.-D. and G.T.; Formal analysis, H.J., I.A.-D., G.T., L.L. and M.O.; Investigation, H.J., I.A.-D., G.T., L.L. and M.O.; Resources, H.J., I.A.-D., G.T. and M.O.; Data curation, I.A.-D., G.T. and M.O.; Writing—original draft, I.A.-D. and H.J.; Writing—review & editing, I.A.-D. and H.J.; Visualization, I.A.-D.; Supervision, H.J., I.A.-D., G.T., L.L. and M.O.; Project administration, H.J., I.A.-D., G.T., L.L. and M.O.; Funding acquisition, H.J., L.L. and M.O. All authors have read and agreed to the published version of the manuscript.

Funding: This research work was funded by Institutional Fund Projects under grant no. (IFPIP 344-135-1443). The authors gratefully acknowledge technical and financial support from the Ministry of Education and King Abdulaziz University, DSRJeddah, Saudi Arabia.

Data Availability Statement: Data are contained within the article.

Conflicts of Interest: The authors declare no conflict of interest.

References

1. Hsu, W.-L.; Chen, C.-Y.; Tsauo, J.-Y.; Yang, R.-S. Balance control in elderly people with osteoporosis. *J. Formos. Med. Assoc.* **2014**, *113*, 334–339. [CrossRef]
2. Hassan, M.; Kadone, H.; Ueno, T.; Suzuki, K.; Sankai, Y. Feasibility study of wearable robot control based on upper and lower limbs synergies. In Proceedings of the 2015 International Symposium on Micro-NanoMechatronics and Human Science (MHS), Nagoya, Japan, 23–25 November 2015; pp. 1–6. [CrossRef]
3. Mohammed, S.; Amirat, Y. Towards intelligent lower limb wearable robots: Challenges and perspectives-State of the art. In Proceedings of the 2008 IEEE International Conference on Robotics and Biomimetics, Bangkok, Thailand, 22–25 February 2009; pp. 312–317. [CrossRef]
4. Accoto, D.; Sergi, F.; Tagliamonte, N.L.; Carpino, G.; Sudano, A.; Guglielmelli, E. Robomorphism: A Nonanthropomorphic Wearable Robot. *IEEE Robot. Autom. Mag.* **2014**, *4*, 45–55. [CrossRef]
5. Bacek, T.; Moltedo, M.; Langlois, K.; Prieto, G.A.; Sanchez-Villamanan, M.C.; Gonzalez-Vargas, J.; Vanderborght, B.; Lefeber, D.; Moreno, J.C. BioMot exoskeleton-Towards a smart wearable robot for symbiotic human-robot interaction. In Proceedings of the 2017 International Conference on Rehabilitation Robotics (ICORR), London, UK, 17–20 July 2017; pp. 1666–1671. [CrossRef]

6. Al-Darraji, I.; Kılıç, A.; Kapucu, S. Mechatronic design and genetic-algorithm-based MIMO fuzzy control of adjustable-stiffness tendon-driven robot finger. *Mech. Sci.* **2018**, *9*, 277–296. [CrossRef]

7. Al-Darraji, I.; Kılıç, A.; Kapucu, S. Optimal control of compliant planar robot for safe impact using steepest descent technique. In Proceedings of the International Conference on Information and Communication Technology (ICICT '19), London, UK, 25–26 February 2019; Association for Computing Machinery: New York, NY, USA, 2019; pp. 233–238. [CrossRef]

8. Choo, J.; Park, J.H. Increasing Payload Capacity of Wearable Robots Using Linear Actuators. *IEEE/ASME Trans. Mechatron.* **2017**, *22*, 1663–1673. [CrossRef]

9. Jafri, S.R.A.; Abbasi, M.B.A.; Shah, S.M.U.A.; Hanif, A. BIPATRON (Bionic parageliatron): A wearable robot for rehabilitation . . . Lets Walk! In Proceedings of the 2017 First International Conference on Latest trends in Electrical Engineering and Computing Technologies (INTELLECT), Karachi, Pakistan, 15–16 November 2017; pp. 1–7. [CrossRef]

10. Nascimento, L.B.P.D.; Eugenio, K.J.S.; Fernandes, D.H.D.S.; Alsina, P.J.; Araujo, M.V.; Pereira, D.D.S.; Sanca, A.S.; Silva, M.R. Safe Path Planning Based on Probabilistic Foam for a Lower Limb Active Orthosis to Overcoming an Obstacle. In Proceedings of the 2018 Latin American Robotic Symposium, 2018 Brazilian Symposium on Robotics (SBR) and 2018 Workshop on Robotics in Education (WRE), Joao Pessoa, Brazil, 6–10 November 2018; pp. 413–419. [CrossRef]

11. Cha, K.-H.; Kang, S.J.; Choi, Y. Knee-wearable Robot System EMG signals. *J. Inst. Control. Robot. Syst.* **2009**, *15*, 286–292. [CrossRef]

12. Belkhier, Y.; Shaw, R.N.; Bures, M.; Islam, M.R.; Bajaj, M.; Albalawi, F.; Alqurashi, A.; Ghoneim, S.S.M. Robust interconnection and damping assignment energy-based control for a permanent magnet synchronous motor using high order sliding mode approach and nonlinear observer. *Energy Rep.* **2022**, *8*, 1731–1740. [CrossRef]

13. Belkhier, Y.; Achour, A. Passivity-based voltage controller for tidal energy conversion system with permanent magnet synchronous generator. *Int. J. Control. Autom. Syst.* **2021**, *19*, 988. [CrossRef]

14. Park, Y.-L.; Santos, J.; Galloway, K.G.; Goldfield, E.C.; Wood, R.J. A soft wearable robotic device for active knee motions using flat pneumatic artificial muscles. In Proceedings of the 2014 IEEE International Conference on Robotics and Automation (ICRA), Hong Kong, China, 31 May–7 June 2014; pp. 4805–4810. [CrossRef]

15. Jeong, M.; Woo, H.; Kong, K. A Study on Weight Support and Balance Control Method for Assisting Squat Movement with a Wearable Robot, Angel-suit. *Int. J. Control. Autom. Syst.* **2020**, *18*, 114–123. [CrossRef]

16. Yuan, Y.; Li, Z.; Zhao, T.; Gan, D. DMP-Based Motion Generation for a Walking Exoskeleton Robot Using Reinforcement Learning. *IEEE Trans. Ind. Electron.* **2020**, *67*, 3830–3839. [CrossRef]

17. Bian, Y.; Shao, J.; Yang, J.; Liang, A. Jumping motion planning for biped robot based on hip and knee joints coordination control. *J. Mech. Sci. Technol.* **2021**, *35*, 1223–1234. [CrossRef]

18. Kagawa, T.; Takahashi, F.; Uno, Y. On-line learning system for gait assistance with wearable robot. In Proceedings of the 2017 56th Annual Conference of the Society of Instrument and Control Engineers of Japan (SICE), Kanazawa, Japan, 19–22 September 2017; pp. 640–643. [CrossRef]

19. Li, Z.; Deng, C.; Zhao, K. Human-Cooperative Control of a Wearable Walking Exoskeleton for Enhancing Climbing Stair Activities. *IEEE Trans. Ind. Electron.* **2020**, *67*, 3086–3095. [CrossRef]

20. Richards, J. *The Comprehensive Textbook of Clinical Biomechanics*; Elsevier: Amsterdam, The Netherlands, 2018; ISBN 9780702064951.

21. Human Body Part Weights. Available online: https://robslink.com/SAS/democd79/body_part_weights.htm (accessed on 2 February 2023).

22. Kirsch, N.A.; Bao, X.; Alibeji, N.A.; Dicianno, B.E.; Sharma, N. Model-Based Dynamic Control Allocation in a Hybrid Neuroprosthesis. *IEEE Trans. Neural Syst. Rehabil. Eng.* **2018**, *26*, 224–232. [CrossRef]

23. Han, H.-G.; Qiao, J.-F. Adaptive Computation Algorithm for RBF Neural Network. *IEEE Trans. Neural Netw. Learn. Syst.* **2012**, *23*, 342–347. [CrossRef]

24. Bazargan-Lari, Y.; Eghtesad, M.; Khoogar, A.; Mohammad-Zadeh, A. Tracking Control of A Human Swing Leg Considering Self-Impact Joint Constraint by Feedback Linearization Method. *Control. Eng. Appl. Inform.* **2015**, *17*, 99–110.

25. Zuo, Q.; Zhao, J.; Mei, X.; Yi, F.; Hu, G. Design and Trajectory Tracking Control of a Magnetorheological Prosthetic Knee Joint. *Appl. Sci.* **2021**, *11*, 8305. [CrossRef]

26. Zhang, X.; Lu, Z.; Yuan, X.; Wang, Y.; Shen, X. L2-Gain Adaptive Robust Control for Hybrid Energy Storage System in Electric Vehicles. *IEEE Trans. Power Electron.* **2021**, *36*, 7319–7332. [CrossRef]

27. Coutinho, D.F.; Fu, M.; Trofino, A.; Danes, P. L2 Gain analysis and control of uncertain nonlinear systems with bounded disturbance inputs. *Int. J. Robust Nonlinear Control. IFAC Affil. J.* **2008**, *18*, 88–110. [CrossRef]

28. van der Schaft, A. L2-Gain Analysis of Nonlinear Systems and Nonlinear State Feedback H∞ Control. *IEEE Trans. Autom. Control.* **1992**, *37*, 770–784. [CrossRef]

29. Hendzel, Z. Hamilton–Jacobi inequality robust neural network control of a mobile wheeled robot. *Math. Mech. Solids* **2019**, *3*, 723–737. [CrossRef]

30. Wang, Y.; Sun, W.; Xiang, Y.; Miao, S. Neural Network-Based Robust Tracking Control for Robots. *Intell. Autom. Soft Comput.* **2009**, *2*, 211–222. [CrossRef]

31. Song, Q.; Li, S.; Bai, Q.; Yang, J.; Zhang, A.; Zhang, X.; Zhe, L. Trajectory Planning of Robot Manipulator Based on RBF Neural Network. *Entropy* **2021**, *23*, 1207. [CrossRef]

32. Al-Darraji, I.; Piromalis, D.; Kakei, A.; Khan, F.; Stojmenovic, M.; Tsaramirsis, G.; Papageorgas, P. Adaptive Robust Controller Design-Based RBF Neural Network for Aerial Robot Arm Model. *Electronics* **2021**, *10*, 831. [CrossRef]

33. Liu, J. *Radial Basis Function (RBF) Neural Network Control for Mechanical Systems*; Springer: Berlin/Heidelberg, Germany, 2013; ISBN 978-3-642-34816-7.

34. Yang, J.; Na, J.; Gao, G.; Zhang, C. Adaptive Neural Tracking Control of Robotic Manipulators with Guaranteed NN Weight Convergence. *Complexity* **2018**, *2018*, 7131562. [CrossRef]

35. Chaouch, H.; Charfeddine, S.; Ben Aoun, S.; Jerbi, H.; Leiva, V. Multiscale Monitoring Using Machine Learning Methods: New Methodology and an Industrial Application to a Photovoltaic System. *Mathematics* **2022**, *10*, 890. [CrossRef]

36. Charfeddine, M.; Jouili, K.; Jerbi, H.; Braiek, N.B. Linearizing control with a robust relative degree based on a Lyapunov function: Case of the ball and beam system. *Int. Rev. Model. Simul.* **2010**, *3*, 219–226.

37. Qiu, S.; Pei, Z.; Wang, C.; Tang, Z. Systematic Review on Wearable Lower Extremity Robotic Exoskeletons for Assisted Locomotion. *J. Bionic. Eng.* **2023**, *20*, 436–469. [CrossRef]

 *mathematics*

Article

An Unfitted Method with Elastic Bed Boundary Conditions for the Analysis of Heterogeneous Arterial Sections

Stephan Gahima [1,2], Pedro Díez [1,2], Marco Stefanati [3], José Félix Rodríguez Matas [3] and Alberto García-González [1,2,*]

[1] Laboratori de Càlcul Numèric, E.T.S. de Ingeniería de Caminos, Universitat Politècnica de Catalunya, 08034 Barcelona, Spain

[2] The International Centre for Numerical Methods in Engineering, CIMNE, 08034 Barcelona, Spain

[3] Laboratory of Biological Structure Mechanics (LaBS), Department of Chemistry, Materials and Chemical Engineering "Giulio Natta", Politecnico di Milano, 20133 Milan, Italy

[*] Correspondence: berto.garcia@upc.edu

Abstract: This manuscript presents a novel formulation for a linear elastic model of a heterogeneous arterial section undergoing uniform pressure in a quasi-static regime. The novelties are twofold. First, an elastic bed support on the external boundary (elastic bed boundary condition) replaces the classical Dirichlet boundary condition (i.e., blocking displacements at arbitrarily selected nodes) for elastic solids to ensure a solvable problem. In addition, this modeling approach can be used to effectively account for the effect of the surrounding material on the vessel. Secondly, to study many geometrical configurations corresponding to different patients, we devise an unfitted strategy based on the Immersed Boundary (IB) framework. It allows using the same (background) mesh for all possible configurations both to describe the geometrical features of the cross-section (using level sets) and to compute the solution of the mechanical problem. Results on coronary arterial sections from realistic segmented images demonstrate that the proposed unfitted IB-based approach provides results equivalent to the standard finite elements (FE) for the same number of active degrees of freedom with an average difference in the displacement field of less than 0.5%. However, the proposed methodology does not require the use of a different mesh for every configuration. Thus, it is paving the way for dimensionality reduction.

Keywords: elastic bed boundary condition; robin boundary condition; immersed boundary method; level set; arterial biomechanics; unfitted method

MSC: 65H99

Citation: Gahima, S.; Díez, P.; Stefanati, M.; Rodríguez Matas, J.F.; García-González, A. An Unfitted Method with Elastic Bed Boundary Conditions for the Analysis of Heterogeneous Arterial Sections. *Mathematics* **2023**, *11*, 1748. https://doi.org/10.3390/math11071748

Academic Editor: Fernando Simoes

Received: 19 February 2023
Revised: 22 March 2023
Accepted: 30 March 2023
Published: 6 April 2023

1. Introduction

Ischemic heart disease is the first cause of death globally, accounting for 27% of fatalities in 2019 [1,2], with coronary atherosclerosis being the cause of most myocardial infarctions [3]. Atherosclerotic plaques (within the coronaries) result from a complex inflammatory process starting with the accumulation and retention of low-density lipoprotein within the intima. The result is a build-up of material (cholesterol and other lipid compositions) within the wall layers, producing stenosis and blood flux reduction in the vessel. Typically, a patient presents either stable or unstable (low or high risk of rupture) plaque. The fast distinction between these two groups is crucial regarding the treatment and disposition of the patient [4]. Thus, the need for patient-specific approaches is self-evident. It is here where computation-supported decision-making processes play a crucial role. This work contributes with a new approach to modeling two-dimensional coronary sections undergoing uniform physiological internal pressure in a quasi-static regime. Holzapfel et al. [5] showed that the pertinence and accuracy of the results depend on the method used to define material properties and to acquire in vivo patient-specific geometries. Typical methods for geometry acquisition include Magnetic Resonance Imaging [6,7], Computer Tomography [8,9], Optical Coherence Tomography [10,11], and Intravascular Ultrasound [12,13], among others. Hyperelastic [14,15], piece-wise homogeneous [16,17], incompressible [18,19], and plane strain [20,21] hypotheses characterize the two-dimensional models. In this work, we have used for the mechanical properties of all plaque components (i.e., normal vessel wall, loose matrix, calcification, and lipid core) a linear approximation of the stress–strain curve up to around 10% deformation reported by [22,23]. In particular, the rational behind using linear properties was to test the proposed approach, since it allows reducing the calculation times. However, recent studies use linear mechanical properties of the arterial tissue to perform clinical predictions based on geometrical and biomechanical markers obtained from finite element simulation [24]. In addition, the proposed methodology can directly accommodate a nonlinear hyper-elastic behavior of the tissues composing the coronary plaque.

One of the main problems when solving a finite element problem is to properly constrain the structure to prevent rigid body motion. The simplest approach to fixing the singularity and suppressing rigid-body modes consists in blocking three degrees of freedom (an isostatic condition with vertical displacements in two arbitrary nodes and horizontal displacement in another, for example). Thus, it is about setting essential (Dirichlet type) boundary conditions (BC) in arbitrarily selected nodes. In addition, blocking axial displacement and allowing free radial expansion [25,26] is another possibility. In addition, fixing two adjacent points [27,28], or the entire external boundary [29], or creating a soft and compressible layer surrounding the section with a zero-displacement BC constraining the latter [30,31] are examples of BCs used to suppress rigid body motions. All these methods struggle, to different degrees, to consider what surrounds the coronary section. Some of them depend on arbitrary choices (e.g., choosing the nodes where to impose BC). Few works have attempted to account for the influence of the surrounding tissue on the artery. In [32], an elastic bed boundary condition was applied along the coronary artery to simulate the effect of the myocardial tissue. In [33], the artery is surrounded by the cardiac wall for half of its circumference to simulate the coronary artery embedded into the myocardium, with the cardiac wall modeled using finite elements. We propose to assume the section to be surrounded by a material along its external boundary. This embedding matrix produces a linear elastic reaction, and it is described with an elastic bed coefficient α to be assessed depending on the stiffness of the surrounding medium.

Additionally, to improve the computational efficiency of the realistic calculation and analysis process, the numerical methodology proposed here implements the aforementioned elastic bed coefficient in an Immersed Boundary (IB) [34] framework with a generic description of the domain based on level sets. IB (combined with elastic bed BC) bases its simulations on a unique (background) mesh supporting the solutions corresponding to different configurations (different patients). It allows comparing solutions and opens the door to reduced-order modeling leading to fast simulations for different patient-specific geometries from initial medical images (with the same degrees of freedom), avoiding individual meshing and preprocessing steps. This methodology is motivated by its potential applicability with voxelized data [35], such as medical images. Via segmentation [36–39], it is possible to identify the different components' contours, and the IB performs the stress analysis on a refined voxelized background mesh to increase accuracy. In general, an IB approach allows seamless integration of structural analysis in a medical image processing pipeline.

The remainder of the paper is structured as follows: Section 2 describes the problem statement (Section 2.1), presents the level set approach (Sections 2.2 and 2.3), and includes the description of the IB framework (Section 2.4), emphasizing the details of the mathematical formulation required in the weak form of the problem (Section 2.5). Section 3 shows the results of the proposed methodology, finishing with a discussion and the main conclusions in Section 4.

2. Materials and Methods

With a patient-specific application in mind, the discretization method to be used has to handle different configurations corresponding to various patients effortlessly. Moreover, input (the diversity of arterial cross-section geometries) and output (the solution in terms of deformation, strains, and stresses) data are to be expressed in homogeneous formats to ease the analysis and the possible application of reduced-order models. Here, level sets defined on a background mesh (discretizing a background domain Ω^B) describe the diversity of the geometric configurations, that is, all the possible instances of the actual computational domain, Ω. It comes naturally to solve the problem with an unfitted approach. Specifically, it uses the background mesh not only to describe the geometry (actually the same background mesh for all the possible geometries) but also to compute the solution, following an IB methodology. Thus, one may prescind the conformal meshes adapted to the geometry that change from case to case. Note that to solve the problem with conformal finite elements, the mesh must be such that it tallies with Ω, matching the boundary $\partial\Omega$. Such an approach requires ad-hoc meshing algorithms, especially for convoluted geometries, and complicates comparing different configurations and their solutions.

2.1. Problem Statement

Let the section occupy a region $\Omega \subset \mathbb{R}^2$ with boundary $\partial\Omega$. The intrinsic heterogeneity of arterial cross-sections is described by dividing Ω into different subdomains $\Omega_1, \Omega_2, \dots$, corresponding to homogeneous regions having different material properties i.e., normal vessel wall, loose matrix, calcification, and lipid core (see Figure 1). Without body forces, the equilibrium is governed by

$$\nabla \cdot \sigma(\mathbf{u}) = \mathbf{0} \quad \text{in } \Omega, \tag{1}$$

with boundary conditions

$$\sigma(\mathbf{u}) \cdot \hat{\mathbf{n}} = \mathbf{t} \qquad\qquad\qquad \text{on } \Gamma_N, \tag{2}$$

$$\sigma(\mathbf{u}) \cdot \hat{\mathbf{n}} = \alpha\mathbf{u} \quad (\alpha < 0) \qquad\qquad \text{on } \Gamma_R, \tag{3}$$

where σ is the Cauchy stress tensor and $\mathbf{u}$ is the displacement field; $\mathbf{t}$ is the surface traction, α is the elastic bed coefficient, and $\hat{\mathbf{n}}$ is the outward unit normal to the boundary. Equation (3) represents the Robin boundary condition, physically corresponding to an elastic bed condition, simulating the surrounding tissue of the artery. The Neumann and elastic bed boundaries cover the whole boundary, i.e., $\partial\Omega = \Gamma_N \cup \Gamma_R$.

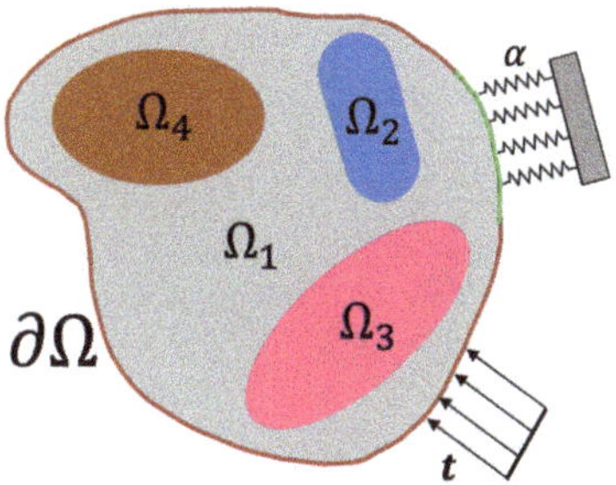

Figure 1. Schematic description of Problem (1) in the Euclidean space. In particular, $\Omega = \bigcup_{k=1}^{4} \Omega_k$ with $\partial\Omega = \Gamma_N \cup \Gamma_R$ where Γ_N and Γ_R are depicted in red and cyan, respectively.

The weak form of Problem 1 (physically corresponding to the principle of virtual work) reads: find $\mathbf{u} \in \left[\mathcal{H}^1(\Omega)\right]^2$ such that

$$\int_\Omega \sigma(\mathbf{u}) : \varepsilon(\mathbf{v}) \, d\Omega - \int_{\Gamma_R} \alpha \mathbf{u} \cdot \mathbf{v} \, d\Gamma = \int_{\Gamma_N} \mathbf{t} \cdot \mathbf{v} \, d\Gamma, \tag{4}$$

for all $\mathbf{v} \in \left[\mathcal{H}^1(\Omega)\right]^2$. $\mathcal{H}^1(\Omega)$ is the Sobolev space of order 1 on Ω; refer to [40] for details. Note that the test function $\mathbf{v}$ is also seen as a virtual infinitesimal displacement (a perturbation from the equilibrium configuration of the body) consistent with the imposed boundary displacements, and $\varepsilon(\mathbf{v}) = 1/2(\nabla\mathbf{v} + \nabla\mathbf{v}^\top)$. The elastic bed BC (3) is an alternative to the standard practice of suppressing rigid-body motions by prescribing displacements at some arbitrarily selected points. As shown in the following, enforcing an isostatic scheme by prescribing point displacements and elastic bed BC produce similar results. We advocate the latter because the elastic bed BC includes physical information about the surrounding medium and does not require selecting arbitrary points to prescribe displacements. This is crucial for model order reduction, where one has to perform operations with the solutions of different configurations, and hence, they need to be comparable.

2.2. Level Set Description of the Domain and Subdomains

As introduced previously, the domain Ω is divided into n subdomains Ω_i, $i = 1, \ldots, n$. The n subdomains cover Ω, that is

$$\Omega = \bigcup_{k=1}^{n} \Omega_k. \tag{5}$$

Level set functions implicitly describe the geometry of Ω and its subdomains in a unique framework. A background domain Ω^B, having a simple geometry (here rectangle or square shape), is introduced to accommodate all possible instances of Ω, resulting in $\Omega \subset \Omega^B$; see Figure 2A.

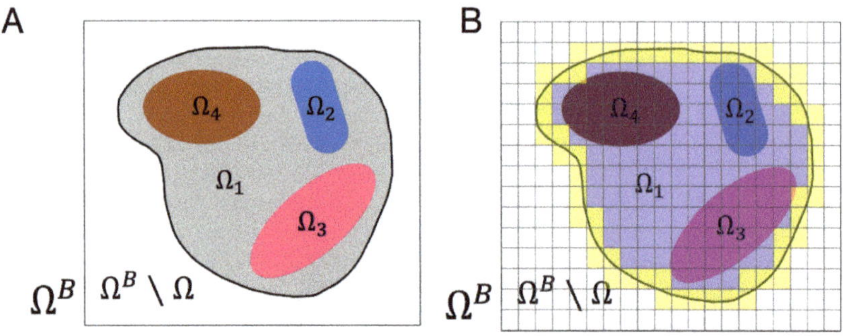

Figure 2. (**A**) The background domain Ω^B and (**B**) (one of its possible) mesh $\mathcal{T}_h(\Omega^B)$ (for more details regarding accurate estimations of the displacement fields at the interfaces, see Section 2.4). Inner and cut (by $\partial\Omega$) elements $T \in \mathcal{T}_h$ are in blue and yellow, respectively.

A standard level set to describe Ω in Ω^B is a continuous function ϕ taking values in Ω^B such that $\phi(\mathbf{x}) > 0$ for $\mathbf{x} \in \Omega$ and negative elsewhere. Thus, $\phi(\mathbf{x}) = 0$ for $\mathbf{x} \in \partial\Omega$. Typically, ϕ is a signed distance to $\partial\Omega$ [41,42]. For a configuration such as the one in Figure 3, with two non-connected parts of the boundary, Γ_N and Γ_R, it is convenient to describe Ω using two level sets to distinguish between the two. Thus, Ω is identified with $\phi^{(1)}$ and $\phi^{(2)}$ such that: $\phi^{(1)}(\mathbf{x}) = 0$ on Γ_N, and $\phi^{(2)}(\mathbf{x}) = 0$ on Γ_R. Both level sets are positive in Ω; see Figure 3B,C for an illustration. Note that one may recover a standard level set for Ω by just taking $\phi = \phi^{(1)}\phi^{(2)}$. Then, following the ideas in [43], new level set functions are introduced to describe the n subdomains. Function $\phi^{(3)}$ provides the information to identify Ω_1 and distinguish it from the remainder subdomains. In particular, $\phi^{(3)}(\mathbf{x}) > 0$ for $\mathbf{x} \in \bigcup_{k=2}^{n} \Omega_k$ and is negative elsewhere in Ω (that is in Ω_1). Similarly, $\phi^{(4)}$ is positive in $\bigcup_{k=3}^{n} \Omega_k$ and negative in the remainder, that is in $\bigcup_{k=1}^{2} \Omega_k$. The last hierarchical level set needed is $\phi^{(n+1)}$ identifying Ω_{n-1} because then Ω_n is precisely the remainder ($\phi^{(n+1)} > 0$). The values of $\phi^{(k)}$ with $k = 3, \ldots, n+1$ outside Ω are not relevant. This is consistent with the hierarchical character of this approach. A visualization of the hierarchical level sets is illustrated in the panels of Figure 3 and summarized in Table 1.

Table 1. Level set-based criteria to classify a point $\mathbf{x}$ in Ω and its subdomains.

Condition	Classification
$\phi^{(1)} > 0$ and $\phi^{(2)} > 0$	$\mathbf{x} \in \Omega$
$\phi^{(1)} > 0$ and $\phi^{(2)} > 0$ and $\phi^{(3)} < 0$	$\mathbf{x} \in \Omega_1$
$\phi^{(1)} > 0$ and $\phi^{(2)} > 0$ and $\phi^{(3)} > 0$ and $\phi^{(4)} < 0$	$\mathbf{x} \in \Omega_2$
$\phi^{(1)} > 0$ and $\phi^{(2)} > 0$ and $\phi^{(3)} > 0$ and $\phi^{(4)} > 0$ and $\phi^{(5)} < 0$	$\mathbf{x} \in \Omega_3$
$\phi^{(1)} > 0$ and $\phi^{(2)} > 0$ and $\phi^{(5)} > 0$	$\mathbf{x} \in \Omega_4$

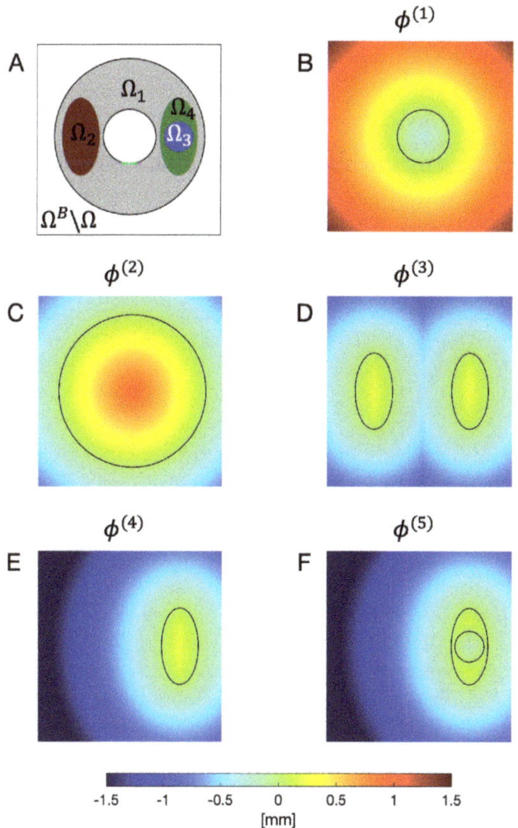

Figure 3. (**A**) Ω embedded in Ω^{B}. Level set (**B**) $\phi^{(1)} = 0$ describes Γ_N, and (**C**) $\phi^{(2)} = 0$ describes Γ_R. (**D**) $\phi^{(3)} = 0$ describes the interfaces between the subset $\bigcup_{k=2}^{4} \Omega_k$ and the background domain. (**E**) $\phi^{(4)} = 0$ describes the interfaces between $\Omega_3 \cup \Omega_4$ and $\Omega^{\mathrm{B}} \setminus \Omega_3 \cup \Omega_4$ and finally, (**F**) $\phi^{(5)} = 0$ describes the interface between Ω_4 and the rest.

The approach described above is similar to the front-tracking method used to simulate multiphase flow with a fixed grid for the flow [44], with the difference that the front does not change with time, and therefore, the level set is computed only once at the beginning of the analysis.

2.3. Discretization of the Level Set Functions in a Background Mesh

With a finite element (FE) discretization of Ω^B (see Figure 2), the level set approach is implemented. A tessellation $\mathcal{T}_h$ of Ω^B consisting of n_e elements T_e, $e = 1, 2, \ldots, n_e$ (h stands for the characteristic size of the elements) is introduced, such that $\Omega^B = \bigcup_{e=1}^{n_e} T_e$. The number of nodes in the mesh is denoted by n_p and the corresponding shape functions are denoted by $N_i(\mathbf{x})$, for $i = 1, 2, \ldots, n_p$ c[cyan]. Thus, each level set $\phi^{(k)}$, $k = 1, \ldots, n+1$, is represented in the background mesh as

$$\phi^{(k)}(\mathbf{x}) \approx \sum_{i=1}^{n_p} [\mathbf{\Phi}^k]_i N_i(\mathbf{x}), \tag{6}$$

with $\mathbf{\Phi}^k \in \mathbb{R}^{n_p}$ being the vector of nodal values describing $\phi^{(k)}$. In this framework, $n+1$ vectors in $\mathbb{R}^{n_p}$ describe any geometrical configuration. This standardized representation allows for dimensionality reduction given the variance in the population of input samples (each corresponding to a different patient). To ease the task of the machine learning algorithms to be used in dimensionality reduction, standard geometric normalizations are performed previously to store the information in the discrete level set format. For instance, all samples are centered (their barycenter is translated to the origin of coordinates) and rotated such that the principal axes of inertia are parallel to the coordinate axes.

2.4. Unfitted Approach: Solving the Problem in the Background Mesh

The framework for approximating the level set over $\mathcal{T}_h(\Omega^B)$ just described is used to solve the original problem (4) using an unfitted approach based on the ideas of the Immersed Boundary Method (IBM). Thus, the displacement field $\mathbf{u}(\mathbf{x})$ is approximated in the background mesh using a standard FE approximation, namely

$$\mathbf{u}(\mathbf{x}) \approx \sum_{i=1}^{n_p} \mathbf{U}_i N_i(\mathbf{x}), \tag{7}$$

with $\mathbf{U}_i \in \mathbb{R}^2$ being the displacement vector in node i. All vectors $\mathbf{U}_i$, $i = 1, 2, \ldots, n_p$, are collected in the standard vector of nodal displacements $\mathbf{U} \in \mathbb{R}^{2n_p}$. Using the Galerkin strategy to solve Equation (4) results in a linear system of equations for $\mathbf{U}$:

$$[\mathbf{K} + \mathbf{M}]\mathbf{U} = \mathbf{F}, \tag{8}$$

where matrices $\mathbf{K}$ and $\mathbf{M}$ in $\mathbb{R}^{2n_p \times 2n_p}$ are the discrete counterparts of the two bilinear forms in the left-hand side of Equation (4) and $\mathbf{F} \in \mathbb{R}^{2n_p}$ is the discretization of the linear form in the right-hand side.

Note that a node i in the mesh is represented by the degrees of freedom $\ell = 2(i-1)+1$ and $\ell+1$ in $\mathbf{U}$, and some other node j is represented by $\tilde{\ell} = 2(j-1)+1$ and $\tilde{\ell}+1$. Assuming these relations, some illustrative examples of the expressions for the corresponding entries in the matrices and the right-hand-side vector are given below

$$[\mathbf{K}]_{\ell\tilde{\ell}} = \int_\Omega \sigma\left(\begin{bmatrix} N_i(\mathbf{x}) \\ 0 \end{bmatrix}\right) : \varepsilon\left(\begin{bmatrix} N_j(\mathbf{x}) \\ 0 \end{bmatrix}\right) d\Omega \ ; \ [\mathbf{K}]_{\ell,\tilde{\ell}+1} = \int_\Omega \sigma\left(\begin{bmatrix} N_i(\mathbf{x}) \\ 0 \end{bmatrix}\right) : \varepsilon\left(\begin{bmatrix} 0 \\ N_j(\mathbf{x}) \end{bmatrix}\right) d\Omega$$

$$[\mathbf{M}]_{\ell\tilde{\ell}} = -\int_{\Gamma_R} \alpha \begin{bmatrix} N_i(\mathbf{x}) \\ 0 \end{bmatrix} \cdot \begin{bmatrix} N_j(\mathbf{x}) \\ 0 \end{bmatrix} d\Gamma \ ; \ [\mathbf{M}]_{\ell,\tilde{\ell}+1} = -\int_{\Gamma_R} \alpha \begin{bmatrix} N_i(\mathbf{x}) \\ 0 \end{bmatrix} \cdot \begin{bmatrix} 0 \\ N_j(\mathbf{x}) \end{bmatrix} d\Gamma$$

$$[\mathbf{F}]_\ell = \int_{\Gamma_N} \mathbf{t} \cdot \begin{bmatrix} N_i(\mathbf{x}) \\ 0 \end{bmatrix} d\Gamma \ ; \ [\mathbf{F}]_{\ell+1} = \int_{\Gamma_N} \mathbf{t} \cdot \begin{bmatrix} 0 \\ N_i(\mathbf{x}) \end{bmatrix} d\Gamma.$$

Note that all the integrals in the expressions above are defined in Ω, Γ_R, and Γ_N, and not in the background domain Ω^B where the FE functions $N_i(\mathbf{x})$ are supported. In particular, evaluating the local contributions (the integrals are restricted to some element T_e) requires identifying whether an element intersects Γ_R or Γ_N. Thus, the main implementation challenge of the unfitted approach is classifying the elements $\mathcal{T}_h$ of Ω^B inside Ω, those outside, and those crossed by the interfaces. For a given configuration, the geometrical information is encoded in the level sets, as described in Section 2.2. This allows elaborating a list of the elements in $\mathcal{T}_h$ that are completely inside Ω, namely $\mathcal{I}_\Omega$ such that if $e \in \mathcal{I}_\Omega$, then $T_e \subset \Omega$. Similarly, lists $\mathcal{I}_{\Gamma_R}$ and $\mathcal{I}_{\Gamma_N}$ are such that if $e \in \mathcal{I}_{\Gamma_R}$ then $T_e \cap \Gamma_R \neq \varnothing$, and if $e \in \mathcal{I}_{\Gamma_N}$ then $T_e \cap \Gamma_N \neq \varnothing$. Figure 4 shows an example of such classification. The elements indexed in these three lists are *active*, meaning that they play a role in the solution for the configuration described by the level sets. Thus, T_e is said to be active if $e \in \mathcal{I}_\Omega \cup \mathcal{I}_{\Gamma_R} \cup \mathcal{I}_{\Gamma_N}$. Accordingly, all the nodes belonging to active elements are active nodes since the corresponding degrees of freedom are the unknowns of (8) (the non-active nodes are to be eliminated from the system).

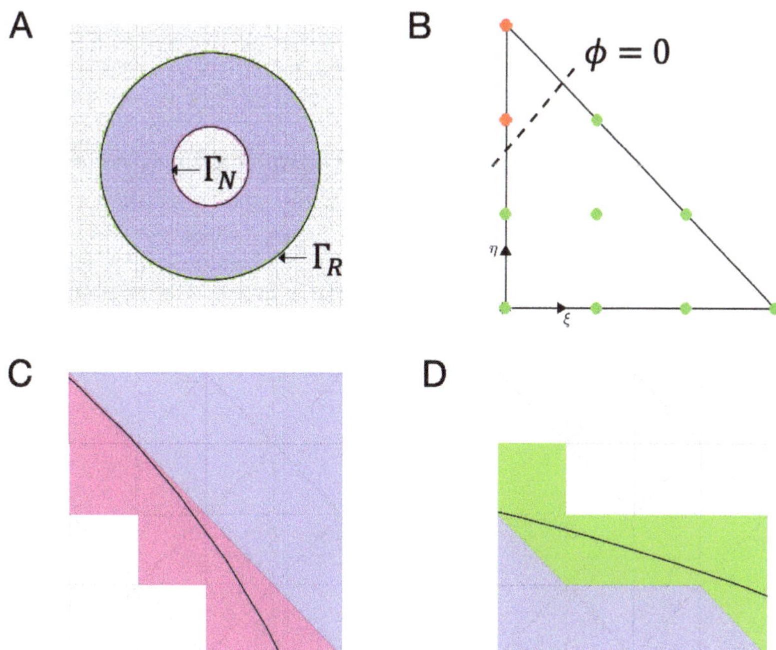

Figure 4. (**A**) Elements T_e for $e \in \mathcal{I}_\Omega \cup \mathcal{I}_{\Gamma_R} \cup \mathcal{I}_{\Gamma_N}$, are colored in violet ($e \in \mathcal{I}_\Omega$), magenta ($e \in \mathcal{I}_{\Gamma_N}$), and cyan ($e \in \mathcal{I}_{\Gamma_R}$), being Γ_N and Γ_R the black lines. The square background domain Ω^B (2.5×2.5 mm^2) is meshed with $n_p = 100^2$ nodes and $n_e = 2 \times 99^2$ elements. Close-ups for better illustration in panels (**C**), and (**D**). Panel (**B**) illustrates that in the elements crossed by the boundary, the quadrature is enriched to avoid having no integration points in the part of the element outside Ω. This suggests using in these elements closed quadratures (as the third-degree closed Newton–Cotes quadrature [45]).

The computation of the elementary contributions to the stiffness matrix $\mathbf{K}$ is standard for the elements completely inside T_e for $e \in \mathcal{I}_\Omega$ (violet elements in Figure 4A). The only particular feature to be accounted for is that the material properties of each Gauss point in the numerical quadrature belong to a subdomain Ω_k. With the values of the level sets interpolated at the Gauss point, following the classification described in Table 1, the material properties are quickly recovered. Note that for the example shown in Figure 4A, only two level sets, $\phi^{(1)}$ and $\phi^{(2)}$, are required. In the elements crossed by the interfaces Γ_R and Γ_N, the integration has to exclude the part of the domain outside Ω. There, a more refined quadrature is used, and null material properties are assigned to the integration points outside Ω. A closed quadrature is preferred to avoid accounting for integration points inside elements with a small portion inside Ω. These geometric checks are performed by setting a tolerance and considering that the distance to the interface is zero when it is below this value. Computing elementary contributions to matrix $\mathbf{M}$ and vector $\mathbf{F}$ requires integrating within the portion of Γ_R or Γ_N in the element T_e. Thus, for $e \in \mathcal{I}_{\Gamma_N}$, T_e intersects Γ_N and contributes to $\mathbf{F}$. Analogously, for $e \in \mathcal{I}_{\Gamma_R}$, T_e intersects Γ_R and contributes to $\mathbf{M}$.

For $e \in \mathcal{I}_{\Gamma_N}$, the elementary contribution from element T_e to $\mathbf{F}$ requires computing

$$\int_{\Gamma_N \cap T_e} [\mathbf{t}]_1 N_i(\mathbf{x}) d\Gamma \quad \text{and} \quad \int_{\Gamma_N \cap T_e} [\mathbf{t}]_2 N_i(\mathbf{x}) d\Gamma, \tag{9}$$

for all the nodes i in element T_e. If the load corresponds to a pressure p applied in the internal wall, then $\mathbf{t} = -p\hat{\mathbf{n}}$, recalling that $\hat{\mathbf{n}} = [n_1 \ n_2]^\top$ is the outward unit normal. Thus, $[\mathbf{t}]_1 = -pn_1$ and $[\mathbf{t}]_2 = -pn_2$. This integral, as it is standard in the FE practice, is computed in a reference element (for linear triangles, it is handy using the triangle with vertices $(0,0)$, $(1,0)$ and $(0,1)$, see Figure 5), where the shape functions are defined (and available in their analytical expressions) in terms of the reference coordinates (ξ, η), namely $\hat{N}_{(i)}(\xi, \eta)$, for $(i) = 1, 2, 3$. Mesh connectivity provides the link between the local numbering of the node inside the element, (i) (from 1 to 3 in the case of linear triangles), and the global numbering i (from 1 to n_{p}). Since T_e is crossed by Γ_N, it is important to identify the entry and exit points in the element, that is the points $\{P_I, P_{II}\} = \Gamma_N \cap \partial T_e$; see Figure 5. This task is performed while identifying the elements in $\mathcal{I}_{\Gamma_N}$, and it is straightforward after the nodal values of $\phi^{(1)}$. Recall that $\phi^{(1)}(\mathbf{x}) = 0$ for $\mathbf{x} \in \Gamma_N$. Same rationale works for Γ_R, using $\phi^{(2)}$. A quadrature is required to integrate along the segment $P_I P_{II}$ (a portion of Γ_N). Here, a Simpson quadrature is adopted and involves computing the values of the function to be integrated on the endpoints of the interval and in the midpoint, P_{m}; see Figure 5. The general expression for Simpson quadrature to approximate the integral of some function ψ reads:

$$\int_{P_I}^{P_{II}} \psi \, d\Gamma \approx \frac{|P_I P_{II}|}{6} (\psi(P_I) + 4\psi(P_{\mathrm{m}}) + \psi(P_{II})), \tag{10}$$

where $|P_I P_{II}|$ is the length of the interval $P_I P_{II}$. Thus, computing the terms in (9) requires obtaining the values of N_i in the three points P_I, P_{II} and P_{m}. These values are easily obtained after their coordinates in the reference element, (ξ_I, η_I), (ξ_{II}, η_{II}) and $(\xi_{\mathrm{m}}, \eta_{\mathrm{m}})$. Then, it suffices using the quadrature given in (10) for $\psi(\mathbf{x}) = -pn_1 N_i(\mathbf{x})$ and $\psi(\mathbf{x}) = -pn_2 N_i(\mathbf{x})$ to obtain the horizontal and vertical components of the nodal forces (on node i from element e).

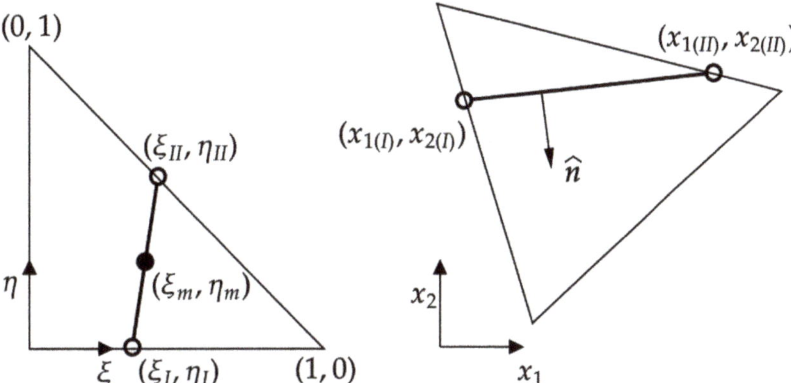

Figure 5. Element T_e (right) described in the Cartesian coordinate system (x_1, x_2) is mapped into reference element (left) described by (ξ, η) coordinates. Outward normal $\hat{n}$ to the portion of Γ_N (respectively, Γ_R) in T_e is to be determined in the Cartesian framework. The two points where the interface meets the boundary of T_e (entry and exit points, here denoted by I and II) are necessary to numerically integrate the coefficients of $\mathbf{F}$ (respectively $\mathbf{M}$) along Γ_N (respectively, Γ_R).

Similarly, for $e \in \mathcal{I}_{\Gamma_R}$, the elementary contribution from T_e to $\mathbf{M}$ requires computing terms of the form

$$\int_{\Gamma_R \cap T_e} \alpha N_i(\mathbf{x}) N_j(\mathbf{x}) \mathrm{d}\Gamma. \tag{11}$$

This is achieved by taking $\psi(\mathbf{x}) = \alpha N_i(\mathbf{x}) N_j(\mathbf{x})$ and using the quadrature (10) accordingly. In the unfitted solution, the degrees of freedom corresponding to nodes that do not belong to any of the active elements (those with index e in $\bigcup \mathcal{I}_{\Gamma_R} \bigcup \mathcal{I}_{\Gamma_N}$) must be removed from system (8).

The number of active nodes (that is the number of nodes in the active elements) is denoted by n_{act} and indicates the measure of the size of the system to be solved. Note that in conformal FE, n_{act} is the number of nodes in the mesh. On the other hand, in an unfitted approach, $n_{\text{act}} < n_{\text{p}}$.

The proposed methodology has been entirely implemented in Matlab R2022b, The MathWorks Inc, and executed in a 3.2 GHz Apple M1 with 8 GB RAM.

2.5. Validating the Methodolgy

The solution of an infinite linear elastic solid with a cylindrical cavity of radius r_{int} subjected to internal pressure p (Figure 6A) is considered.

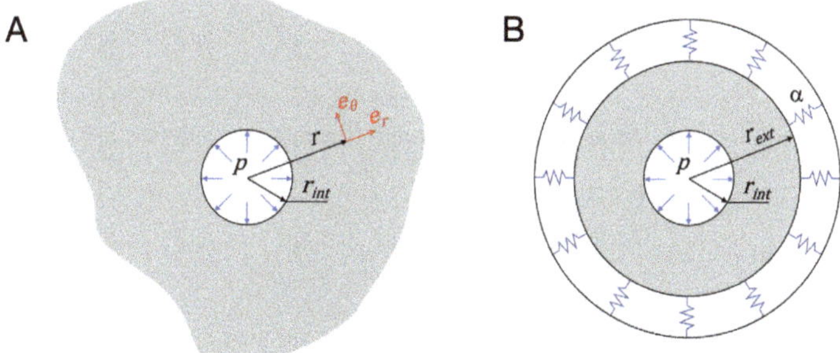

Figure 6. Verification problem. (**A**) Infinite solid with a cylindrical cavity of radius r_{int} subjected to internal pressure p; (**B**) Infinite cylinder of inner radius r_{int} and external radius r_{ext} subjected to an internal pressure p and elastic bed boundary conditions at r_{ext}.

Assuming cylindrical coordinates, the analytical solution of the problem is [46]

$$u_r = p\frac{r_{int}^2}{2\mu r}, \quad u_\theta = u_z = 0, \tag{12}$$

where $\mu = \mu(E, v)$ is the shear modulus of the material, E and v are its Young's modulus and Poisson ratio, respectively, and $r \geq r_{int}$ is the radial coordinate. The strain and stress fields are obtained with the constitutive equations for a linear elastic solid as

$$\varepsilon_{rr} = \frac{\partial u_r}{\partial r} = -\frac{p}{2\mu}\left(\frac{r_{int}}{r}\right)^2, \tag{13}$$

$$\varepsilon_{\theta\theta} = \frac{u_r}{r} = \frac{p}{2\mu}\left(\frac{r_{int}}{r}\right)^2, \tag{14}$$

$$\varepsilon_{zz} = 0, \tag{15}$$

$$\sigma_{rr} = -p\left(\frac{r_{int}}{r}\right)^2, \tag{16}$$

$$\sigma_{\theta\theta} = p\left(\frac{r_{int}}{r}\right)^2, \tag{17}$$

$$\sigma_{zz} = 0. \tag{18}$$

This problem is equivalent to that of an infinite cylinder of internal radius r_{int} and external radius, r_{ext}, subjected to internal pressure, p (applied at $r = r_{int}$), and elastic bed BC on $r = r_{ext}$ (see Figure 6B), with the *ballast* coefficient α given as

$$\alpha = -\frac{2\mu}{r_{ext}}, \tag{19}$$

obtained from (3) together with (12), and (16)–(18).

With this problem at hand, the accuracy of the methodology is quantified in terms of local and global quantities. The displacement field is a local quantity of accuracy, and the total deformation energy (TDE) is used as a global metric to assess the convergence for the numerical solution. The TDE for the infinite cylinder in Figure 6B is given by

$$\text{TDE} = \frac{1}{2}\int_0^{2\pi}\int_{r_{int}}^{r_{ext}}(\sigma_{rr}\varepsilon_{rr} + \sigma_{\theta\theta}\varepsilon_{\theta\theta})r\,\mathrm{d}r\mathrm{d}\theta =$$
$$= \pi\int_{r_{int}}^{r_{ext}}\left[\left(\frac{p^2 r_{int}^4}{2\mu r^4}\right) + \left(\frac{p^2 r_{int}^4}{2\mu r^4}\right)\right]r\,\mathrm{d}r = \frac{\pi p^2 r_{int}^4}{2\mu}\left(\frac{1}{r_{int}^2} - \frac{1}{r_{ext}^2}\right) > 0. \tag{20}$$

Note that (20) only accounts for the elastic energy stored in the cylinder and not in the elastic bed. Numerically, the TDE is calculated as

$$\text{TDE}_{\text{num}} = \frac{1}{2}\mathbf{U}^\top\mathbf{K}\,\mathbf{U}, \tag{21}$$

where $\mathbf{U}$ corresponds to the displacement vector of the active nodes in the background mesh, and $\mathbf{K}$ is the stiffness matrix associated with the active elements in the background mesh.

3. Results

3.1. Covergence Analysis

To show the accuracy of the methodology, problem (1) is solved using the proposed Immersed Boundary Robin-based (IBR) model, on the idealized geometry domain of Figure 6B, for an internal pressure of $p = 10^{-2}$ MPa. Consequently, the analytical solution of the problem is calculated for the infinite solid with a cylindrical cavity at r_{int} (Figure 6A). Recall that, to make equivalent both the analytical results with the Robin-based approach, the *ballast* coefficient α of Equation(19) is used. The different model parameters used for this study are reported in Table 2.

Table 2. Model parameters: Young's modulus E, Poisson ratio ν and the *ballast* coefficient α.

E (MPa)	ν	α (MPa/mm)	r_{int} (mm)	r_{ext} (mm)
7.3313×10^{-1}	0.475	-1.2426×10^{-1}	1.5	4

Figure 7A represents the relative error of the displacements at an arbitrary point with radius $r \in \{r_{int}, r_{ext}\}$ (compared to the analytical solution obtained using Equation (12)) of both the proposed Immersed Boundary Robin-based approach (in black) and the classical FE approach in red (which is calculated for completeness in the analysis).

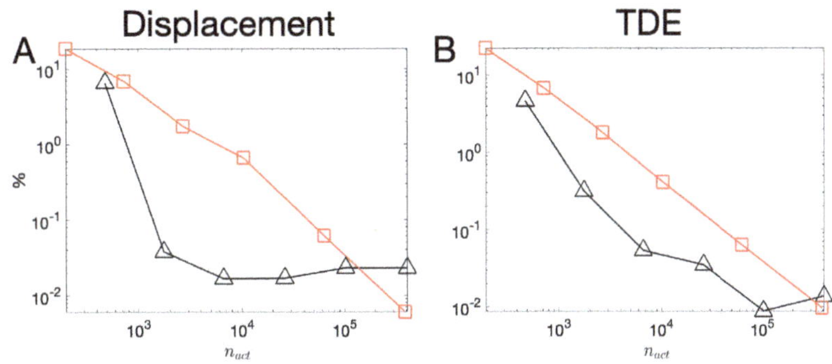

Figure 7. Relative error of (**A**) displacements magnitude and (**B**) TDE obtained for the IBR (black) and classical FE (red) formulations.

In addition, recalling Equations (20) and (21), the relative error (with respect to the analytical value) of the total deformation energy (TDE) is also calculated for both the IBR and the classical FE models and represented in Figure 7B in black and red, respectively. It is worth noting the convergence behavior by increasing the number of active nodes, which means refining the conformal mesh for the classical FE approach and the unfitted background mesh for the IBR solution; the solution improves faster for a smaller number of active nodes using our IBR methodology until a stabilization plateau, where the relative error of TDE stagnates at a value close to 10^{-4}. This is likely due to truncation errors affecting quantities computed in the unfitted procedure, mainly integrals in elements divided by boundaries. This level of accuracy is perfectly acceptable in this type of model.

3.2. Elastic Bed Coeficient α: Sensitivity Analysis

As mentioned in Section 2.1, the elastic bed coefficient α represents an elastic bed boundary condition that simulates the interaction of the body with its surroundings. Therefore, it is also possible to tune α so that it only avoids rigid body motions (and not influencing nodal displacements and stresses). For this study, Figure 8A shows a realistic arterial section where the proposed α analysis is applied under an internal pressure $p = 2.6267 \times 10^{-2}$ MPa. Homogeneous linear elastic material properties were used, being the material parameters reported in Table 3.

Table 3. Material parameters of the piece-wise homogeneous domain $\Omega = \bigcup_{k=1}^{4} \Omega_k$ (i.e., normal vessel wall, loose matrix, calcification, and lipid core), E (Young's modulus) and ν (Poisson ratio).

Subdomain	Material	E [MPa]	ν	Ref.
Ω_1	Normal vessel-wall	0.73	0.475	[22]
Ω_2	Loose matrix	0.431	0.475	[22]
Ω_3	Calcification	1.5×10^4	0.3	[23]
Ω_4	Lipid core	1.8841×10^{-1}	0.475	[22]

According to this, the graph in Figure 8B shows that by decreasing the elastic bed coefficient α, the average displacements at the external boundary ($\overline{\mathbf{u}}_{|\Gamma_R}$) becomes unaffected by the surroundings while eliminating rigid body motion. That is to say, the left part of the graph corresponds to the limit value of a floating object with no surrounding stiffness: the small values of α do only suppress rigid-body modes. The limit case for large values of α has zero displacements in the external boundary, as it is reflected in the plot.

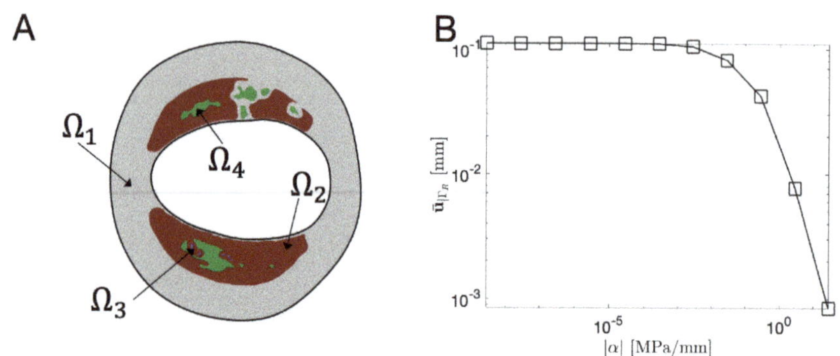

Figure 8. (**A**) Realistic arterial (coronary) section domain $\Omega = \bigcup_{k=1}^{4} \Omega_k$. (**B**) Elastic bed coefficient against average displacement on Γ_R for the coronary section under an internal pressure $p = 2.6267 \times 10^{-2}$ MPa.

3.3. Characteristic Length h of the Background Mesh $\mathcal{T}_h(\Omega^B)$: Sensitivity Analysis

A relevant parameter in the IBR methodology is the characteristic length h of the background mesh. It is strictly related to the number of active nodes (n_{act}) by inverse proportionality, meaning the smaller the h value, the greater the n_{act} value. It is possible to verify how the IBR solution, in terms of displacements and TDE, converges to those obtained with the classical FE method using a very fine mesh. For this, six background meshes $\mathcal{T}_{h(i)}(\Omega^B)$ $i = 1, \ldots, 6$ are used with a decreasing value of h as i increases.

Figure 9 shows the results of such analysis for the section depicted in Figure 8. Figure 9A reports the error for both displacements and TDE associated with the six background meshes with respect to the conformal mesh solution, where it is shown the clear error decreases by increasing the number of active nodes. As an example, Figure 9B shows a plot of the local displacements difference error associated with the finest background mesh developed for this analysis. Table 4 shows the maximum differences in TDE and the displacement magnitude for the different background meshes considered in the analysis.

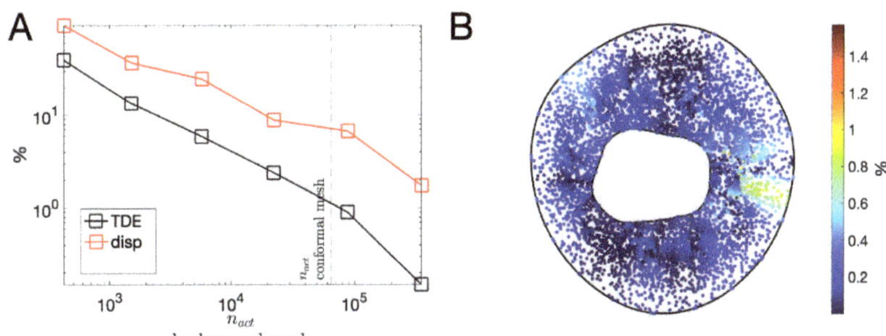

Figure 9. (**A**) Convergence profile for the maximum difference for TDE and displacement magnitude. (**B**) A plot of the local difference in displacement magnitude for a subset of the nodes in the conformal mesh. In particular, B refers to mesh 6 from Table 4.

Table 4. Results of the sensitivity analysis about the characteristic length h for TDE and displacement magnitude. From mesh 5, there are more active nodes (n_{act}) in the background than nodes in the conformal mesh (n_{act} conformal mesh $= 64{,}552$).

Mesh #	n_{act}	Diff. for TDE	max Diff. for Displacement Magnitude
1	414	39.3%	93.6%
2	1507	13.4%	36.7%
3	5734	5.9%	24.6%
4	22,270	2.4%	8.8%
5	87,776	0.9%	6.7%
6	348,550	0.15%	1.75%

3.4. Realistic Immersed Boundary Robin-Based approach

Figure 10 shows a comparison between the IBR methods and the classical FE approach for a realistic coronary section, described in Figure 10A, subjected to an internal pressure of $p = 10^{-2}$ MPa. The model parameters used for this study are shown in Table 5, being the material properties reported in Table 3.

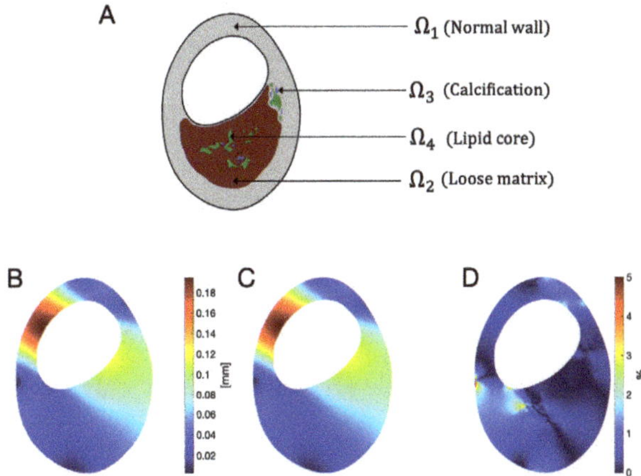

Figure 10. Realistic arterial (coronary) section (**A**) domain $\Omega = \bigcup_{k=1}^{4} \Omega_k$ displacements distribution (**B**) using IBR and (**C**) classical framework. Panel (**D**) depicts relative local error for displacements.

Table 5. Background mesh $\mathcal{T}_h(\Omega^B)$ and conformal mesh $\mathcal{T}_h(\Omega)$ parameters.

	n_{act}	**Active Elements**	α **(MPa/mm)**
$\mathcal{T}_h(\Omega^B)$	42,397	83,378	-1.7736×10^{-4}
$\mathcal{T}_h(\Omega)$	47,505	94,210	

Figure 10B shows the displacement field obtained with the proposed Immersed Boundary method (obtaining a total deformation energy TDE = 6.8226×10^{-3}), while Figure 10C corresponds to the solution obtained with the classical FE method (TDE = 6.8248×10^{-3}). Differences in the total deformation energy are found to be less than 0.05%, with a maximum difference in the displacement magnitude of less than 5% (Figure 10D), being an average error of less than 0.5% in the section. Figure 11A,B show the distribution of the Von Misses stress obtained for both methodologies, IBR method, and the classical FE method, respectively, showing a highly similar stress distribution.

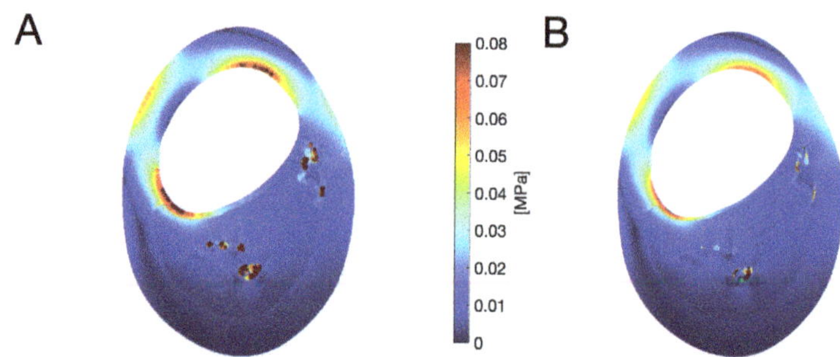

Figure 11. Realistic arterial (coronary) section domain $\Omega = \bigcup_{k=1}^{4} \Omega_k$ Von Misses stress distribution using (**A**) IBR methods and (**B**) classical framework.

4. Discussion

Computational engineering is considered a potentially powerful tool for biomedical sciences. Nevertheless, one of the most critical drawbacks is the time required to develop credible and accurate numerical solutions where fast patient-specific decision making is necessary, i.e., in the vascular biomechanics field. Advancing in this direction, the article presents a novel formulation that combines hierarchical level sets (from a 2D arterial segmentation) with an Immerse Boundary-Robin-based (IBR) formulation to obtain stress and strain distributions in arterial sections under physiological conditions of blood pressure.

The hierarchical level sets allow us to describe the arterial geometries (segmentations), including internal materials distributions (healthy tissue, lipid core, calcified core, among others) and their properties in a highly standardized format. The use of level sets also allows for using a single background mesh to simulate different patient-specific arterial segmentations, all with the same number of degrees of freedom, thus avoiding the preprocessing stages for developing a different conformal finite element mesh per geometry. Furthermore, having the same number of degrees of freedom allows developing so-called "a posteri" reduced-order models (ROMs) or dimensionality reduction methodologies toward fast (even real-time) simulations. It is worth mentioning that despite vascular tissues exhibiting nonlinear behavior, here, a simple linear elastic model is adopted to demonstrate the performance of the proposed approach in the benchmarks selected. Note that the proposed formulation is straightforwardly generalizable to any type of material model. As expected, the accuracy of the IBR depends on the size of the background mesh relative to the minimum size of the heterogeneity. One way to overcome this limitation is to consider an adaptive background mesh with a higher element density near the heterogeneities.

Using the elastic bed boundary conditions (instead of classical Dirichlet) allows us to effectively remove the rigid body motion without altering the natural deformation of the arterial section due to the internal pressure. However, this boundary condition may also be used to account for the effect of the surrounding tissue on the artery in case the information is known. For instance, a uniform ballast coefficient, α, may represent an artery entirely surrounded by tissue as could be the case of the middle cerebral artery or a penetrating myocardial coronary artery. It could also be used to model a partially surrounded artery by specifying a non-uniform ballast coefficient along the external contour of the section. The results obtained with the proposed methodology in terms of displacement and stress fields were very similar compared to those imposing isostatic Dirichlet-type boundary conditions, which are widely accepted in the scientific community. To the best knowledge of the authors, this is the first attempt to propose Immersed Boundary methods with elastic bed boundary conditions for this type of simulation, showing the strong potential of the methodology for biomechanical applications.

Author Contributions: Conceptualization, P.D., J.F.R.M. and A.G.-G.; methodology, P.D., J.F.R.M. and A.G-.G.; software, S.G. and M.S.; validation, S.G. and M.S.; formal analysis, P.D.; investigation, J.F.R.M. and A.G.-G.; resources, P.D. and A.G.-G.; data curation, M.S. and S.G.; writing—original draft preparation, S.G.; writing—review and editing, M.S., P.D., J.F.R.M. and A.G.-G.; visualization, S.G. and J.F.R.M.; supervision, P.D., J.F.R.M. and A.G.-G.; project administration, P.D. and A.G.-G.; funding acquisition, P.D. All authors have read and agreed to the published version of the manuscript.

Funding: The authors acknowledge the financial support from the Ministerio de Ciencia e Innovación (MCIN/AEI/10.13039/501100011033) through the grants PID2020-113463RB-C33 and CEX2018-000797-S and the Italian Ministry of Education, University and Research (Grant number 1613 FISR2019_03221, CECOMES).

Acknowledgments: The authors acknowledge Zhongzhao Teng of the University of Cambridge for providing the realistic 2D geometries used for this numerical development.

Conflicts of Interest: The authors declare no conflict of interest.

Abbreviations

The following abbreviations are used in this manuscript:

BC	Boundary conditions
FE	Finite element
IB	Immersed Boundary
TDE	Total Deformation Energy

References

1. World Health Statistics 2020: Monitoring Health for the SDGs, Sustainable Development Goals. Available online: https://www.who.int/publications/i/item/9789240005105 (accessed on 20 September 2022).
2. The Top 10 Causes of Death. Available online: https://www.who.int/news-room/fact-sheets/detail/the-top-10-causes-of-death (accessed on 20 September 2022).
3. Thygesen, K.; Alpert, J.S.; White, H.D. Universal Definition of Myocardial Infarction. *Circulation* **2007**, *116*, 2634–2653. [CrossRef] [PubMed]
4. Goyal, A.; Zeltser, R. Unstable Angina. Available online: https://www.ncbi.nlm.nih.gov/books/NBK442000/ (accessed on 27 September 2022).
5. Holzapfel, G.A.; Mulvihill, J.J.; Cunnane, E.M.; Walsh, M.T. Computational approaches for analyzing the mechanics of atherosclerotic plaques: A review. *J. Biomech.* **2014**, *47*, 859–869. [CrossRef]
6. Leiner, T.; Gerretsen, S.; Botnar, R.; Lutgens, E.; Cappendijk, V.; Kooi, E.; van Engelshoven, J. Magnetic resonance imaging of atherosclerosis. *Eur. Radiol.* **2005**, *15*, 1087–1099. [CrossRef]
7. Corti, R.; Fuster, V. Imaging of atherosclerosis: Magnetic resonance imaging. *Eur. Heart J.* **2011**, *32*, 1709–1719. [CrossRef] [PubMed]
8. Ostrom, M.P.; Gopal, A.; Ahmadi, N.; Nasir, K.; Yang, E.; Kakadiaris, I.; Flores, F.; Mao, S.S.; Budoff, M.J. Mortality Incidence and the Severity of Coronary Atherosclerosis Assessed by Computed Tomography Angiography. *J. Am. Coll. Cardiol.* **2008**, *52*, 1335–1343. [CrossRef]
9. Achenbach, S.; Raggi, P. Imaging of coronary atherosclerosis by computed tomography. *Eur. Heart J.* **2010**, *31*, 1442–1448. [CrossRef]

10. Yabushita, H.; Bouma, B.E.; Houser, S.L.; Aretz, H.T.; Jang, I.K.; Schlendorf, K.H.; Kauffman, C.R.; Shishkov, M.; Kang, D.H.; Halpern, E.F.; et al. Characterization of Human Atherosclerosis by Optical Coherence Tomography. *Circulation* **2002**, *106*, 1640–1645. [CrossRef] [PubMed]

11. Araki, M.; Park, S.J.; Dauerman, H.L.; Uemura, S.; Kim, J.S.; Mario, C.D.; Johnson, T.W.; Guagliumi, G.; Kastrati, A.; Joner, M.; et al. Optical coherence tomography in coronary atherosclerosis assessment and intervention. *Nat. Rev. Cardiol.* **2022**, *19*, 684–703. [CrossRef]

12. Böse, D.; von Birgelen, C.; Erbel, R. Intravascular Ultrasound for the Evaluation of Therapies Targeting Coronary Atherosclerosis. *J. Am. Coll. Cardiol.* **2007**, *49*, 925–932. [CrossRef]

13. Garcia-Garcia, H.M.; Costa, M.A.; Serruys, P.W. Imaging of coronary atherosclerosis: Intravascular ultrasound. *Eur. Heart J.* **2010**, *31*, 2456–2469. [CrossRef]

14. Akyildiz, A.C.; Speelman, L.; Nieuwstadt, H.A.; van Brummelen, H.; Virmani, R.; van der Lugt, A.; van der Steen, A.F.; Wentzel, J.J.; Gijsen, F.J. The effects of plaque morphology and material properties on peak cap stress in human coronary arteries. *Comput. Methods Biomech. Biomed. Eng.* **2016**, *19*, 771–779. [CrossRef]

15. Kok, A.M.; van der Lugt, A.; Verhagen, H.J.; van der Steen, A.F.; Wentzel, J.J.; Gijsen, F.J. Model-based cap thickness and peak cap stress prediction for carotid MRI. *J. Biomech.* **2017**, *60*, 175–180. [CrossRef]

16. Li, Z.Y.; Howarth, S.; Trivedi, R.A.; U-King-Im, J.M.; Graves, M.J.; Brown, A.; Wang, L.; Gillard, J.H. Stress analysis of carotid plaque rupture based on in vivo high resolution MRI. *J. Biomech.* **2006**, *39*, 2611–2622. [CrossRef]

17. Sadat, U.; Teng, Z.; Young, V.; Graves, M.; Gaunt, M.; Gillard, J. High-resolution Magnetic Resonance Imaging-based Biomechanical Stress Analysis of Carotid Atheroma: A Comparison of Single Transient Ischaemic Attack, Recurrent Transient Ischaemic Attacks, Non-disabling Stroke and Asymptomatic Patient Groups. *Eur. J. Vasc. Endovasc. Surg.* **2011**, *41*, 83–90. [CrossRef]

18. Gijsen, F.J.; Nieuwstadt, H.A.; Wentzel, J.J.; Verhagen, H.J.; van der Lugt, A.; van der Steen, A.F. Carotid Plaque Morphological Classification Compared With Biomechanical Cap Stress. *Stroke* **2015**, *46*, 2124–2128. [CrossRef]

19. Nieuwstadt, H.A.; Kassar, Z.A.M.; van der Lugt, A.; Breeuwer, M.; van der Steen, A.F.W.; Wentzel, J.J.; Gijsen, F.J.H. A Computer-Simulation Study on the Effects of MRI Voxel Dimensions on Carotid Plaque Lipid-Core and Fibrous Cap Segmentation and Stress Modeling. *PLoS ONE* **2015**, *10*, e0123031. [CrossRef]

20. Akyildiz, A.C.; Speelman, L.; van Brummelen, H.; Gutiérrez, M.A.; Virmani, R.; van der Lugt, A.; van der Steen, A.F.; Wentzel, J.J.; Gijsen, F.J. Effects of intima stiffness and plaque morphology on peak cap stress. *BioMed. Eng. Online* **2011**, *10*, 25. [CrossRef]

21. Sadat, U.; Teng, Z.; Gillard, J.H. Biomechanical structural stresses of atherosclerotic plaques. *Expert Rev. Cardiovasc. Ther.* **2010**, *8*, 1469–1481. [CrossRef]

22. Teng, Z.; Zhang, Y.; Huang, Y.; Feng, J.; Yuan, J.; Lu, Q.; Sutcliffe, M.P.; Brown, A.J.; Jing, Z.; Gillard, J.H. Material properties of components in human carotid atherosclerotic plaques: A uniaxial extension study. *Acta Biomater.* **2014**, *10*, 5055–5063. [CrossRef]

23. Ebenstein, D.M.; Coughlin, D.; Chapman, J.; Li, C.; Pruitt, L.A. Nanomechanical properties of calcification, fibrous tissue, and hematoma from atherosclerotic plaques. *J. Biomed. Mater. Res. Part A* **2009**, *91A*, 1028–1037. [CrossRef]

24. Milzi, A.; Lemma, E.D.; Dettori, R.; Burgmaier, K.; Marx, N.; Reith, S.; Burgmaier, M. Coronary plaque composition influences biomechanical stress and predicts plaque rupture in a morpho-mechanic OCT analysis. *eLife* **2021**, *10*, e64020. [CrossRef] [PubMed]

25. Neumann, E.E.; Young, M.; Erdemir, A. A pragmatic approach to understand peripheral artery lumen surface stiffness due to plaque heterogeneity. *Comput. Methods Biomech. Biomed. Eng.* **2019**, *22*, 396–408. [CrossRef] [PubMed]

26. Noble, C.; Carlson, K.D.; Neumann, E.; Lewis, B.; Dragomir-Daescu, D.; Lerman, A.; Erdemir, A.; Young, M.D. Finite element analysis in clinical patients with atherosclerosis. *J. Mech. Behav. Biomed. Mater.* **2022**, *125*, 104927. [CrossRef]

27. Costopoulos, C.; Huang, Y.; Brown, A.J.; Calvert, P.A.; Hoole, S.P.; West, N.E.; Gillard, J.H.; Teng, Z.; Bennett, M.R. Plaque Rupture in Coronary Atherosclerosis Is Associated With Increased Plaque Structural Stress. *JACC* **2017**, *10*, 1472–1483. [CrossRef]

28. Teng, Z.; Brown, A.J.; Calvert, P.A.; Parker, R.A.; Obaid, D.R.; Huang, Y.; Hoole, S.P.; West, N.E.; Gillard, J.H.; Bennett, M.R. Coronary Plaque Structural Stress Is Associated with Plaque Composition and Subtype and Higher in Acute Coronary Syndrome. *Circulation* **2014**, *7*, 461–470. [CrossRef]

29. Madani, A.; Bakhaty, A.; Kim, J.; Mubarak, Y.; Mofrad, M.R.K. Bridging Finite Element and Machine Learning Modeling: Stress Prediction of Arterial Walls in Atherosclerosis. *J. Biomech. Eng.* **2019**, *141*, 84502. [CrossRef]

30. Akyildiz, A.C.; Hansen, H.H.G.; Nieuwstadt, H.A.; Speelman, L.; Korte, C.L.D.; van der Steen, A.F.W.; Gijsen, F.J.H. A Framework for Local Mechanical Characterization of Atherosclerotic Plaques: Combination of Ultrasound Displacement Imaging and Inverse Finite Element Analysis. *Ann. Biomed. Eng.* **2016**, *44*, 968–979. [CrossRef]

31. Akyildiz, A.C.; Speelman, L.; Velzen, B.V.; Stevens, R.R.; Steen, A.F.V.D.; Huberts, W.; Gijsen, F.J. Intima heterogeneity in stress assessment of atherosclerotic plaques. *Interface Focus* **2018**, *8*, 20170008. [CrossRef] [PubMed]

32. Ohayon, J.; Gharib, A.M.; Garcia, A.; Heroux, J.; Yazdani, S.K.; Malvè, M.; Tracqui, P.; Martinez, M.A.; Doblare, M.; Finet, G.; et al. Is arterial wall-strain stiffening an additional process responsible for atherosclerosis in coronary bifurcations: An in vivo study based on dynamic CT and MRI. *Am. J. Physiol. Heart Circ. Physiol.* **2011**, *301*, H1097–H1106. [CrossRef] [PubMed]

33. Morlacchi, S.; Pennati, G.; Petrini, L.; Dubini, G.; Migliavacca, F. Influence of plaque calcifications on coronary stent fracture: A numerical fatigue life analysis including cardiac wall movement. *J. Biomech.* **2014**, *47*, 899–907. [CrossRef]

34. Peskin, C.S. The immersed boundary method. *Acta Numer.* **2002**, *11*, 479–517. [CrossRef]

35. Aleksandrov, M.; Zlatanova, S.; Heslop, D.J. Voxelisation Algorithms and Data Structures: A Review. *Sensors* **2021**, *21*, 8241. [CrossRef] [PubMed]

36. Tsai, A.; Yezzi, A.; Wells, W.; Tempany, C.; Tucker, D.; Fan, A.; Grimson, W.; Willsky, A. A shape-based approach to the segmentation of medical imagery using level sets. *IEEE Trans. Med. Imaging* **2003**, *22*, 137–154. [CrossRef]

37. Ramesh, K.; Kumar, G.; Swapna, K.; Datta, D.; Rajest, S. A Review of Medical Image Segmentation Algorithms. *EAI Endorsed Trans. Pervasive Health Technol.* **2018**, *7*, e6. [CrossRef]

38. Liu, X.; Song, L.; Liu, S.; Zhang, Y. A Review of Deep-Learning-Based Medical Image Segmentation Methods. *Sustainability* **2021**, *13*, 1224. [CrossRef]

39. Taghanaki, S.A.; Abhishek, K.; Cohen, J.P.; Cohen-Adad, J.; Hamarneh, G. Deep semantic segmentation of natural and medical images: A review. *Artif. Intell. Rev.* **2021**, *54*, 137–178. [CrossRef]

40. Quarteroni, A. *Numerical Models for Differential Problems*; Springer: Milan, Italy, 2014. [CrossRef]

41. Gomes, J.; Faugeras, O. Reconciling Distance Functions and Level Sets. *J. Vis. Commun. Image Represent.* **2000**, *11*, 209–223. [CrossRef]

42. Osher, S.; Fedkiw, R. *Signed Distance Functions*; Springer: Berlin/Heidelberg, Germany, 2003; pp. 17–22. [CrossRef]

43. Zlotnik, S.; Díez, P. Hierarchical X-FEM for n-phase flow. *Comput. Methods Appl. Mech. Eng.* **2009**, *198*, 2329–2338. [CrossRef]

44. Bayareh, M.; Mortazavi, S. Equilibrium Position of a Buoyant Drop in Couette and Poiseuille Flows at Finite Reynolds Numbers. *J. Mech.* **2012**, *29*, 53–58. [CrossRef]

45. Silvester, P. Symmetric Quadrature Formulae for Simplexes. *Math. Comput.* **1970**, *24*, 95. [CrossRef]

46. Mal, A.K.; Singh, S.J. *Deformation of Elastic Solids*; Prentice-Hall: Hoboken, NJ, USA, 1990.

 *mathematics*

Article

Multiple Myeloma Cell Simulation Using an Agent-Based Framework Coupled with a Continuous Fluid Model

Pau Urdeitx [1,2,3], Sandra Clara-Trujillo [4,5], Jose Luis Gomez Ribelles [4,5] and Mohamed H. Doweidar [1,2,3,*]

[1] Mechanical Engineering Department, School of Engineering and Architecture (EINA), University of Zaragoza, 50018 Zaragoza, Spain
[2] Aragon Institute of Engineering Research (I3A), University of Zaragoza, 50018 Zaragoza, Spain
[3] Biomedical Research Networking Center in Bioengineering, Biomaterials and Nanomedicine (CIBER-BBN), 50018 Zaragoza, Spain
[4] Centre for Biomaterials and Tissue Engineering (CBIT), Universitat Politecnica de Valencia, 46022 Valencia, Spain
[5] Biomedical Research Networking Center in Bioengineering, Biomaterials and Nanomedicine (CIBER-BBN), 46022 Valencia, Spain
* Correspondence: mohamed@unizar.es

Abstract: Bone marrow mechanical conditions play a key role in multiple myeloma cancer. The complex mechanical and chemical conditions, as well as the interactions with other resident cells, hinder the development of effective treatments. Agent-based computational models, capable of defining the specific conditions for every single cell, can be a useful tool to identify the specific tumor microenvironment. In this sense, we have developed a novel hybrid 3D agent-based model with coupled fluid and particle dynamics to study multiple myeloma cells' growth. The model, which considers cell–cell interactions, cell maturation, and cell proliferation, has been implemented by employing user-defined functions in the commercial software Fluent. To validate and calibrate the model, cell sedimentation velocity and cell proliferation rates have been compared with in vitro results, as well as with another previously in-house developed model. The results show that cell proliferation increased as cell–cell, and cell–extracellular matrix interactions increased, as a result of the reduction n maturation time. Cells in contact form cell aggregates, increasing cell–cell interactions and thus cell proliferation. Saturation in cell proliferation was observed when cell aggregates increased in size and the lack of space inhibited internal cells' proliferation. Compared with the previous model, a huge reduction in computational costs was obtained, allowing for an increase in the number of simulated cells.

Keywords: in silico; 3D model; multiple myeloma; tumor aggregate; dense discrete particle model

MSC: 92-08

Citation: Urdeitx, P.; Clara-Trujillo, S.; Gomez Ribelles, J.L.; Doweidar, M.H. Multiple Myeloma Cell Simulation Using an Agent-Based Framework Coupled with a Continuous Fluid Model. *Mathematics* 2023, 11, 1824. https://doi.org/10.3390/math11081824

Academic Editor: Fernando Simoes

Received: 27 February 2023
Revised: 5 April 2023
Accepted: 6 April 2023
Published: 12 April 2023

1. Introduction

Multiple myeloma (MM) is a heterogeneous cancer in the bone marrow, in which malignant cells grow from the resident plasma cells. These malignant cells can spread via the circulatory system and peripheral soft tissues [1]. MM is currently the second-most common hematological cancer in the world, being the cause of 15–20% blood cancers, and is incurable in most cases [2,3]. MM's mortality rate is remarkably high, with less than 10% of patients achieving complete remission [1,2,4]. Appropriate diagnostic techniques can be key to establish patient-specific therapies, such as autologous hematopoietic stem cell transplantation—that may improve survival [1]. As this is a heterogeneous tumor, with multiple conditions related to malignant cells, high rates of drug resistance are common [1]. Likewise, the interaction of myeloma cells with the extracellular matrix (ECM) and other resident cells can also increase the risk of metastasis [4–6]. Indeed, the mechanical properties of the ECM have a high impact on both cell motility and the multiple myeloma cell (MMC) proliferation rate [4–6]. Feng et al. (2010) reported an increase in MMC stiffness, associated with the activation of focal adhesion kinases (FAKs) in hydrogel matrices as the collagen concentration increased [6]. In turn, this increased stiffness resulted in an increase in cell motility and tumor growth. A similar phenomenon was also observed when MMCs interacted with different cells in their microenvironment, such as mesenchymal stem cells (MSC), which increased tumor growth [5–7]. As such, the mechanical properties of the cellular environment, the ECM's stiffness and cell–cell interactions, for instance, seemed to play a key role in tumor development, thereby affecting treatment efficacy and metastasis risk [5,8].

A better understanding of this heterogeneous tumor should lead to the development of new and more efficient treatments and better diagnostic techniques. In this regard, microfluidic chips and hydrogel matrices are widely used to study the cell response to different stimuli produced by the culture medium's mechanical properties and mechanical interactions with other cells in the tumor environment [9,10]. However, given the complex nature of the bone marrow microenvironment, adjusting in vitro culture conditions to mimic in vivo conditions is not always straightforward [11,12]. In this sense, computational models can play a key role in conducting preliminary studies of these conditions, thereby allowing lower costs, as well as saving time [13,14]. In this context, there are two broad categories of computational models: 1—Continuous computational models, which are widely used to analyze the cell culture conditions—such as fluid dynamics, nutrients, chemical species' concentrations [15–17]—and cells' distribution in the medium [18–20], but which cannot analyze cell–cell and cell–ECM interactions, are critical for understanding MMC behavior; 2—Single-agent models, which can provide detailed information about each single cell considered—including their response to the cell-specific mechanical, chemical, electrical, and thermal stimuli received from their surrounding microenvironment [21–30]—but these are limited to small cell populations and require a detailed characterization of the cell environment.

Herein, we present a novel computational multiphase model designed for the advanced analysis of cell–cell, and cell–ECM interactions during MM tumor evolution. This model defines cells using the discrete particle model (DPM) in a fluidic environment, in which fluid and particle dynamics are coupled, thereby combining the concepts of the previously mentioned models (continuous and single-agent models) to enhance the performance of single-agent models while maintaining the advantages of continuous models. The cell processes are implemented via user-defined functions (UDF), taking into account fluid dynamics and cell-specific conditions.

This approach offers a valuable tool to support in vitro experiments by providing detailed information about the specific MMC environment and opening novel perspectives on the cell–ECM and cell–cell interactions.

2. Methods

The present computational model was developed to analyze the interactions between MMCs and their ECM. It couples particle dynamics with fluid dynamics, considering cell–cell interactions by analyzing the cells using the discrete element method (DPM-DEM), implemented in Fluent (Ansys [31]) (ANS). The results were compared with those of a model developed previously in-house, using the commercial software Abaqus (ABQ) [32], which was adapted and calibrated with appropriate parameters to simulate MMCs [33,34]. Cell processes in both models were modulated by the specific cell conditions and implemented via UDFs. Both models were used to analyze the progression and development of MM tumor aggregation.

2.1. Multiphase Coupled Fluid-Particle Model (ANS)

In this article, we present a hybrid computational model, which combines the formulation of two classical methods: continuous and discrete methods. Indeed, due to the mathematical incompatibility of both models, each one calculates the corresponding phase in parallel, and the bidirectional interactive coupling of the methods is achieved through the intercommunication of the properties and conditions of each phase. Fluid dynamics are included in the cell through the inclusion of velocity, density, and viscosity values of the fluid in the motion equation of the cell. Besides, the perturbation of the continuous phase, due to the motion of the cells, is included as an external contribution to the momentum balance equation. Therefore, although both models are calculated in parallel, consistency and stability are guaranteed through common variables such as cell and fluid velocities.

In the discrete phase, our objective was to determine the cell velocity, $\mathbf{v}_c$, at every time step, t. In such a case, cell motion was transferred as a moment applied to the particle corresponding to the drag forces applied during the cell–ECM interaction. Other contributions, such as the gravity or collision forces, were also taken into account, with the trajectory of the discrete phase being obtained using the force balance acting on the cell as [31]:

$$m\frac{d\mathbf{v}_c}{dt} = \mathbf{F}_{drag} + \mathbf{F}_r + \mathbf{F}_{grav} + \mathbf{F}_{ij}, \tag{1}$$

where m is the cell mass, and $\mathbf{F}_{drag}$, $\mathbf{F}_r$, and $\mathbf{F}_{grav}$ are the corresponding contributions of the drag force, domain motion, and gravity, respectively. As the considered experiments have been developed in static conditions, in this paper, there is no contribution of $\mathbf{F}_r$ component. Besides, the contribution of the gravity forces is defined by considering differences in the cell and media mass density as:

$$\mathbf{F}_{grav} = m\frac{\mathbf{g}(\rho_c - \rho_f)}{\rho_c}, \tag{2}$$

where $\mathbf{g}$ is the gravity acceleration, ρ_c is the cell mass density, and ρ_f is the fluid mass density.

Other forces acting on the cell, such as cell–cell, and cell–wall collision forces, are referred to as $\mathbf{F}_{ij}$.

The drag force on cells, as calculated using the cell constant spherical shape, is proportional to the difference between the cell and fluid velocity, as follows:

$$\mathbf{F}_{drag} = \frac{1}{2}C_D\rho_f A_c \parallel \mathbf{v}_f - \mathbf{v}_c \parallel (\mathbf{v}_f - \mathbf{v}_c), \tag{3}$$

where C_D is the drag coefficient, A_c is the cell cross-section, and $\mathbf{v}_f$ is the fluid velocity.

In the event of overlap, cell–cell and cell–wall collision forces were taken into account using the discrete element method (DEM) (Figure 1). These collision forces were described by defining the stiffness of the contributing parts (cell–cell or cell–wall) and the overlap distance, δ_{ij}, which was calculated as follows [31]:

$$\delta_{ij} = \parallel \mathbf{x}_i - \mathbf{x}_j \parallel - (r_i + r_j), \tag{4}$$

where $\mathbf{x}_i$ and $\mathbf{x}_j$ are the positions of the participating cells, while r_i and r_j are their radii. The cell–cell contact direction, $\mathbf{e}_{ij}$, can be defined as:

$$\mathbf{e}_{ij} = \frac{(\mathbf{x}_i - \mathbf{x}_j)}{\parallel \mathbf{x}_i - \mathbf{x}_j \parallel}. \tag{5}$$

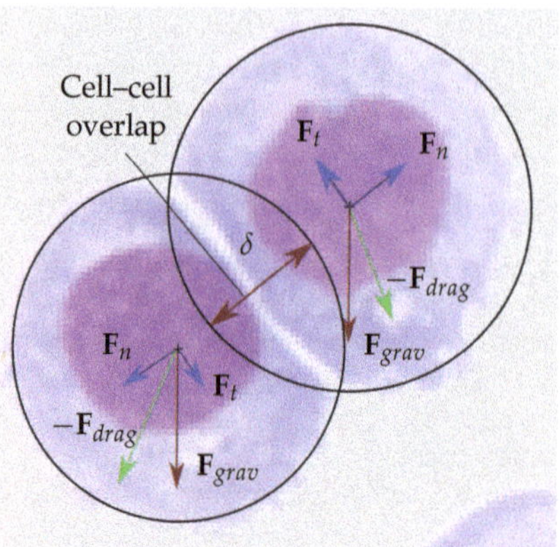

Figure 1. Forces acting on the cell. The cell–cell overlap distance and cell stiffness were considered when computing contact forces. Drag forces were determined using the momentum exchange between the cell and the fluidic ECM.

The Hertzian–Dashpot Collision Law was then used to establish the collision normal force, $\mathbf{F}_{n,}$:

$$\mathbf{F}_n = (K_H \delta_{ij}^{3/2} + \gamma(\mathbf{v}_{ij}\mathbf{e}_{ij}))\mathbf{e}_{ij}, \tag{6}$$

where γ is the damping coefficient, and $\mathbf{v}_{ij}$ represents the relative velocity of the collision, defined as $\mathbf{v}_{ij} = \mathbf{v}_j - \mathbf{v}_i$. Finally, K_H is the stiffness coefficient of the collision, which was calculated using the stiffness of the participating cells, as:

$$K_H = \frac{4}{3} \frac{E_i E_j}{E_j(1 - v_i^2) + E_i(1 - v_j^2)} \sqrt{\frac{r_i r_j}{r_i + r_j}}, \tag{7}$$

where E_i, E_j and v_i, v_j are the stiffness and the Poisson coefficient of the cells involved, respectively.

The tangential force, $\mathbf{F}_t$, in the contact area described by the friction coefficient, μ, is defined as:

$$\mathbf{F}_t = \mathbf{F}_n \mu. \tag{8}$$

The contribution of the collision force, $\mathbf{F}_{ij}$, to cell motion (Equation (1)) was then calculated for each cell using tangential and normal forces as follows:

$$\mathbf{F}_{ij} = \sum_{i!=j}^{n} \left(\mathbf{F}_n^{ij} + \mathbf{F}_t^{ij}\right). \tag{9}$$

For its part, in the continuous phase, the objective is to determine the fluid velocity, $\mathbf{v}_f$, taking into account its active interaction with the cells. In such a case, the momentum exchange between the continuous and discrete phases, $\mathbf{F}_{ex}$, is the change in the particle's momentum as it goes through each control volume at each flow time, Δt. During the momentum balance, this momentum exchange is regarded as a momentum source in the continuous phase and is defined as [31]:

$$\mathbf{F}_{ex} = \sum \left[\frac{18\mu C_D Re}{\rho_c(2r_i)^2 24}(\mathbf{v}_c - \mathbf{v}_f)\right] \dot{m}_c \Delta t, \tag{10}$$

where μ is the fluid viscosity, Re is the relative Reynolds number, ρ_c is the cell density, and $\dot{m}_c$ is the mass flow rate of the discrete phase.

Thus, the general equation for the momentum conservation of the continuous phase results in [31]:

$$\frac{\partial}{\partial t}(\rho_f \mathbf{v}_f) + \nabla(\rho_f \mathbf{v}_f \mathbf{v}_f) = -\nabla p + \nabla \tau + \rho_f \mathbf{g} + \mathbf{F}_{ex}, \tag{11}$$

where ∇ is the divergence operator, p is the static pressure, $\mathbf{g}$ is the gravitational accelera-tion, and τ is the contribution of the stress tensor given by:

$$\tau = \mu\left[(\nabla\mathbf{v}_f + \nabla\mathbf{v}_f^T) - \frac{2}{3}\nabla\mathbf{v}_f I\right], \tag{12}$$

where I is the unit tensor.

Cells evaluate and respond to the specific mechanical conditions to which they are sub-jected during cellular processes, such as proliferation and differentiation [35,36]. Thus, we defined the cell-cycle progression as the maturation index (MI), which is a time-dependent process, as:

$$\mathrm{MI} = \begin{cases} \frac{t_c}{t_{mat}} & t_c < t_{mat}, \\[2mm] 1 & t_c \geq t_{mat}, \end{cases} \tag{13}$$

where t_c is the cell-cycle progression time, and t_{mat} is the time needed for the cell to undergo cellular processes, such as proliferation and differentiation [30]. In addition, the maturation time is dependent on the mechanical conditions of the cell. In this model, due to the consideration of the fluidic environment, the mechanical conditions perceived by the cell were defined in terms of the number of cell–cell (N_c) and cell–wall (N_w) contacts at each time step increment as:

$$t_{mat} = \frac{t_n}{1 + \gamma_w N_w + \gamma_c N_c}, \tag{14}$$

where γ_w and γ_c are the proportional time factors for the cell–wall and cell–cell interactions, respectively, and t_n is the natural maturation time, which is the time needed by the cell to complete the cell cycle without stimulus. As a cell progresses through the cell cycle, it becomes ready to divide into two new cells, thereby implying that the cell expands in size before proliferating. Thus, we assumed that the cell radius increases proportionally to the state of maturation as follows:

$$r_i = r_0(1 + \frac{\mathrm{MI}}{2}), \tag{15}$$

where r_i is the cell radius for each ith cell at each time step, and r_0 is the nominal cell radius obtained from the literature [37].

2.2. Design of In Vitro Experiments and Data Collection

Experimental data from MMC lines were used to validate the model. The MMC line RPMI8226 was cultured in vitro in multi-well plates. RPMI8226 was purchased from the American Type Culture Collection (ATCC, Rockville, MD, USA) and cultured in RPMI1640 medium supplemented with 10% fetal bovine serum (FBS), 2 mM L-glutamine and 100 µg/mL penicillin, and 100 µg/mL streptomycin. Cells were cultured at 37 °C un-der a 5% CO_2 atmosphere in a Galaxy S incubator (Eppendorf New Brunswick, Hamburg, Germany). A total of 100,000 cells were seeded in 500 µL of medium in p24 multi-well plates. Every 24 h, 50% of the volume of culture medium was renewed. After 24, 72, and 96 h of culture, cell proliferation was determined using the colorimetric MTS assay, according to the manufacturer's instructions, pipetting the supernatant onto a 96-well plate and reading at 490 nm (Victor 3 microplate reader, Perkin Elmer, Waltham, MA, USA). The results obtained are presented in [38].

2.3. Model Assumptions

In order to reduce the computational costs of the model, we made some assumptions. Thus, the cell geometry is considered to be spherical with no changes in its morphology [14,22]. We also considered a reduced, but significant, portion of the whole in vitro experiment to reduce the number of cells simulated. Cell properties were obtained, when possible, for the RPMI-8226 cell line to adequately compare results with the in vitro experiments (Table 1). In general, these properties were considered homogeneous for all participating cells, although slight discrepancies are possible due to the non-homogeneous nature of MM tumors [9]. Finally, neither oxygen nor nutrient consumption were considered.

Table 1. Mechanical parameters considered in the model.

Parameter	Description	Value	Refs.
r_0	Nominal cell radius	5 ± 1 μm	[37]
ρ_c	Cell mass density	1077 kg/m^3	[39,40]
ρ_f	ECM mass density	940 kg/m^3	[41]
ν_f	ECM viscosity	0.05 Pa s	[42,43]
E	Cell stiffness	0.4 ± 0.2 kPa	[5,6]
E_{ECM}	ECM stiffness	0.3 ± 0.1 kPa	[42]
μ	Friction coefficient for cell contacts	0.9	[6,44]
t_n	Natural maturation time without stimulus	60 h	[45]
γ_w	Proportional time factor for cell–wall contacts	1.60	–
γ_c	Proportional time factor for cell–cell contacts	0.14	–

3. Results

We proposed two experiments to validate the in silico model. In the first experiment, we compared cell sedimentation velocity to validate the mechanical model as well as the ECM and cell interactions considering a few cells in a large-scale ECM. The duration of the experiment was adapted to the time required for cell sedimentation. In the second experiment, we compared cell proliferation in a reduced slice of the ECM, starting with the same cell concentration as described in Section 2.2, considering four days of cell–cell and cell–ECM interactions. In this experiment, we increased the number of cells and the geometry to study tumor aggregates' growth.

3.1. Cell Sedimentation

The velocity of cell sedimentation was compared to that of in vitro models found in the literature [44]. We compared the evolution of cell sedimentation velocity using a reference portion of the ECM with a base of 30×30 μm and a height of 4.5 mm. The results were compared using the mean values from five cases with different initial cell distributions. In all cases, cells were seeded in the ECM with a random distribution, with an equivalent initial cell concentration in the in vitro experiment used as a reference (5×10^5 cells/mL) [44]. The time step increment was decoupled for cell position tracking (0.1 s) and fluid time step (5 s), and the cell sedimentation velocity was obtained for all cells over a simulation time of 15 min.

The mean cell sedimentation velocity for both computational models was around 13 mm/h (Figure 2), which is consistent with the experimental results [44]. The obtained curves with both models have been compared with the reference data, through the value of the mean square error (MSE), obtaining values of 0.0026, and 0.0029 for the ABQ and ANS models, respectively. As the objective of the case was the calibration and validation of the models, it is possible to conclude that both models give a valid approximation. In addition, the difference between them is not significant. Higher velocities were obtained using the ANS model, which considered the inertial effects of cells, as well as the changes in cell diameter, which is related to the cell maturation state. The ABQ model was unable to reproduce the maximum velocity (16 mm/h), and the cell distribution velocity observed was more homogeneous.

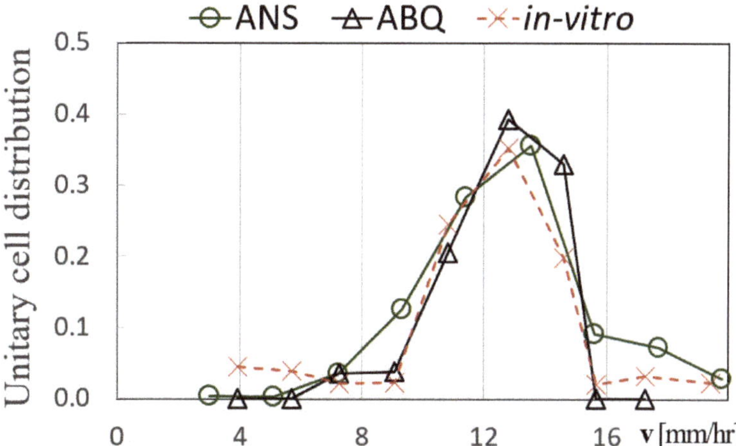

Figure 2. Distribution of mean cell sedimentation velocity per unit cell. The results obtained were compared to the in vitro results (red). The results of the ANS model showed a wider cell sedimentation velocity due to variational inertial forces and a non-homogeneous cell diameter (MSE = 0.0029), while the ABQ model revealed a lower distribution of the cell velocity due to the more homogeneous conditions considered (MSE = 0.0026).

3.2. Cell Proliferation

Cell proliferation was compared to the experimental data detailed in Section 2.2 [38]. Similarly to the first case (cell sedimentation) the results were compared with five different initial cell distributions. Based on the initial cell concentration described in Section 2.2, 20 MMCs were randomly seeded in an ECM of 200 × 200 × 200 μm, with their initial state of maturation being randomly assigned. Due to differences in the timescale of the various events, different time step increments were considered for the fluid domain (0.5 h), cell position tracking (1 s), and cell processes (0.5 h), for a total time of 96 h.

The cell distribution and initial state of maturation were randomly assigned for both computational models (Figure 3 top). After 10–30 h, depending on the initial state of maturation considered in each case, a few cells proliferated. The proliferated cells were randomly scattered around the mother cells. After four days of simulation, several cell distributions were observed in all cases (Figure 3 bottom), although the final number of cells was always comparable (240–260 cells). The results of cell proliferation were normalized with respect to the initial cell number and compared with the in vitro experiments (Figure 4a). In general, cell proliferation was consistent with the experimental data, and the ANS model exhibited greater homogeneity in terms of cell proliferation. An increase in the variability of cell proliferation was observed at later times, with a higher number of cells and higher variability in terms of the specific conditions of each cell (Figure 4b).

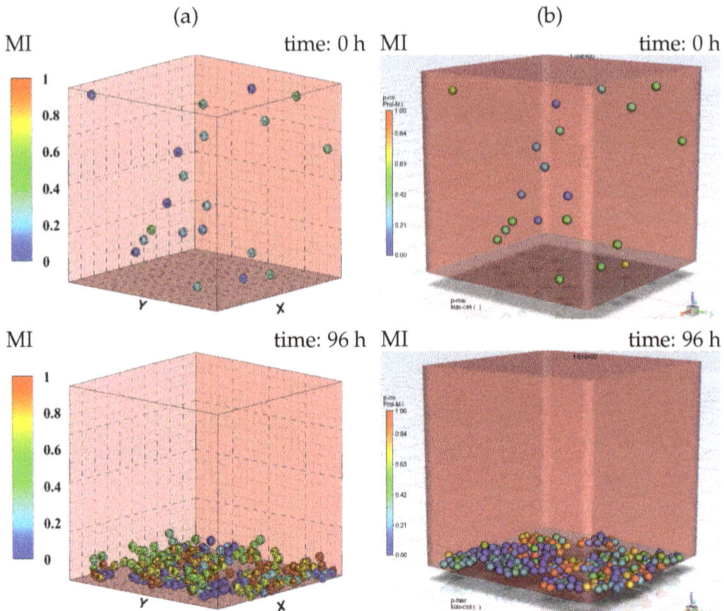

Figure 3. MI during 96 h of simulation using the ABQ (**a**) and ANS (**b**) models. MMCs were seeded in a random initial distribution, and at a random initial state of maturation (**top**). After 96 h of cell-in-culture simulation, MMC aggregates started to form (**bottom**) (see also Supplementary Material Videos S1 and S2).

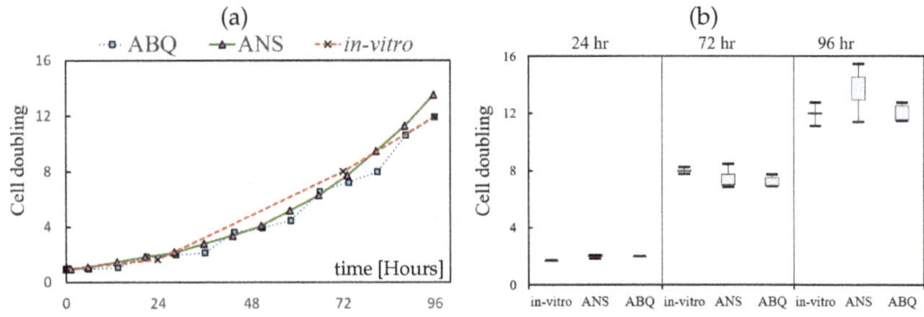

Figure 4. Cell doubling. (**a**) Comparison of the mean values after 96 h in a cells-in-culture simulation using the ANS and ABQ models, with results of the in vitro model (Section 2.2). (**b**) Dispersion of the results due to the random nature of the experiment.

3.3. Tumor Aggregation Growth

After calibration and validation of the initial model, the formation of tumor aggregates, due to the proliferation of individual cells, was studied in a long-term case. Cells were initially distributed into groups to form cell aggregates in an ECM of $400 \times 400 \times 50$ μm, then were simulated for 360 h.

The number of cells in each cluster increased due to cell proliferation and, after 30 h, all clusters were considered to be tumor aggregates (minimum of 30 cells). Given the increase in the number of cell–cell contacts ($\sim$6.1% per hour), the internal cells in these aggregates showed the highest increase in maturation rate ($\sim$30.3%). However, once the cell was completely surrounded, cell proliferation was inhibited due to the lack of space (Figure 5). After 60 h, tumor aggregates started to merge with each other, and by the end of the simulation, all cells had joined together to form a single cluster. The cell proliferation rate was higher in the first 48 h, when exponential cell growth was observed, and then progressively reduced as more cells became completely surrounded (Figure 5d).

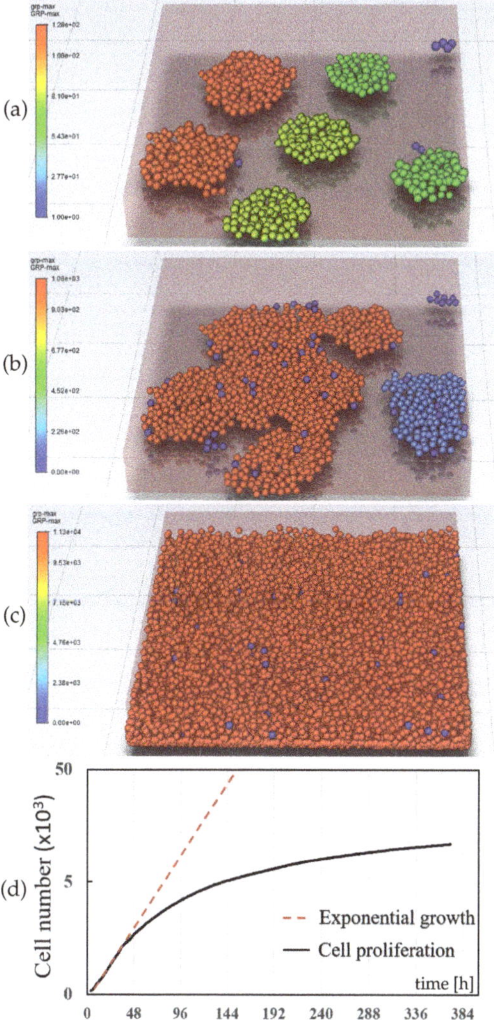

Figure 5. Tumor aggregate growth. (**a**) Initially, cells were randomly distributed into groups of 20–30 cells. (**b**) After 48 h, cell groups merged, and the inner cells of the aggregates lost their ability to proliferate. (**c**) Surface cells continued to proliferate up until 360 h of simulation. (**d**) Cell growth for 360 h of simulation (log scale) (see also Supplementary Material Video S3).

4. Discussion

The results obtained for the two models studied are qualitatively consistent with the in vitro results found in the literature [9,10,38,44,46,47]. In general, when compared to the results for the ABQ model, the ANS model appeared to be better. Thus, in the first experiment, we observed a wider range of cell sedimentation velocities for the ANS model when compared to the ABQ model. This disparity was caused by two major factors: firstly, the ANS model considers inertial contributions in the force balance, although the effect may be very small due to the model's scale; secondly, in the ANS model, the cell diameter was considered to be proportional to the state of maturation, which results in an increase in both the cell weight and cross-section and, consequently, in the drag coefficient, which, in turn, may cause higher variability in the distribution of cell velocities, as well as higher maximum values. As such, the ANS model appears to be more realistic with regard to the non-homogeneity of cellular properties.

The ABQ model was able to reproduce the mechanical environment more precisely by considering the internal cell deformations, and cell-active forces that lead to cell–cell adhesion (Figure 3a) [48,49]. Thus, cells in the ABQ model tended to stay attached, while cells in the ANS model tended to form layers at the bottom. The results of cell proliferation for both models were consistent with the in vitro results (Figure 4). The presence of newly proliferated cells and the formation of new cell–cell contacts accelerated cell maturation. Due to the reduced number of MMCs simulated, we observed abrupt initial stages of proliferation, which became more homogeneous as the number of cells increased and the cells responded to their specific conditions. This increase in homogeneity was higher for the ANS model, with a continuous progression in cell proliferation after 48 h, ($\sim$9.9% per hour). In the ABQ model, a complete smoothing of the curve was not achieved (Figure 4), due to the limited ability of the model to reproduce cell mechanical conditions in fluidic environments. As the main factor that modulates maturation in the ABQ model was related to the ECM stiffness and internal cell deformations, the results did not fit well in fluidic environment simulations. The ANS model appears to be more suitable for defining the cellular microenvironment of a liquid medium, as it can capture a wide range of conditions for each individual cell.

The growth of MMCs, which form tumor aggregates, increased as cell–cell and cell–ECM interactions increased. However, in vitro, once the cell aggregate reached a certain size, the increase in proliferation rate was reversed, thus indicating that cell proliferation can be inhibited by a variety of factors. Likewise, in the numerical models, the lack of available space to generate new cells caused a saturation effect, thus preventing internal cells from proliferating further. This lack of space was exacerbated in in vitro and in vivo experiments by the cells' difficulty in accessing necessary resources, such as O_2, nutrients, and growth factors. At this point, the growth of the aggregate requires the development of new blood vessels to allow the internal cells to be supplied with the necessary resources.

From a computational point of view, higher computational costs were observed with the ABQ model, which was discretized using elements with a size proportional to that of the cell (3 μm), since the definition of the cells was based on the nodes of the mesh. As a reference, for the second experiment, a mesh of 343,000 trilinear hexahedral elements was generated. The time cost for the first and second experiments was approximately 30 and 60 h, respectively, in a computing cluster with four cores and 16 GB of RAM. The ANS model, in turn, can be discretized with a much larger mesh size (10 μm), and use of the decoupled cell-tracking time allowed us to significantly reduce computational costs. In this case, the required time cost was approximately 5 h on a personal computer with an i5-650 processor and 16 GB of random-access memory (RAM). We also observed a significant limitation in the maximum number of cells for the ABQ model, which increased the memory requirement exponentially. Thus, it was not possible to simulate the third experiment in the ABQ model due to the computational costs.

As a result, there are several advantages to using the ANS model rather than the ABQ model, including the ability to define, more accurately, the conditions of the experiments conducted. In this sense, when mechanical conditions are the most important factor in cell behavior and a precise description of cellular forces is required, the ABQ model appears to be more appropriate, as it provides a more comprehensive control of the mechanical conditions of the cell microenvironment. However, when these conditions were less important (for example, in fluidic environments) and a large number of cells (and, thus, a large control volume) was required, the coupled fluid-particle model appears to be the best choice.

5. Conclusions

We have presented a new computational model to study MMC behavior and tumor growth. The model was compared to a previous in-house ABQ computational model [29], as wel as with the experimental results obtained by the authors and from the literature [9,10,38,44,46,47]. The results obtained are qualitatively consistent with the literature. The obtained results show that the initial cell concentration has a clear effect on MMC growth, with cell proliferation increasing as the numbers of cell–cell and cell–ECM interactions increase. As such, we observed faster cell maturation for cells with more cell–cell and cell–ECM interactions (reduction in maturation time of up to 69.7%), while cell proliferation was inhibited when a lack of space was observed.

From a computational viewpoint, the ANS model offers several advantages when compared with the model developed previously. Better results for the MMC sedimentation velocity and tumor growth were obtained due to the more representative definition of the fluidic environment and the consideration of a non-homogeneous cell volume. Compared with the ABQ model, we observed reduced computational costs for adequately simulating cells and the cell microenvironment. The maximum number of cells in the ABQ model was limited and proportional to the number of discretized elements in the ECM. In this case, increasing the ECM implied increasing the number of elements, which dramatically increased computational costs. As the ANS model simulation time step seems to be the main responsible of these computational costs, using a decoupled calculation time step for the cells and fluid reduced them. Thus, the model developed herein has significant advantages compared to the previous model in computation, as it results in a significant reduction (84–91%) in computational costs, which allows a more realistic simulation. Although not considered in this model, additional advantages can be added to the ANS model, such as the consideration of nutrient diffusion and consumption, non-stationary fluid flow and pressure gradients, or the presence of blood vessels in an in vivo analysis.

Supplementary Materials: The following supporting information can be downloaded at: https://www.mdpi.com/article/10.3390/math11081824/s1, Video S1: Video_Fig_3(a); Video S2: Video_Fig_3(b); Video S3: Video_Fig_5.

Author Contributions: Conceptualization, P.U. and M.H.D.; methodology, P.U. and M.H.D.; in vitro analysis, S.C.-T. and J.L.G.R., software, P.U. and M.H.D.; validation, P.U., S.C.-T., J.L.G.R. and M.H.D.; formal analysis, P.U. and M.H.D.; investigation, P.U. and M.H.D.; resources, J.L.G.R. and M.H.D.; data curation, P.U., S.C.-T., J.L.G.R. and M.H.D.; writing—original draft preparation, P.U. and S.C.-T.; writing—review and editing, J.L.G.R. and M.H.D.; visualization, P.U.; supervision, J.L.G.R. and M.H.D.; project administration, J.L.G.R. and M.H.D.; funding acquisition, J.L.G.R. and M.H.D. All authors have read and agreed to the published version of the manuscript.

Funding: This research was funded by the Spanish State Research Agency (AEI/10.13039/501100011033) through the projects PID2019-106099RB-C44 and C41, and the Government of Aragon (DGA-T24 20R). This work was also supported by the Spanish Ministry of Science, Innovation and Universities through Grant N- FPU17/05810 awarded to Sandra Clara-Trujillo.

Data Availability Statement: All the data and results are included within the manuscript.

Acknowledgments: The authors would like to thank the anonymous reviewers for their thorough reading and professional comments that helped improve the manuscript.

Conflicts of Interest: The authors declare no conflict of interest.

References

1. Kumar, S.K.; Rajkumar, V.; Kyle, R.A.; van Duin, M.; Sonneveld, P.; Mateos, M.V.; Gay, F.; Anderson, K.C. Multiple myeloma. *Nat. Rev. Dis. Prim.* **2017**, *3*, 17046. [CrossRef] [PubMed]
2. Perez-Amill, L.; Suñe, G.; Antoñana-Vildosola, A.; Castella, M.; Najjar, A.; Bonet, J.; Fernández-Fuentes, N.; Inogés, S.; López, A.; Bueno, C.; et al. Preclinical development of a humanized chimeric antigen receptor against B cell maturation antigen for multiple myeloma. *Haematologica* **2020**, *106*, 173–184. [CrossRef]

3. Sun, J.; Muz, B.; Alhallak, K.; Markovic, M.; Gurley, S.; Wang, Z.; Guenthner, N.; Wasden, K.; Fiala, M.; King, J.; et al. Targeting CD47 as a Novel Immunotherapy for Multiple Myeloma. *Cancers* **2020**, *12*, 305. [CrossRef] [PubMed]

4. Qiang, Y.W.; Walsh, K.; Yao, L.; Kedei, N.; Blumberg, P.M.; Rubin, J.S.; Shaughnessy, J.; Rudikoff, S. Wnts induce migration and invasion of myeloma plasma cells. *Blood* **2005**, *106*, 1786–1793. [CrossRef] [PubMed]

5. Wu, D.; Guo, X.; Su, J.; Chen, R.; Berenzon, D.; Guthold, M.; Bonin, K.; Zhao, W.; Zhou, X. CD138-negative myeloma cells regulate mechanical properties of bone marrow stromal cells through SDF-1/CXCR4/AKT signaling pathway. *Biochim. Biophys. Acta-Mol. Cell Res.* **2015**, *1853*, 338–347. [CrossRef]

6. Feng, Y.; Ofek, G.; Choi, D.S.; Wen, J.; Hu, J.; Zhao, H.; Zu, Y.; Athanasiou, K.A.; Chang, C.C. Unique biomechanical interactions between myeloma cells and bone marrow stroma cells. *Prog. Biophys. Mol. Biol.* **2010**, *103*, 148–156. [CrossRef]

7. Clara-Trujillo, S.; Ferrer, G.G.; Ribelles, J.L.G. In Vitro Modeling of Non-Solid Tumors: How Far Can Tissue Engineering Go? *Int. J. Mol. Sci.* **2020**, *21*, 5747. [CrossRef]

8. Podar, K.; Tai, Y.T.; Lin, B.K.; Narsimhan, R.P.; Sattler, M.; Kijima, T.; Salgia, R.; Gupta, D.; Chauhan, D.; Anderson, K.C. Vascular Endothelial Growth Factor-induced Migration of Multiple Myeloma Cells Is Associated with $\beta1$ Integrin- and Phosphatidylinositol 3-Kinase-dependent PKCα Activation. *J. Biol. Chem.* **2002**, *277*, 7875–7881. [CrossRef]

9. Zlei, M.; Egert, S.; Wider, D.; Ihorst, G.; Wäsch, R.; Engelhardt, M. Characterization of in vitro growth of multiple myeloma cells. *Exp. Hematol.* **2007**, *35*, 1550–1561. [CrossRef]

10. Jin, J.; Wang, T.; Wang, Y.; Chen, S.; Li, Z.; Li, X.; Zhang, J.; Wang, J. SRC3 expressed in BMSCs promotes growth and migration of multiple myeloma cells by regulating the expression of Cx43. *Int. J. Oncol.* **2017**, *51*, 1694–1704. [CrossRef]

11. Huh, D.; Hamilton, G.A.; Ingber, D.E. From 3D cell culture to organs-on-chips. *Trends Cell Biol.* **2011**, *21*, 745–754. [CrossRef] [PubMed]

12. Bhatia, S.N.; Ingber, D.E. Microfluidic organs-on-chips. *Nat. Biotechnol.* **2014**, *32*, 760–772. [CrossRef]

13. Mogilner, A. Mathematics of cell motility: Have we got its number? *J. Math. Biol.* **2009**, *58*, 105–134. [CrossRef] [PubMed]

14. Rodriguez, M.L.; McGarry, P.J.; Sniadecki, N.J. Review on cell mechanics: Experimental and modeling approaches. *Appl. Mech. Rev.* **2013**, *65*, 060801. . [CrossRef]

15. Ayensa-Jiménez, J.; Pérez-Aliacar, M.; Randelovic, T.; Oliván, S.; Fernández, L.; Sanz-Herrera, J.A.; Ochoa, I.; Doweidar, M.H.; Doblaré, M. Mathematical formulation and parametric analysis of in vitro cell models in microfluidic devices: Application to different stages of glioblastoma evolution. *Sci. Rep.* **2020**, *10*, 21193. [CrossRef]

16. Cioffi, M.; Küffer, J.; Ströbel, S.; Dubini, G.; Martin, I.; Wendt, D. Computational evaluation of oxygen and shear stress distributions in 3D perfusion culture systems: Macro-scale and micro-structured models. *J. Biomech.* **2008**, *41*, 2918–2925. [CrossRef]

17. Soleimani, S.; Shamsi, M.; Ghazani, M.A.; Modarres, H.P.; Valente, K.P.; Saghafian, M.; Ashani, M.M.; Akbari, M.; Sanati-Nezhad, A. Translational models of tumor angiogenesis: A nexus of in silico and in vitro models. *Biotechnol. Adv.* **2018**, *36*, 880–893. [CrossRef]

18. Carlier, A.; Skvortsov, G.A.; Hafezi, F.; Ferraris, E.; Patterson, J.; Koc, B.; Van Oosterwyck, H. Computational model-informed design and bioprinting of cell-patterned constructs for bone tissue engineering. *Biofabrication* **2016**, *8*, 025009. [CrossRef]

19. Kang, K.T.; Park, J.H.; Kim, H.J.; Lee, H.Y.H.M.; Lee, K.I.; Jung, H.H.; Lee, H.Y.H.M.; Jang, J.W. Study of Tissue Differentiation of Mesenchymal Stem Cells by Mechanical Stimuli and an Algorithm for Bone Fracture Healing. *Tissue Eng. Regen. Med.* **2011**, *8*, 359–370.

20. Fouliard, S.; Benhamida, S.; Lenuzza, N.; Xavier, F. Modeling and simulation of cell populations interaction. *Math. Comput. Model.* **2009**, *49*, 2104–2108. [CrossRef]

21. Bissell, M.J.; Rizki, A.; Mian, I.S. Tissue architecture: The ultimate regulator of breast epithelial function. *Curr. Opin. Cell Biol.* **2003**, *15*, 753. [CrossRef] [PubMed]

22. te Boekhorst, V.; Preziosi, L.; Friedl, P. Plasticity of Cell Migration In Vivo and In Silico. *Annu. Rev. Cell Dev. Biol.* **2016**, *32*, 491–526. [CrossRef] [PubMed]

23. Kim, M.C.; Silberberg, Y.R.; Abeyaratne, R.; Kamm, R.D.; Asada, H.H. Computational modeling of three-dimensional ECM-rigidity sensing to guide directed cell migration. *Proc. Natl. Acad. Sci. USA* **2018**, *115*, E390–E399. [CrossRef]

24. Mousavi, S.J.; Doweidar, M.H. Encapsulated piezoelectric nanoparticle–hydrogel smart material to remotely regulate cell differentiation and proliferation: A finite element model. *Comput. Mech.* **2019**, *63*, 471–489. [CrossRef]

25. Farsad, M.; Vernerey, F.J. An XFEM-based numerical strategy to model mechanical interactions between biological cells and a deformable substrate. *Int. J. Numer. Methods Eng.* **2012**, *92*, 238–267. [CrossRef]

26. Katti, D.R.; Katti, K.S. Cancer cell mechanics with altered cytoskeletal behavior and substrate effects: A 3D finite element modeling study. *J. Mech. Behav. Biomed. Mater.* **2017**, *76*, 125–134. [CrossRef] [PubMed]

27. Urdeitx, P.; Farzaneh, S.; Mousavi, S.J.; Doweidar, M.H. Role of oxygen concentration in the osteoblasts behavior: A finite element model. *J. Mech. Med. Biol.* **2020**, *20*, 1950064. [CrossRef]

28. Malekian, N.; Habibi, J.; Zangooei, M.H.; Aghakhani, H. Integrating evolutionary game theory into an agent-based model of ductal carcinoma in situ: Role of gap junctions in cancer progression. *Comput. Methods Programs Biomed.* **2016**, *136*, 107–117. [CrossRef] [PubMed]

29. Urdeitx, P.; Doweidar, M.H. Mechanical stimulation of cell microenvironment for cardiac muscle tissue regeneration: A 3D in-silico model. *Comput. Mech.* **2020**, *66*, 1003–1023. [CrossRef]

30. Urdeitx, P.; Doweidar, M.H. A Computational Model for Cardiomyocytes Mechano-Electric Stimulation to Enhance Cardiac Tissue Regeneration. *Mathematics* **2020**, *8*, 1875. [CrossRef]
31. ANSYS Inc. *Fluent Theory Guide*, 15th ed.; ANSYS, Inc.: Canonsburg, PA, USA, 2013; pp. 724–746.
32. Abaqus (ABQ). *Abaqus 6.14*; Dassault Systemes: Vélizy-Villacoublay, France, 2014.
33. Urdeitx, P.; Doweidar, M.H. Enhanced Piezoelectric Fibered Extracellular Matrix to Promote Cardiomyocyte Maturation and Tissue Formation: A 3D Computational Model. *Biology* **2021**, *10*, 135. [CrossRef] [PubMed]
34. Urdeitx, P.; Mousavi, S.J.; Avril, S.; Doweidar, M.H. Computational modeling of multiple myeloma interactions with resident bone marrow cells. *Comput. Biol. Med.* **2023**, *153*, 106458. [CrossRef] [PubMed]
35. Wu, Q.Q.; Chen, Q. Mechanoregulation of chondrocyte proliferation, maturation, and hypertrophy: Ion-channel dependent transduction of matrix deformation signals. *Exp. Cell Res.* **2000**, *256*, 383–391. [CrossRef] [PubMed]
36. Cheng, G.; Tse, J.; Jain, R.K.; Munn, L.L. Micro-environmental mechanical stress controls tumor spheroid size and morphology by suppressing proliferation and inducing apoptosis in cancer cells. *PLoS ONE* **2009**, *4*, e4632. [CrossRef]
37. Isobe, T.; Ikeda, Y.; Ohta, H. Comparison of sizes and shapes of tumor cells in plasma cell leukemia and plasma cell myeloma. *Blood* **1979**, *53*, 1028–1030. [CrossRef]
38. Clara-Trujillo, S.; Tolosa, L.; Cordón, L.; Sempere, A.; Ferrer, G.G.; Luis, J.; Ribelles, G.; Ribelles, J.L.G. Novel microgel culture system as semi-solid three-dimensional in vitro model for the study of multiple myeloma proliferation and drug resistance. *Biomater. Adv.* **2022**, *135*, 212749. [CrossRef]
39. Bam, R.; Ling, W.; Khan, S.; Pennisi, A.; Venkateshaiah, S.U.; Li, X.; van Rhee, F.; Usmani, S.; Barlogie, B.; Shaughnessy, J.; et al. Role of Bruton's tyrosine kinase in myeloma cell migration and induction of bone disease. *Am. J. Hematol.* **2013**, *88*, 463–471. [CrossRef]
40. Zipursky, A.; Bow, E.; Seshadri, R.; Brown, E. Leukocyte density and volume in normal subjects and in patients with acute lymphoblastic leukemia. *Blood* **1976**, *48*, 361–371. [CrossRef]
41. Aranda-Lara, L.; Torres-García, E.; Oros-Pantoja, R. Biological Tissue Modeling with Agar Gel Phantom for Radiation Dosimetry of 99mTc. *Open J. Radiol.* **2014**, *4*, 44–52. [CrossRef]
42. Thompson, B.R.; Horozov, T.S.; Stoyanov, S.D.; Paunov, V.N. An ultra melt-resistant hydrogel from food grade carbohydrates. *RSC Adv.* **2017**, *7*, 45535–45544. [CrossRef]
43. WATASE, M.; ARAKAWA, K. Rheological Properties of Hydrogels of Agar-agar. *Nippon Kagaku Zassi* **1971**, *92*, 37–42. [CrossRef]
44. Hamburger, A.; Salmon, S.E. Primary Bioassay of Human Myeloma Stem Cells. *J. Clin. Investig.* **1977**, *60*, 846–854. [CrossRef] [PubMed]
45. Cowley, G.S.; Weir, B.A.; Vazquez, F.; Tamayo, P.; Scott, J.A.; Rusin, S.; East-Seletsky, A.; Ali, L.D.; Gerath, W.F.; Pantel, S.E.; et al. Parallel genome-scale loss of function screens in 216 cancer cell lines for the identification of context-specific genetic dependencies. *Sci. Data* **2014**, *1*, 140035. [CrossRef] [PubMed]
46. Lambert, K.E.; Huang, H.; Mythreye, K.; Blobe, G.C. The type III transforming growth factor-β receptor inhibits proliferation, migration, and adhesion in human myeloma cells. *Mol. Biol. Cell* **2011**, *22*, 1463–1472. [CrossRef] [PubMed]
47. Peacock, C.D.; Wang, Q.; Gesell, G.S.; Corcoran-Schwartz, I.M.; Jones, E.; Kim, J.; Devereux, W.L.; Rhodes, J.T.; Huff, C.A.; Beachy, P.A.; et al. Hedgehog signaling maintains a tumor stem cell compartment in multiple myeloma. *Proc. Natl. Acad. Sci. USA* **2007**, *104*, 4048–4053. [CrossRef]
48. Mousavi, S.J.; Doblaré, M.; Doweidar, M.H. Computational modelling of multi-cell migration in a multi-signalling substrate. *Phys. Biol.* **2014**, *11*, 026002. [CrossRef]
49. Mousavi, S.J.; Doweidar, M.H. Numerical modeling of cell differentiation and proliferation in force-induced substrates via encapsulated magnetic nanoparticles. *Comput. Methods Programs Biomed.* **2016**, *130*, 106–117. [CrossRef]

 *mathematics*

Article

Biomechanical Effects of Medializing Calcaneal Osteotomy on Bones and the Tissues Related to Adult-Acquired Flatfoot Deformity: A Computational Study

Javier Bayod [1,*], Ricardo Larrainzar-Garijo [2], Brayan David Solórzano [3] and Christian Cifuentes-De la Portilla [3]

[1] Applied Mechanics and Bioengineering Group (AMB), Aragón Institute of Engineering Research (I3A), Universidad de Zaragoza, 50018 Zaragoza, Spain
[2] Orthopaedics and Trauma Department, Medicine School, Hospital Universitario Infanta Leonor, Universidad Complutense, 28031 Madrid, Spain
[3] Biomedical Engineering Department, Universidad de los Andes, Bogotá 111711, Colombia
* Correspondence: jbayod@unizar.es

Abstract: Medializing calcaneal osteotomy (MCO) is a flatfoot treatment in stages IIa–IIb. It is true that structural correction is well known, but stress changes in foot tissues have not been sufficiently studied to date. Our objective was to evaluate the stress generated by MCO in both hindfoot and forefoot bones and in some soft tissues that support the arch. A finite element foot model was employed, simulating some situations related to flatfoot development. Results show a higher stress concentration around the osteotomy region when MCO is used in patients with plantar fascia weakness. Additionally, the stress increase found in lateral metatarsals would be the explanation for the long-term pain reported by patients.

Keywords: finite elements; flatfoot; stress redistribution; osteotomy; pes planus

MSC: 92-08

Citation: Bayod, J.; Larrainzar-Garijo, R.; Solórzano, B.D.; Cifuentes-De la Portilla, C. Biomechanical Effects of Medializing Calcaneal Osteotomy on Bones and the Tissues Related to Adult-Acquired Flatfoot Deformity: A Computational Study. *Mathematics* 2023, 11, 2243. https://doi.org/10.3390/math11102243

Academic Editors: Fernando Simoes and Mauro Malvè

Received: 22 February 2023
Revised: 19 April 2023
Accepted: 29 April 2023
Published: 10 May 2023

1. Introduction

Adult acquired flatfoot deformity (AAFD) is a pathology that causes a progressive flattening of the foot arch, which has been traditionally related to a tibialis posterior tendon (TPT) dysfunction. However, some clinical studies found that a failure/rupture of the plantar fascia (PF) or the calcaneonavicular ligament (also spring ligament (SL)) could also generate the arch collapse and the forefoot abduction [1–5]. Treatment options depend on the injury stage. In the first stages, AAFD treatments are related to reinforcing the TPT [6]. Nevertheless, sometimes the foot deformation reappears over time, forcing surgeons to use more aggressive techniques, intervening directly over the foot's bone structure. If the foot deformity is still flexible (stages IIa and IIb), the most habitual procedure is medializing calcaneal osteotomy (MCO) [7,8], which allows both the progressive foot arch flattening and the foot pronation caused by the flatfoot deformity to be corrected [4,6,9]. This procedure provokes a supination momentum in the foot to compensate the pronation [1,10]. In this way, the foot's structural correction is achieved by MCO and its results are normally satisfactory. Nevertheless, some clinical studies have shown that this procedure generates long-term side-effects related to stress distribution changes in forefoot and metatarsals [11–13], which could increase the risk of bone fractures, as has been reported with Evans' osteotomy [14].

In a recent study published by our research group using a previous version of our foot model, we showed that MCO can reduce foot pronation on its own [9]. However, changes in the biomechanical stress caused in bones and the main soft tissues that support the arch remained unstudied. Even in the literature, these stress changes have not been sufficiently studied, because of the difficulty of measuring tissue stresses in cadavers.

Some cadaver-based models have been used to study the structural correction of the foot, evaluating changes in both the plantar footprint using force platforms, and foot arch falling using radiographic (Rx) images. For example, Patrick et al. [15] measured the subtalar joint pressure produced by MCO using a cadaveric model suffering with flatfoot. They introduced a pressure sensor in the posterior facet joint, obtaining some, but limited, information about the effects of MCO on hindfoot joint pressures. As can be noted, these kinds of studies require high economic investment in measurement equipment, as well as meticulous control over the tested tissues to guarantee their biomechanical characteristics [16].

An alternative now accepted by clinicians and biomechanical researchers for evaluating the complex biomechanics of the human foot is finite element modelling (FEM) [17–19]. There are many models that study foot biomechanics and the effects produced by some surgical techniques. However, none of them have been used to study the stress effects of MCO on foot tissues. This kind of model specifically evaluates foot structure deformation and plantar pressure measurement [13,19]. Thus, these models greatly simplify the tissue anatomy and do not take into account important aspects such as the biomechanical difference between cortical and trabecular bones (which is very important when tissue stresses are evaluated [20], nor the geometry of some soft tissues such as the plantar fascia, the spring ligament, ligaments, or tendons, which are habitually modelled as bar elements. Thus, previously reported models cannot measure and locate the stresses around the foot anatomy.

The objective of this research was to investigate the biomechanical effects in terms of stress concentrations and displacements that an MCO provokes in both foot bones and the main foot arch stabilizers (TPT, PF and SL), using an enhanced version of the model used in [9]. This analysis was performed by simulating different pathological scenarios related to AAFD development.

2. Materials and Methods

This study was based on the foot model (segmentation and tissue properties) proposed by Cifuentes-De la Portilla et al [3], which has been used for the flatfoot evaluation of some other surgical procedures [21]. However, for this study, the entire model was reconstructed to simulate the MCO procedure, maintaining both tissue characteristics and loading conditions but including both the tibia and fibula bones to better represent the anatomical tendons' trajectories. The model used reconstructed a healthy human unloaded foot, based on CT images (radiographs of 0.6 mm/slide) acquired from the right foot of a 49-year-old man (weight = 75 Kg, height = 1.70 m).

2.1. FE Foot Model and Modifications

Tissue segmentation and 3D reconstruction (bones, PF, TPT, Achilles' tendon, Peroneus Longus tendon (PLT) and Peroneus Brevis tendon (PBT)) were performed using MIMICS V. 10 (Materialize, Leuven, Belgium). The spring ligament (SL) and both plantar ligaments (short plantar ligament and long plantar ligament) were added following atlas images, following the surgeons' guidance due to the difficulty of segmenting these from the CT images. The tibia and fibula were reoriented with tools available in MIMICS from the scan position to correspond to the orientation during the stance phase of gait. The previous finite element model [9] was enhanced by adding the TPT, Achilles' tendon, PLT, and PBT.

To simulate the MCO, calcaneus bone was modified, performing a 45-degrees transversal cut and translating the segment medially by 10-mm (See Figure 1) [15,22]. This modification was performed following the guidance of a specialist in foot surgeries. Elements allowing internal fixation, such as plates, screws, and bone graft, were not simulated because a complete joint fusion was supposed. The complete FE model is shown in Figure 1. allowing internal fixation, such as plates, screws, and bone graft, were not simulated because a complete joint fusion was supposed. The complete FE model is shown in Figure 1.

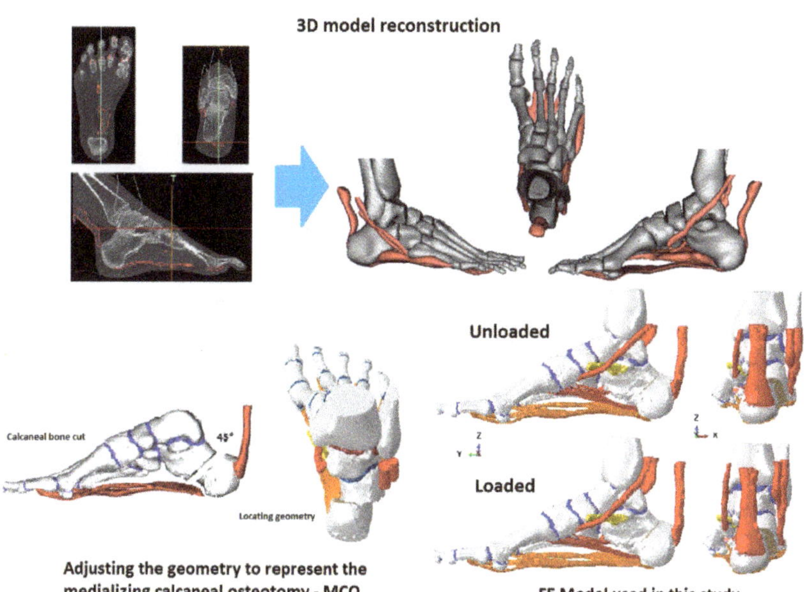

Figure 1. Reconstruction and modifications in the model to simulate a medializing calcaneal osteotomy. The Achilles' tendon, tibialis posterior tendon, and both Peroneus tendons' geometries and the pieces of Tibia and Fibula bones were reoriented to obtain a vertical position.

2.2. Meshing

The model's meshing was developed by means of the software ANSYS V.15 (Canonsburg, PA, USA). In summary, the model include 28 cortical bone pieces, 24 trabecular bone pieces, 26 cartilage segments, 4 tendons, 3 ligaments, and the plantar fascia. A trial–error approach was used to optimize the mesh size of each segment [17]. These authors suggested that the number of inaccurate elements must be less than 5% in all the measured parameters. All simulations and post-processing were developed in Abaqus/CAE 6.14 (Dassault Systèmes, Vélizy-Villacoublay, France) using the available nonlinear geometry solver.

Some of the conditions considered in order to achieve a reasonable mesh size without compromising the calculation time included having a minimum mesh size sufficiently small to fit into the tightest segments, a mesh accuracy of more than 99% of the elements being better than 0.2 mesh quality (Jacobians) and checking that the poor elements were located away from the region of greatest interest (hindfoot bones, metatarsals, PF, and SL) (see Figure 2). The convergence analysis was performed for 265,547 linear tetrahedral elements (C3D4). All parameters exhibited good mesh quality ratios (see Table 1).

Table 1. Mesh quality metrics based on Burkhart et al. (2013) recommendations [17].

Quality Metric	Assessment Criteria	Accurate Elements	Inaccurate Elements
Element Jacobians	>0.2	99.96%	0.04%
Aspect ratio	>0.3	96.9%	3.1%
Min. angles	>30°	95.3%	4.7%
Max. angles	<120°	99.46%	0.84%

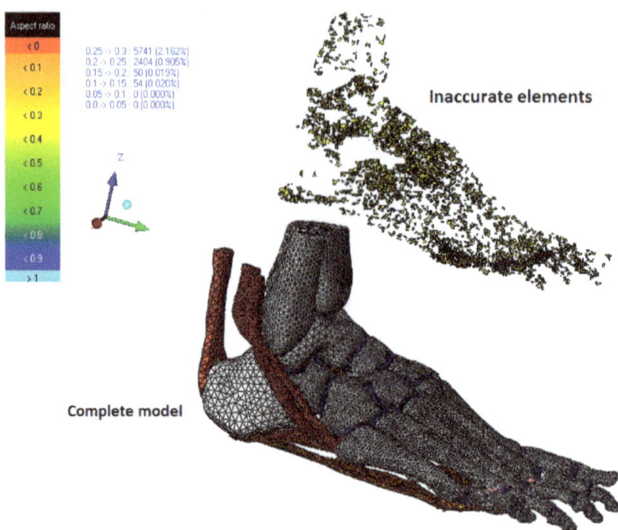

Figure 2. Location of the inaccurate elements, applying the Jacobians as the quality mesh criteria for evaluation.

2.3. Tissue Properties

Two kinds of behavior were considered in this finite element model: linear elastic behavior and hyper-elastic behavior.

Tissues with elastic linear behavior were the cortical bone, trabecular bone, ligaments, and plantar fascia.

Tissues with hyper-elastic behavior were tendons and cartilages.

The numerical values for each of these tissues were as follows:

The material properties (Young's modulus (E) and Poisson's ratio (v)) of the cortical bone, trabecular bone, ligaments, and plantar fascia were assigned in accordance with published data: cortical bone (E = 17,000 MPa, v = 0.3), trabecular bone (E = 700 Mpa, v = 0.3), ligaments (E = 250 Mpa, v = 0.28), and plantar fascia (E = 240 MPa, v = 0.28) [16,20,23].

Tendons and cartilages were modelled as hyper-elastic materials (Ogden model), using the parameters taken from specialized articles [24]. The strain energy density function U is:

$$U = \frac{\mu}{\alpha^2}(\lambda_1^\alpha + \lambda_2^\alpha + \lambda_3^\alpha - 3) + \frac{1}{D}(J-1)^2, \tag{1}$$

where the initial shear modulus μ = 4.4 MPa (cartilage)/33.16 MPa (tendons), the strain hardening exponent α = 2 (cartilage)/24.89 (tendons), and the compressibility parameter D = 0.45 (cartilage)/0.0001207 (tendons) [16,24,25]. The plantar fascia and spring ligament failures were simulated, applying the isotropic hardening theory that generates a progressive reduction of the tissue's stiffness, resulting in a very flexible material. The initial parameters were a Young's modulus of 240 and a Poisson ratio of 0.3. This strategy allowed us to improve the convergence of the model. Tibialis posterior tendon failure was simulated by removing the traction force of this tendon. Additionally, we considered that this characterization could be more realistic than simply changing the properties of the tissues because it approximates the viscoelastic behavior of these tissues, where stiffness depends on the loading application. The model used for this study maintains the differences of the bone characterization (cortical and trabecular) presented in Cifuentes-De la Portilla, C. et al. [3], where the internal parts of all the foot bones were modeled as trabecular. However, fibula and tibia bones were entirely simulated as cortical because a stress evaluation was not performed on these parts.

2.4. Loading and Boundary Conditions

The FE model was reconstructed from CT images of an unloaded foot. First of all, a standing load position was created (midstance phase) that was used as a reference case to compare against all the pathological cases. In Figure 3, load and boundary conditions are shown. The value of 720 N for the load corresponds to the full weight of an adult of about 70 Kg leaning on one foot. This condition represents a traditional scenario of an AAFD diagnosis assessment. Both loading conditions and boundaries were kept unaltered for all the MCO simulations.

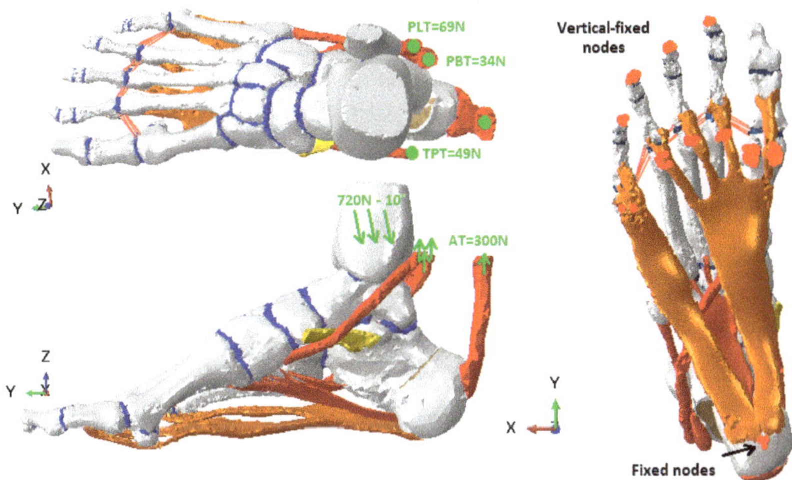

Figure 3. Boundary and loading values applied in the model. Loading values correspond to the weight of a person of 70 Kg.

The direction of the load exhibits an inclination of 10 degrees (descending vertical). This load was distributed over the tibia–talus joint (90%) and fibula–talus (10%) [26]. The tendon traction forces were included as reported by Arangio et al. [27]. To simulate the contact with the floor and to avoid the foot structure displacement under loading tests, some nodes located at the lower part of the calcaneus were fixed, while the Z-axis displacement (vertical) of the lower nodes of the first and fifth metatarsals was constrained to 0, using boundary tools available in Abaqus. The nodes remained unaltered for all the simulations performed. To avoid the tendon geometries crossing through the bones, we used the contact surfaces method, using the surfaces of the bones and tendons in contact during simulations.

2.5. About the Model Validation

This study used a model that has been previously validated for other studies related to AAFD [9]. They followed the recommendations of Tao et al. [5], measuring the vertical displacements of some anatomical points: the highest point of the talus (TAL) and of the navicular (NAV), the midpoint of the first cuneiform (CUN), and the highest point of the first metatarsal head (1MT), in two different loading conditions: light loading (minimal contact with the ground) and normal stance loading, using lateral Rx images (sagittal plane) (see Figure 4). The light loading condition was the position before starting simulations.

2.6. Model Analysis and Evaluation Criteria

To quantify the structural deformation of the foot and to evaluate biomechanical stress changes generated by MCO in foot tissues, we simulated the weakness/failure of the plantar fascia (PF), spring ligament (SL) and the tibialis posterior tendon (TPT) in isolation, but also combining these three elements. The weakness/failure was simulated by applying the isotropic hardening theory. This method was implemented using the function

"Parameter" in Abaqus, which allows modification of the stiffness of a material, reducing the Young's modulus from its initial value until obtaining a stiffness reduction by about 86%. The stresses in hindfoot bones, forefoot bones, and in all the soft tissues included in the model were calculated.

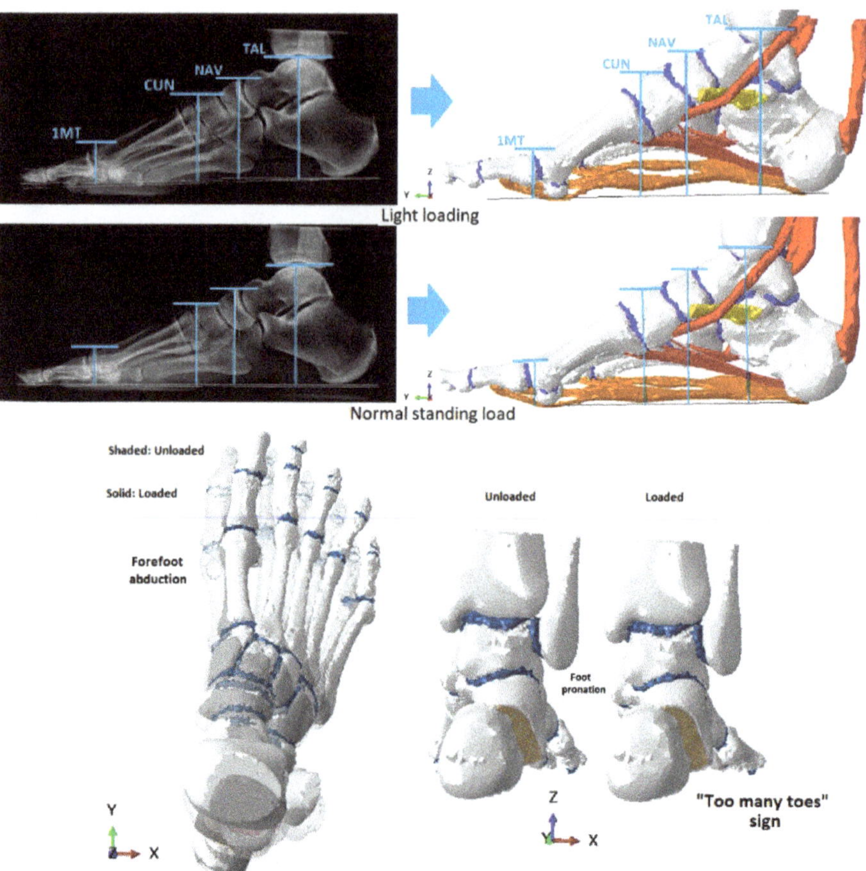

Figure 4. (**Up**) Validation strategy which compares the foot deformation in two different loading values. (**Bottom**) Signs of adult acquired flatfoot deformity achieved with our model.

For measuring stress on tissues, the maximum principal stress (S. Max) was used. This magnitude is closely related to the tensile stress that is generated in foot tissues [28]. Structural deformation was quantified measuring the vertical displacement of the entire structure (in millimeters).

3. Results

3.1. About Model Validation

Deformation found in the model is very approximated for a patient in a loading test. Found values can be included in the inter-subject variability under healthy conditions, represented as light loading (minimal contact with the ground) and a normal standing load (Table 2). These were compared with the average deformation of all the evaluated points measured in Rx-images in the sagittal view of 12 healthy patients.

Table 2. Results of the validation process. The values correspond to the difference between the measured distance from each point to the ground, under two different loading values: Light loading (minimal contact with the ground) and normal standing load [9].

Reference Point	Model Prediction (mm)	Patient Average (mm)	Patient Std. Deviation
TAL	−0.33	−0.32	0.14
NAV	−0.27	−0.26	0.04
CUN	−0.26	−0.19	0.08
1MT	−0.07	0.08	0.003

3.2. Flatfoot Simulation and MCO Structural Correction

Quantification of structural changes generated by both the simulated pathological scenarios and the MCO cases was carried out by measuring the vertical displacement of the entire foot skeletal structure (in millimeters). The values show by how much each foot region falls or rises (see Figure 5). Blue represents falling while red represents elevation. As can be seen, in simulations performed before applying MCO, the medial foot region suffers higher displacements downwards. These changes are also related to arch lengthening and foot pronation. After applying MCO, the tendency towards supination of the foot structure can be seen clearly, also causing fewer vertical displacements and less arch lengthening because of this (please see the Hallux fingertip—the tip of the big toe).

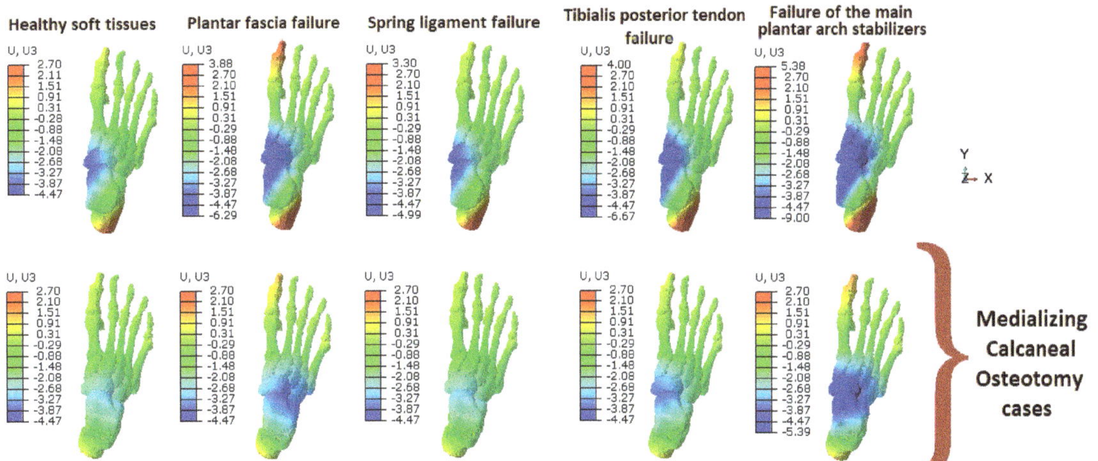

Figure 5. Structural changes obtained after simulating the model in different pathological cases. Values measure vertical displacement. Blue represents negative displacement (downwards). The color scale was normalized to 2.70, which is the maximum displacement reported in the healthy case. The maximum displacement values when these are higher than the reference is shown at the top of the color scale.

The effectiveness of MCO in the correction of the forefoot pronation and hindfoot valgus can be seen. Additionally, it can be observed how well MCO can compensate for the failure of both the spring ligament and the tibialis posterior tendon, reducing the effects expected by these failures (mainly forefoot pronation) [3,9].

The medialization of the calcaneus, through an osteotomy, manages to reduce the typical pronator moment of the subtalar joint. In this situation, the functional demand of the native structures to preserve the plantar arch, spring ligament, and tibiabis posterioris is significantly lower. If the pronator moment is maintained, the mechanical demands are very high and this may explain the known failure in isolated arthrodeses of the talocuneiform joint.

3.3. Stress in Forefoot Bones Generated by MCO

The first analyzed results were stresses on metatarsals. This magnitude was measured in Megapascals (MPa). The color scale was normalized to 60 MPa to better visualize the results (Figure 6). However, for each color scale, we also show the highest value obtained from each case and the stress measured in lateral metatarsals. Meaning of the scale: red means high stress, green represents medium stress, while blue represents low stress. It is observed that MCO increases the stress around the fourth metatarsal, and it increases even more when the PF or TPT fails. Additionally, the MCO reduces the forefoot maximum stress by approximately 35% (from 240,04 MPa to 155,06 MPa). These results are relevant because they show how the MCO by itself caused this stress reduction in metatarsals.

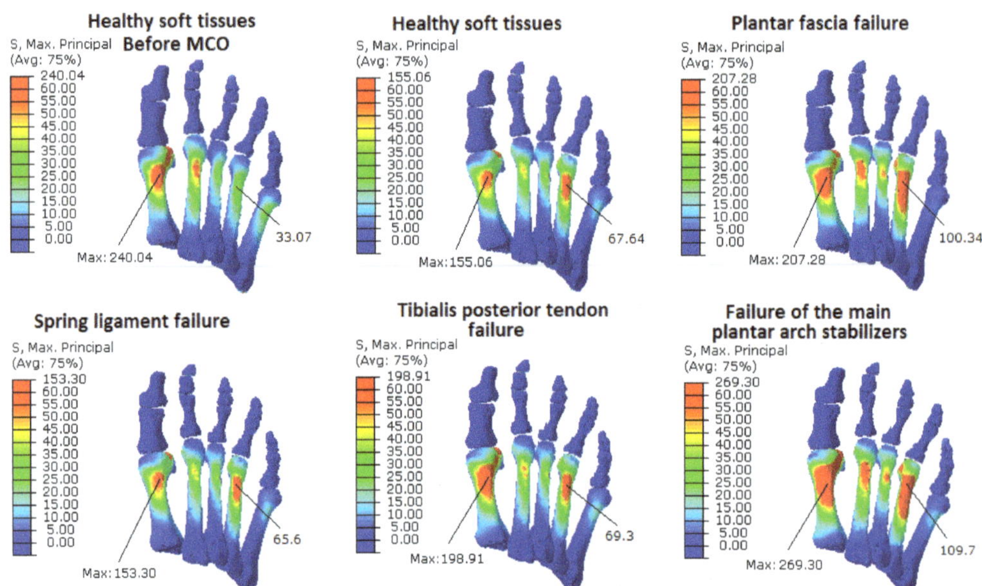

Figure 6. Stress in metatarsals (MPa). All simulations were performed after applying MCO, except for the first case which was included for comparison. The maximum stress values obtained are shown at the top of the color scale.

3.4. Stress in Hindfoot Bones Generated by MCO

In the second place, stresses in hindfoot bones were also evaluated for all the above-mentioned pathological cases for MCO (Figure 7). Now, the color scale for stress values was normalized to 20 MPa to better visualize the results. It can be noted that there is an increase in the stress concentration around the osteotomized region, when one of the main stabilizers fails (PF, TPT or SL).

A summary of bones stress changes is shown in Table 3, including the relative changes obtained from each case against simulating an MCO with healthy soft tissues. The results in Table 3 come from the stress results shown in Figures 6 and 7.

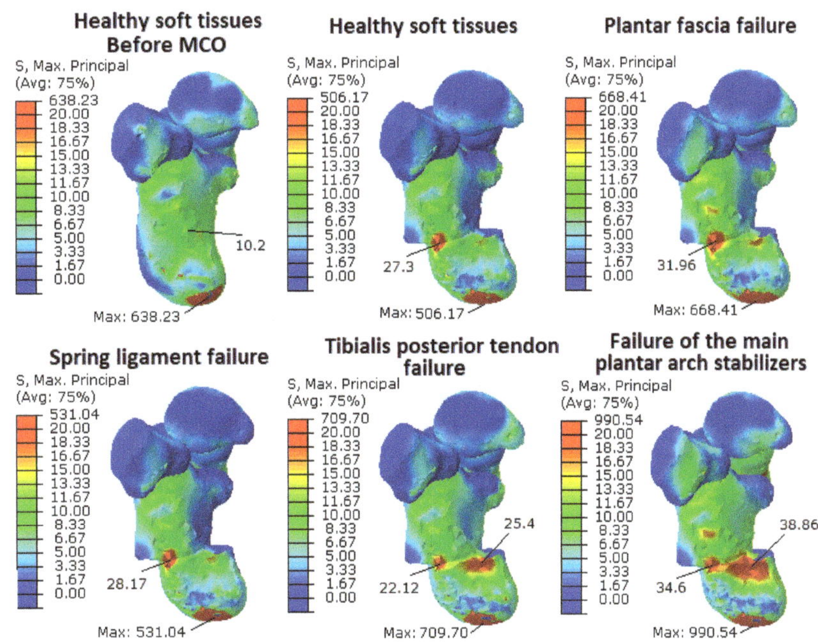

Figure 7. Stress in hindfoot bones (MPa). All simulations were performed after applying MCO, except for the first case which was included for comparison. The maximum stress values obtained are shown at the top of the color scale.

Table 3. Relative differences obtained from simulations, considering the simulation of MCO with healthy soft tissues as reference. This table is related to Figures 6 and 7. Please see these figures to find the maximum stress locations. All stress values are in megapascals (MPa).

	Forefoot Bones Stress		Hindfoot Bones Stress	
	Max. Stress Values (MPa)	Relative Difference (%)	Max. Stress Values (MPa)	Relative Difference (%)
Healthy after MCO	155.06	0	507.17	0
MCO and PF failure	207.28	34	668.41	32
MCO and SL failure	153.3	−1	531.04	5
MCO and TPT failure	198.91	28	709.7	40
MCO and main arch-stabilizers failure	269.3	74	990.54	95

3.5. Stress in Soft Tissues after MCO

We compared the soft tissue stress before and after applying MCO, simulating a failure of one or two of the main stabilizers of the plantar arch (TPT, PF and SL) (Figure 8). The results were normalized using values that allow the differences to be easily identified: 100 MPa for the tibialis posterior tendon, 33 MPa for the plantar fascia and 29 MPa for the spring ligament. The color scale is organized in the same way as described above for the bone stress evaluation.

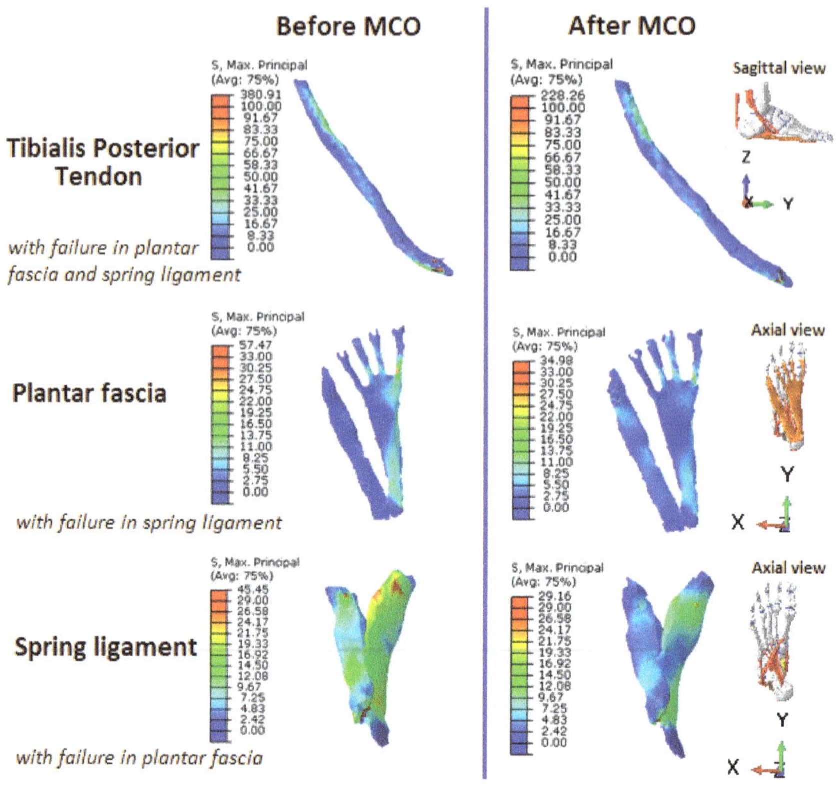

Figure 8. Comparison between stresses generated before and after applying MCO in the main soft tissues that support the plantar arch, in some pathological scenarios. The maximum stress values (MPa) obtained are shown at the top of the color scale.

The maximum stresses in the rest of the soft tissues included in the model were also quantified. The results are summarized in Figure 9.

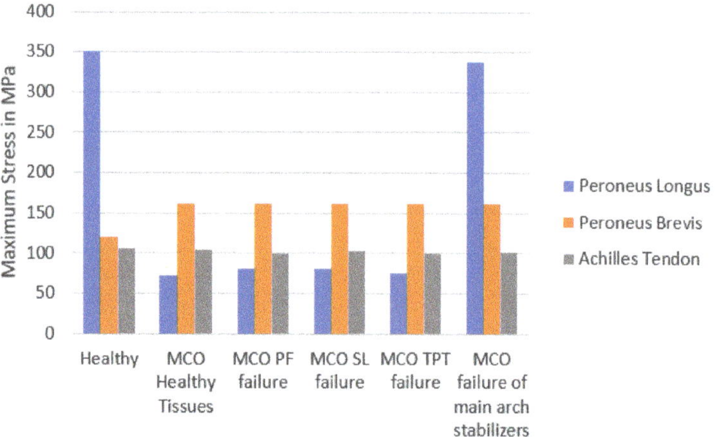

Figure 9. Comparison of the maximum stress obtained in the rest of the tendons included in the model.

4. Discussion

There are many treatment options for AAFD used by surgeons depending on the disease stage. For cases with a flexible deformity (IIa–IIb), tendon reinforcements are insufficient, requiring intervention in the bone structure. One of the most widely used options is medializing calcaneal osteotomy (MCO). Recently, Zanolli et al. and Patrick et al. [15,29] published experimental studies focused on comparing the foot's structural correction achieved with MCO with some other strategies such as Z-osteotomy or lateral column lengthening. Other authors have evaluated the outcomes of MCO and both its effect on the Achilles tendon and its contribution to correcting signs of AAFD [11]. Although the structural correction of the foot that can be achieved with MCO is widely known, some clinical studies have shown that MCO generates some long-term consequences such as stress distribution changes in the forefoot and probable risk of stress fractures, as has been reported with Evans' osteotomy [11,12,14]. These findings may be very relevant for the surgeon's decision-making process [30]. However, the biomechanical side-effects generated by MCO in both foot bones and the main soft tissues that support the plantar arch have not been analyzed sufficiently, mainly because of the difficulty of measuring tissue stress in cadaveric models [5,9].

Traditionally, the performance of an isolated arthrodesis of the cuneometatarsal joint for the treatment of flat feet in adults has been associated with a high rate of malunion. This fact has displaced the technique towards more aggressive ones, such as triple arthrodesis. However, associating an OCM reduces the pronator moment during gait and thus potentially allows it to act only on the cuneometatarsal joint. Selective arthrodesis of this joint would result in a clear clinical benefit.

In view of the above, a computational foot model was used to evaluate the changes on the biomechanics in foot tissue stresses when performing an MCO. This research alternative is used nowadays in clinical biomechanics studies. Thus, some examples such as these can be found in the literature. Smith et al. [31] designed a computational model to evaluate the structural effect of the Evans osteotomy. Wang et al. [13] proposed an FE study to evaluate different variables in MCO application, such as the angle and the medializing displacement distance. Normally, all these studies simplify the anatomy of the soft tissues and the biomechanical properties of bones. These simplifications penalize their use for analyzing the stress changes generated by MCO.

The proposed model can be reproduced, on the one hand, for loaded foot deformities and, on the other hand, for the main signs of AAFD such as foot pronation and the "too many toes" sign [4] (please see Figure 4 (bottom)). It is important to remark on two items: this model differentiates trabecular bone and cortical bone and this model contains the main soft tissues related to AAFD development. These make it possible to localize stress concentrations and evaluate stresses on both hard and soft tissues [3].

The results show that MCO effectively reduces the pronation of the hindfoot typically observed in patients with AAFD (see intense blue color around both astragalus and navicular bones). As can be seen in Figure 5, in most of the cases simulated with MCO, foot pronation was not generated. Moreover, these results indicate a good compensation of the MCO when the spring ligament fails. This is due to the supination momentum caused by the MCO over the foot structure. Nevertheless, if all the main soft tissues support the plantar arch failure, the MCO by itself cannot stop foot pronation. This means that, when performing an MCO, surgeons should add another strategy, such as artrodeses or tendon reinforcement, for example.

Additionally, the results suggest that MCO compensates very well for the spring ligament failure. The supination momentum that MCO causes in the foot structure can explain these results. However, it is noticeable that, when all the main soft tissues that support the plantar arch fail, MCO cannot prevent foot pronation on its own. This means that MCO should be applied in combination with other strategies, such as tendon reinforcement or arthrodeses.

The results of the stress analysis on forefoot bones show that MCO reduces the bone stresses by approximately 35%, considering the healthy case as a reference and comparing the maximum stress values obtained from both healthy cases (without tissue weakness) before and after MCO (see Table 3 and Figure 6). However, when the foot arch stabilizers fail, our simulations found a stress increase in all the metatarsals, mainly in the first, second, and fourth. This increase is much more important if the plantar fascia or the tibialis posterior tendon fail, increasing the stress by approximately 34% and 28% respectively, compared to the maximum stress values obtained from MCO with healthy soft tissues (see Table 3). When the main arch stabilizers fail, the stress increases by 74%. This increase in the stress values and in its redistribution could explain the pain in the toes reported in patients treated with MCO [30] and the findings of Iaquinto [12], who concluded that corrective osteotomies shifted loads from the medial forefoot to the lateral forefoot, with greater impact for combination lateral column lengthening and MCO procedures. The stress on the third and fifth metatarsals increases less than the others, probably because of some differences in the tissue insertion on the phalanges. Despite these differences, it is important to note how the metatarsal stresses change in different scenarios.

The results of the soft tissue analysis show that MCO noticeably reduces the stress in the main foot arch stabilizers, especially when the plantar fascia (mainly) or tibialis posterior tendons fail (Figure 8). Additionally, the stress reduction generated in the peroneus longus tendon is considerable, except in the case simulated with failure of all the main foot arch stabilizers (Figure 9). This result is consistent with the structural analysis, which shows that MCO cannot correct foot pronation when these tissues fail. As expected, no significant stress changes were found in the Achilles' tendon. These results are close to those obtained by Hadfield et al. [11] and Kongsgaard et al. [32] in their study performed using cadaver models.

Finally, if some of the stabilizers of the main arch fail, such as the plantar fascia or tibialis posterior tendon, an important stress concentration around the osteotomy region appears (Table 3 and Figure 7). As is shown in Table 3, the maximum stress in these cases increased by 32% and 40%, respectively. When the foot arch stabilizers fail, the maximum stress concentration increases by about 95% (from 507 to 990 MPa). In this work, fixation methods were not evaluated since complete bone healing is assumed after calcaneal translation.

A limitation of this study is that the analysis was based on a static simulation. Thus, patients' variability in tissues and loading was not considered, because one case study was simulated. However, the relations and differences (in percentages) obtained could be useful for evaluating the MCO effects in all the scenarios simulated. Our results cannot be generalized because only one anatomy was investigated, but the relative differences obtained could help with the study of MCO effects on the foot structure. Additionally, our model does not include the plantar pad, the flexor hallux longus, or flexor digitorum longus tendons. However, clinical studies have shown that these tissues have a minor role in AAFD development and in the foot arch support, compared to the tissues included in the model used [27,33]. Additionally, our model does not include any artificial restriction for the tibialis posterior tendon motion, so the pathway generated after traction forces may not be anatomically correct. Additionally, we used an isotropic characterization for plantar fascia and ligament tissues, which could lead to non-real calculations of stress in the tissues. It is necessary to perform a parametric study to show how sensitive the model predictions are to the material properties chosen. Moreover, our study was based on small displacements and deformations, so a linear elastic behavior for these tissues does not greatly falsify the results. This model also does not allow for error predictions, since statistics or deviations on the model characteristics are not included. One way to be able to make error predictions could be using probabilistic finite elements. Finally, it is important to remark that the values of biomechanical stress found cannot be assumed to be true stress values for all people (because of inter-subject variability). Nevertheless, we can analyze the relative differences generated in each case. The smaller increase in the third and fifth metatarsal stresses could be caused by differences in tissue insertion in the phalanges.

5. Conclusions

As a conclusion, we can say that an MCO is a good option to perform in patients suffering with AAFD developed mainly due to a failure of the spring ligament. Nevertheless, if patients exhibit plantar fascia weakness and/or tibialis posterior tendon dysfunction, this technique could also be applied, but carefully. In this case, we found that an MCO causes an important increase in the lateral metatarsals (a possible reason for pain in the long term) and around the osteotomized area. These factors create a risk of calcaneal fracture.

Author Contributions: Conceptualization, J.B. and R.L.-G.; methodology, C.C.-D.l.P. and J.B.; software, B.D.S. and C.C.-D.l.P.; validation, C.C.-D.l.P. and R.L.-G.; formal analysis, B.D.S. and J.B.; investigation, C.C.-D.l.P. and J.B.; resources, B.D.S. and C.C.-D.l.P.; data curation, R.L.-G.; writing—original draft preparation, C.C.-D.l.P. and J.B.; writing—review and editing, J.B. and R.L.-G.; visualization, B.D.S.; supervision, C.C.-D.l.P.; project administration, J.B.; funding acquisition, J.B. All authors have read and agreed to the published version of the manuscript.

Funding: This research was funded by the Ministry of Economy and competitiveness of the Government of Spain through the project DPI2019-108009RB-100.

Data Availability Statement: All the data and results are included within the manuscript.

Conflicts of Interest: The authors declare no conflict of interest.

References

1. Abousayed, M.M.; Alley, M.C.; Shakked, R.; Rosenbaum, A.J. Adult-Acquired Flatfoot Deformity: Ethology, Diagnosis, and Management. *JBJS Rev.* **2017**, *5*, e7. [CrossRef] [PubMed]
2. Bluman, E.M.; Title, C.I.; Myerson, M.S. Posterior Tibial Tendon Rupture: A Refined Classification System. *Foot Ankle Clin.* **2007**, *12*, 233–249. [CrossRef] [PubMed]
3. Cifuentes-De la Portilla, C.; Larrainzar-Garijo, R.; Bayod, J. Biomechanical stress analysis of the main soft tissues associated with the development of adult acquired flatfoot deformity. *Clin. Biomech.* **2018**, *61*, 163–171. [CrossRef] [PubMed]
4. Deland, J.T. The adult acquired flatfoot and spring ligament complex: Pathology and implications for treatment. *Foot Ankle Clin.* **2001**, *6*, 129–135. [CrossRef] [PubMed]
5. Tao, K.; Ji, W.T.; Wang, D.M.; Wang, C.T.; Wang, X. Relative contributions of plantar fascia and ligaments on the arch static stability: A finite element study. *Biomed. Technol. Biomed. Eng.* **2010**, *55*, 265–271. [CrossRef] [PubMed]
6. Toullec, E. Adult flatfoot. *Orthop. Traumatol. Surg. Res.* **2015**, *101*, S11–S17. [CrossRef]
7. Fowble, V.A.; Sands, A.K. Treatment of adult acquired pes plano abductovalgus (flatfoot deformity): Procedures that preserve complex hindfoot motion. *Oper. Tech. Orthop.* **2004**, *14*, 13–20. [CrossRef]
8. Feibel, J.B.; Donley, B.G. Calcaneal Osteotomy and Flexor Digitorum Longus Transfer for Stage II Posterior Tibial Tendon Insufficiency. *Oper. Technol. Orthop.* **2006**, *16*, 53–59. [CrossRef]
9. Larrainzar-Garijo, R.; Cifuentes de la Portilla, C.; Gutiérrez-Narvate, E.; Díez-Nicolás, E.; Bayod, J. Efecto de la osteotomía medializante de calcáneo sobre tejidos blandos de soporte del arco plantar: Un estudio computacional. *Rev. Esp. Cir. Ortop. Traumatol.* **2018**, *63*, 15–163. [CrossRef]
10. Vora, A.M.; Tien, T.R.; Parks, B.G.; Schon, L.C. Correction of moderate and severe acquired flexible flatfoot with medializing calcaneal osteotomy and flexor digitorum longus transfer. *JBJS* **2006**, *88*, 1726–1734. [CrossRef]
11. Hadfield, M.H.; Snyder, J.W.; Liacouras, P.C.; Owen, J.R.; Wayne, J.S.; Adelaar, R.S. Effects of medializing calcaneal osteotomy on Achilles tendon lengthening and plantar foot pressures. *Foot Ankle Int.* **2003**, *24*, 523–529. [CrossRef] [PubMed]
12. Iaquinto, J.M.; Wayne, J.S. Effects of surgical correction for the treatment of adult acquired flatfoot deformity: A computational investigation. *J. Orthop. Res.* **2011**, *29*, 1047–1054. [CrossRef] [PubMed]
13. Wang, Z.; Kido, M.; Imai, K.; Ikoma, K.; Hirai, S. Towards patient-specific medializing calcaneal osteotomy for adult flatfoot: A finite element study. *Comput. Methods Biomech. Biomed. Eng.* **2018**, *21*, 332–343. [CrossRef]
14. Davitt, J.S.; Morgan, J.M. Stress fracture of the fifth metatarsal after Evans' calcaneal osteotomy: A report of two cases. *Foot Ankle Int.* **1998**, *19*, 710–712. [CrossRef]
15. Patrick, N.; Lewis, G.S.; Roush, E.P.; Kunselman, A.R.; Cain, J.D. Effects of Medial Displacement Calcaneal Osteotomy and Calcaneal Z Osteotomy on Subtalar Joint Pressures: A Cadaveric Flatfoot Model. *J. Foot Ankle Surg.* **2016**, *55*, 1175–1179. [CrossRef]
16. Tao, K.; Wang, D.; Wang, C.; Wang, X.; Liu, A.; Nester, C.J.; Howard, D. An in vivo experimental validation of a computational model of human foot. *J. Bionic Eng.* **2009**, *6*, 387–397. [CrossRef]

17. Burkhart, T.A.; Andrews, D.M.; Dunning, C.E. Finite element modeling mesh quality, energy balance and validation methods: A review with recommendations associated with the modeling of bone tissue. *J. Biomech.* **2013**, *46*, 1477–1488. [CrossRef] [PubMed]

18. Viceconti, M.; Olsen, S.; Nolte, L.-P.; Burton, K. Extracting clinically relevant data from finite element simulations. *Clin. Biomech.* **2005**, *20*, 451–454. [CrossRef]

19. Wang, Y.; Wong, D.W.-C.; Zhang, M. Computational models of the foot and ankle for pathomechanics and clinical applications: A review. *Ann. Biomed. Eng.* **2016**, *44*, 213–221. [CrossRef]

20. García-Aznar, J.M.; Bayod, J.; Rosas, A.; Larrainzar, R.; García-Bógalo, R.; Doblaré, M.; Llanos, L.F. Load transfer mechanism for different metatarsal geometries: A finite element study. *J. Biomech. Eng.* **2009**, *131*, 021011. [CrossRef]

21. Cifuentes-De la Portilla, C.; Larrainzar-Garijo, R.; Bayod, J. Analysis of the main passive soft tissues associated with adult acquired flatfoot deformity development: A computational modeling approach. *J. Biomech.* **2019**, *84*, 183–190. [CrossRef]

22. Chan, J.Y.; Williams, B.R.; Nair, P.; Young, E.; Sofka, C.; Deland, J.T.; Ellis, S.J. The contribution of medializing calcaneal osteotomy on hindfoot alignment in the reconstruction of the stage II adult acquired flatfoot deformity. *Foot Ankle Int.* **2013**, *34*, 159–166. [CrossRef]

23. Wright, D.G.; Rennels, D.C. A Study of the Elastic Properties of Plantar Fascia. *JBJS* **1964**, *46*, 482–492. [CrossRef]

24. Morales-Orcajo, E.; Souza, T.R.; Bayod, J.; de Las Casas, E.B. Non-linear finite element model to assess the effect of tendon forces on the foot-ankle complex. *Med. Eng. Phys.* **2017**, *49*, 71–78. [CrossRef] [PubMed]

25. Wu, L. Nonlinear finite element analysis for musculoskeletal biomechanics of medial and lateral plantar longitudinal arch of Virtual Chinese Human after plantar ligamentous structure failures. *Clin. Biomech.* **2007**, *22*, 221–229. [CrossRef] [PubMed]

26. Bayod, J.; Becerro-de-Bengoa-Vallejo, R.; Losa-Iglesias, M.E.; Doblaré, M. Mechanical stress redistribution in the calcaneus after autologous bone harvesting. *J. Biomech.* **2012**, *45*, 1219–1226. [CrossRef]

27. Arangio, G.A.; Salathe, E.P. A biomechanical analysis of posterior tibial tendon dysfunction, medial displacement calcaneal osteotomy and flexor digitorum longus transfer in adult acquired flat foot. *Clin. Biomech.* **2009**, *24*, 385–390. [CrossRef]

28. Gefen, A. Stress analysis of the standing foot following surgical plantar fascia release. *J. Biomech.* **2002**, *35*, 629–637. [CrossRef] [PubMed]

29. Zanolli, D.H.; Glisson, R.R.; Nunley, J.A.; Easley, M.E. Biomechanical assessment of flexible flatfoot correction: Comparison of techniques in a cadaver model. *JBJS* **2014**, *96*, 45. [CrossRef]

30. Guha, A.R.; Perera, A.M. Calcaneal osteotomy in the treatment of adult acquired flatfoot deformity. *Foot Ankle Clin.* **2012**, *17*, 247–258. [CrossRef]

31. Smith, B.A.; Adelaar, R.S.; Wayne, J.S. Patient specific computational models to optimize surgical correction for flatfoot deformity. *J. Orthop. Res.* **2016**, *35*, 1523–1531. [CrossRef] [PubMed]

32. Kongsgaard, M.; Aagaard, P.; Kjaer, M.; Magnusson, S.P. Structural Achilles tendon properties in athletes subjected to different exercise modes and in Achilles tendon rupture patients. *J. Appl. Physiol.* **2005**, *99*, 1965–1971. [CrossRef] [PubMed]

33. Riddiford-Harland, D.; Steele, J.; Baur, L. Medial midfoot fat pad thickness and plantar pressures: Are these related in children? *Int. J. Paediatr. Obes.* **2011**, *6*, 261–266. [CrossRef] [PubMed]

MDPI

St. Alban-Anlage 66

4052 Basel

Switzerland

Tel. +41 61 683 77 34

Fax +41 61 302 89 18

www.mdpi.com

Mathematics Editorial Office

E-mail: mathematics@mdpi.com

www.mdpi.com/journal/mathematics